Fundamental and Applied Aspects of Nonionizing Radiation

Fundamental and Applied Aspects of Nonionizing Radiation

Edited by

Solomon M. Michaelson
Morton W. Miller
Richard Magin
Edwin L. Carstensen

The University of Rochester

PLENUM PRESS · NEW YORK AND LONDON

Library of Congress Cataloging in Publication Data

Rochester International Conference on Environmental Toxicity, 7th, 1974.
Fundamental and applied aspects of nonionizing radiation.

"Proceedings of a conference organized by the Department of Radiation Biology and Biophysics, the School of Medicine and Dentistry of the University of Rochester, held at Rochester, New York, June 5-7, 1974."
Includes indexes
1. Radiation–Physiological effect–Congresses. 2. Radiobiology–Congresses. I. Michaelson, Sol M. II. Rochester, N. Y. University. Dept. of Radiation Biology and Biophysics. III. Title.
QP82.2.R3R58 1974   574.1'915   75-33698
ISBN 0-306-30901-7

Proceedings of a conference organized by the Department of Radiation Biology and Biophysics, the School of Medicine and Dentistry of The University of Rochester, held at Rochester, New York, June 5-7, 1974

© 1975 Plenum Press, New York
A Division of Plenum Publishing Corporation
227 West 17th Street, New York, N.Y. 10011

United Kingdom edition published by Plenum Press, London
A Division of Plenum Publishing Company, Ltd.
Davis House (4th Floor), 8 Scrubs Lane, Harlesden, London, NW10 6SE, England

All rights reserved

No part of this book may be reproduced, stored in a retrieval system, or transmitted in any form or by any means, electronic, mechanical, photocopying, microfilming, recording, or otherwise, without written permission from the Publisher

Printed in the United States of America

We gratefully acknowledge the support provided for this conference by the Environmental Protection Agency, the Department of the Air Force, the University of Rochester, and the U. S. Atomic Energy Commission. The conference committee wishes to express its appreciation to Judy Havalack, the conference secretary, for her efficient secretarial service and coordination of arrangements for the conference. Secretarial assistance of Margaret Bush and the excellent operation of the projection equipment by Dr. Shin-Tsu Lu are also gratefully acknowledged.

                      The Conference Committee
                          Sol M. Michaelson,
                              Chairman
                          George G. Berg
                          Edwin L. Carstensen
                          Richard Magin
                          Morton W. Miller

Preface

During the last 30 years, there has been a remarkable development and increase in the number of processes and devices that utilize or emit non-ionizing radiant energies such as microwaves, a form of electromagnetic wave energy and ultrasound representative of mechanical vibration. These energies are used in all sectors of our society for military, industrial, telecommunications, medical, and consumer applications. More recently, the use of ultrasound in biology and medicine has been considerably expanded. These increases in sources of non-ionizing radiant energy have resulted in growing interest on the part of government regulatory agencies, industrial and military physicians, research workers, clinicians, and even environmentalists. Although there is information on biologic effects and potential hazards to man from exposure to microwaves or ultrasound, considerable confusion and misinformation has permeated not only the public press but also some scientific and technical publications.

Interest in the biologic effects of high frequency currents developed in the beginning of the present century. This was followed by the introduction of "ultrashortwave" therapy. During the latter part of World War II, the U. S. military services became interested in the possible hazards to personnel working around microwave sources, and the Office of Naval Research of the U.S. Navy began to sponsor research on the biologic effects of microwaves in 1948. In 1956, the U.S. Department of Defense assigned the responsibility for tri-service coordination of studies related to the biologic effects and potential hazards of microwave exposure to the U. S. Air Force. These studies contributed greatly to a better understanding of the biologic effects of microwaves.

In 1968, the U.S. Bureau of Radiological Health sponsored a symposium on the Biological Effects and Health Implications of Microwave Radiation in Richmond, Virginia. This Symposium was held to provide an indication of the state of knowledge in the area of microwave health effects at that time. Subsequently, several symposia have been held both in the U.S. and the U.S.S.R. on the general topic of the biological effects of microwaves. In October, 1973, the first truly international Symposium on Biologic Effects and Health Hazards of Microwave Radiation was held in Warsaw, Poland.

Ultrasound has not been studied as a naturally occurring phenomenon except for low-frequency, low-intensity emanations of animal origin. Interest in the possible harmful effects of ultrasound on man became highlighted when ultrasonic devices came into more general use.

To provide a perspective on the uses of microwaves in the civilian sector of the U.S., it has been noted that in 1972 about 425,000 microwave ovens; an estimated 15,000 shortwave and 15,000 microwave diathermy devices; approximately 120,000 microwave communications towers, each with several separate sources, were in use. About 2 million people are treated annually with radiofrequency (microwave) diathermy. In regard to ultrasonic devices, there are an estimated 50,000 cleaning units now in use in the U.S. These are used in a variety of industries and other non-home applications. Approximately 50,000 other commercial/industrial applications were in use at the end of 1972. In 1970, there were approximately 3,000 medical diagnostic devices in use; however, the use-growth pattern suggests this may have increased to nearly 10,000 by 1972. Industry sales projections and surveys indicate that about 33,000 ultrasonic diathermy units existed at the end of 1972. The very rapid increases in sales of diagnostic ultrasonic devices indicate that approximately 175,000 may be in use by 1976. Population at risk is not known with any degree of accuracy. Based on extrapolations of a 1970 equipment survey, an estimated two million people are treated annually with ultrasonic diathermy. Mixer and other industrial commercial applications accounted for approximately 50,000 units in use by the end of 1972. The number of ultrasonic devices projected for use in 1980 include: cleaning equipment, 200,000 units; medical diagnostic, approximately 175,000; diathermy, 100,000; commerical/industrial, 180,000.

Although thermal effects of microwave absorption have been well demonstrated and documented, some investigators suggest non-thermal or specific effects due to microwave exposure. When animals or man are exposed to microwaves of high intensity for significant periods of time, the absorbed energy may induce physiologic responses as a reaction to the increased body temperature or subtle alterations in thermal gradients in the body. Although there have been reports of functional changes of the neuroendocrine, cardiovascular, or central nervous

PREFACE

system as a result of microwave exposure, these responses are consistent with the pattern of physiologic adjustment to thermal inputs into the body.

The interactions of ultrasound in tissue have been studied; however, these interactions are quite complex, and much further study is needed to understand completely the biologic effects of ultrasound. Presently, much of the reported effects of ultrasound may be explained in light of existing theory. When ultrasound is absorbed by tissue, heat is produced. On the other hand, some investigators have reported biological effects from exposure to ultrasound that cannot be explained on the basis of heating alone.

Because of the complexity of the interactions of non-ionizing radiation in biological systems, an inter-disciplinary approach is necessary to assess and elucidate the problems that evolve as this field advances and as the use of these energies expands. It is important to maintain a proper perspective and assess realistically the biomedical effects of these radiant energies so that the worker or general public will not be unduly exposed nor will research, development and beneficial utilization of these energies be hampered or restricted by an undue concern for effects which may be non-existent or minimal in comparison to other environmental hazards. The goal of this conference is to review and place the available information and concepts in proper perspective to understand and encourage the full potential for the beneficial use of these radiant energies, at the same time preventing adverse effects to individuals exposed to these energies.

Sol M. Michaelson
Conference Chairman

Contents

Welcoming Remarks xv

SESSION I: BIOPHYSICS AND DOSIMETRY
Dietrich E. Beischer, Chairman

Dielectric Properties of Biological Materials and
Interaction of Microwave Fields at the Cellular
and Molecular Level   3
    Herman P. Schwan

Acoustic Properties of Biological Materials 21
    Floyd Dunn

Synthesis of Frequency Response of Electric
Field Probes 41
    Tadeusz M. Babij

Ultrasonic Measuring Techniques 59
    Harold F. Stewart

SESSION II: ENERGY ABSORPTION
James D. Hardy, Chairman

Transient Effects of Low-Level Microwave Irradiation
on Bioelectric and Muscle Cell Properties and on
Water Permeability and Its Distribution 93
    Adolfo Portela, Osvaldo Llobera, Solomon M. Michaelson,
    P. A. Stewart, Juan C. Perez, Ariel H. Guerrero,
    Carlos A. Rodriguez, and Roberto J. Perez

Thermal Factors in Ultrasonic Focal
Destruction in Organized Tissues 129
    Padmakar P. Lele

Physiological Responses to Heat 143
    John Bligh

SESSION III: MICROWAVES - BIOLOGICAL EFFECTS
Karl Lowy, Chairman

Electrophysiological Effects of Electromagnetic
Fields on Animals 167
    Arthur W. Guy, James Lin, and C. K. Chou

Sensation and Perception of Microwave Energy 213
    Sol M. Michaelson

SESSION IV: ULTRASOUND - BIOLOGICAL EFFECTS
George W. Casarett, Chairman

Are Chromosomal Aberrations Reliable
Indicators of Environmental Hazards? 233
    John R. K. Savage

Action of Ultrasound on Isolated Cells
and Cell Cultures 249
    C. R. Hill

Non-Thermal Effects of Ultrasound on
Intact Animal Tissues 263
    K. J. W. Taylor and M. Dyson

Physical Consequences of Ultrasound on
Plant Tissues and Other Bio-Systems 277
    Wesley Nyborg, Douglas Miller, and
    Alexander Gershoy

SESSION V: MEDICAL APPLICATIONS
Raymond Gramiak, Chairman

The Use of Nonionizing Radiation for
Therapeutic Heating 303
    Justus F. Lehmann and C. Gerald Warren

Ultrasound in Surgery 325
    Padmakar P. Lele

Safety of Ultrasound in Diagnosis 341
    C. R. Hill

EMC Design Effectiveness in Electronic
Medical Prosthetic Devices   351
    John C. Mitchell, William D. Hurt,
    and Terry O. Steiner

              SESSION VI: OCCUPATIONAL ASPECTS
                  Paul E. Tyler, Jr., Chairman

Analysis of Occupational Exposure
to Microwave Radiation   367
    P. Czerski and Maksymilian Siekierzynski

Control of Occupational Exposure to
Nonionizing Radiation   379
    Thomas Ely

Military Role in Safe Use of Microwaves   389
    Lawrence T. Odland

          SESSION VII: FUTURE APPLICATIONS AND CONTROLS
             Edythalena Tompkins, Chairwoman

Prospects for Expansion of Industrial
and Consumer Uses of Microwaves   411
    John M. Osepchuk

Solar Power via Satellite   433
    Peter E. Glaser and Owen Maynard

International Cooperation on Nonionizing
Radiation Protection   447
    Michael J. Suess

Participants   459

Index   465

WELCOMING REMARKS

It is a pleasure to welcome you to the University of Rochester, and its Medical Center, and more particularly, to welcome you to this conference. This is the Seventh International Conference on Environmental Toxicity. In this particular instance, we are talking about the toxicity of non-ionizing radiation. In view of the fact that these conferences have proved to be an important part of our educational program and our research effort here, we hope that this one will be equally effective, for in the past we have found that they have given a special kind of insight into environmental problems of man. They identified problems; helped in elucidating them, and helped toward an eventual solution, and we anticipate that this one will have a similar impact in the field of non-ionizing radiation.

During the last quarter of a century or more, there has been a very rapid increase in the number of devices that emit non-ionizing radiant energy. These include lasers as sources of optical radiation (ultraviolet, infrared, and visible light) microwaves, radiowaves, sound and ultrasound. Furthermore, these energies are used in almost all parts of our society: industrial, military, medical, to say nothing of the entertainment field. These increases in sources of non-ionizing radiant energy have resulted in the growing interest of research workers, clinicians and particularly of the government, in the exposures of humans to these sources of energy. As a result there has been considerable legislation, after a series of Congressional hearings, that relates to this particular field. These have resulted in the Radiation Control for Health and Safety Act of 1968, and the Occupational Safety and Health Act of 1970. The former act requires that the Secretary of Health, Education, and Welfare prescribe standards to control man-made radiation from electronic instruments, if it is determined that such standards are necessary for the protection of the public health and safety. The Occupational Safety and Health Act provides an even broader authority for the Departments of

Labor and HEW, to establish occupational safety and health standards for workers exposed to all kinds of potential hazards, including radiation. But the interest and concern about the biologic effects of radiant energy is not just national, but international in scope.

In this context, the Regional Office for Europe of the World Health Organization has under development a program concerned with the health effects of non-ionizing radiation. It is worthy of note that the European Regional Officer for Environmental Health for the World Health Organization, as well as the American Advisor and the United Kingdom Representative to the Regional Office for Europe of the World Health Organization are participating in this symposium.

Because of the complexity of the interactions of non-ionizing radiation in biological systems, an interdisciplinary approach is necessary to the problems that evolve, as this field advances and as the use of these energies expands. The goal, of course, and therefore the goal of this conference, is to understand and encourage the beneficial uses of these radiant energies, at the same time keeping the adverse effects within acceptable limits. The international representation and the interdisciplinary composition of this conference, I believe, augur well for meeting that goal.

And so, let me again welcome you to Rochester. I hope you will find the conference stimulating and productive for each of you, and that you enjoy Rochester and your stay with us. Thank you.

> J. Lowell Orbison
> Dean, School of Medicine
>   and Dentistry
> The University of Rochester
> Rochester, New York

Biophysics and Dosimetry

DIELECTRIC PROPERTIES OF BIOLOGICAL MATERIALS AND INTERACTION OF MICROWAVE FIELDS AT THE CELLULAR AND MOLECULAR LEVEL

Herman P. Schwan

Dept. of Bioengineering, University of Pennsylvania

Philadelphia, Pennsylvania

The propagation of electromagnetic waves in tissues is determined by their electrical properties. In addition, these properties tell us much about the mechanism of interaction of electromagnetic fields with various biological systems, including biopolymers, membranes and cells. Our present day knowledge of the electrical properties is rather advanced, and I shall first summarize the state of our present knowledge of such properties. Then I shall draw some conclusions about possible mechanisms which may or may not give cause to subtle nonthermal effects.

No consideration will be given to magnetic properties since the latter are, for our purposes, identical to those of free space. Hence, we will concern ourselves with the two electrical properties which totally define the electrical characteristics, namely, the dielectric constant relative to free space ε and the conductivity η. Both properties are for all our purposes independent of field strength, since the electrical characteristics of body electrolytes and biopolymers change at first at field strength values of many KV/cm. However, they change with temperature and strongly with frequency. As a matter of fact, as the frequency increases from a few Hertz to gigahertz the dielectric constant decreases from several million to only a few units. Concurrently the conductivity increases from mMho/cm to nearly a thousand.

This work was supported by NIH Grant HE-01253 and ONR Contract N00014-67-A-0216-0015.

Table 1

ELECTRICAL RELAXATION MECHANISM

Inhomogeneous Structure (Maxwell-Wagner)	β
Permanent Dipole Rotation (Debye)	γ, βtail
Counter Ion Relaxation	
Electrophoretic Relaxation	α

ELECTRICAL PROPERTIES

The first figure indicates the dielectric behavior of practically all tissues. Three relaxation regions α, β, γ of the dielectric constant exist at low, medium and very high frequencies. Each of these relaxation regions is in its simplest form characterized by equations of the type

$$\varepsilon = a + \frac{b}{1+x^2} \ ; \ \eta = c + d \frac{x^2}{1+x^2}$$

where x is a multiple of the frequency and the constants are determined by the values at the beginning and end of the dispersion change. However, biological variability may cause the actual data to change with frequency somewhat more smoothly than indicated by the equations.

The mechanisms responsible for these three relaxation regions are indicated in Table 1. Inhomogeneous structure is responsible for the β-dispersion, i.e., the polarization resulting from the charging of interfaces, i.e., membranes through intra- and extracellular fluids (Maxwell-Wagner effect). Rotation of molecules having a permanent dipole moment such as water and proteins is responsible for the γ-dispersion (water) and a small addition to the tail of the β-dispersion resulting from a corresponding β_1-dispersion of proteins. The tissue proteins only slightly elevate the high frequency tail of the tissue's β-dispersion since the addition of the β_1-effect caused by tissue proteins is small compared to the Maxwell-Wagner effect and since it occurs at somewhat higher frequencies. Another contribution to the β-dispersion is caused by smaller subcellular structures, such as mitochondria, cell nuclei and other subcellular organelles. Since these structures are

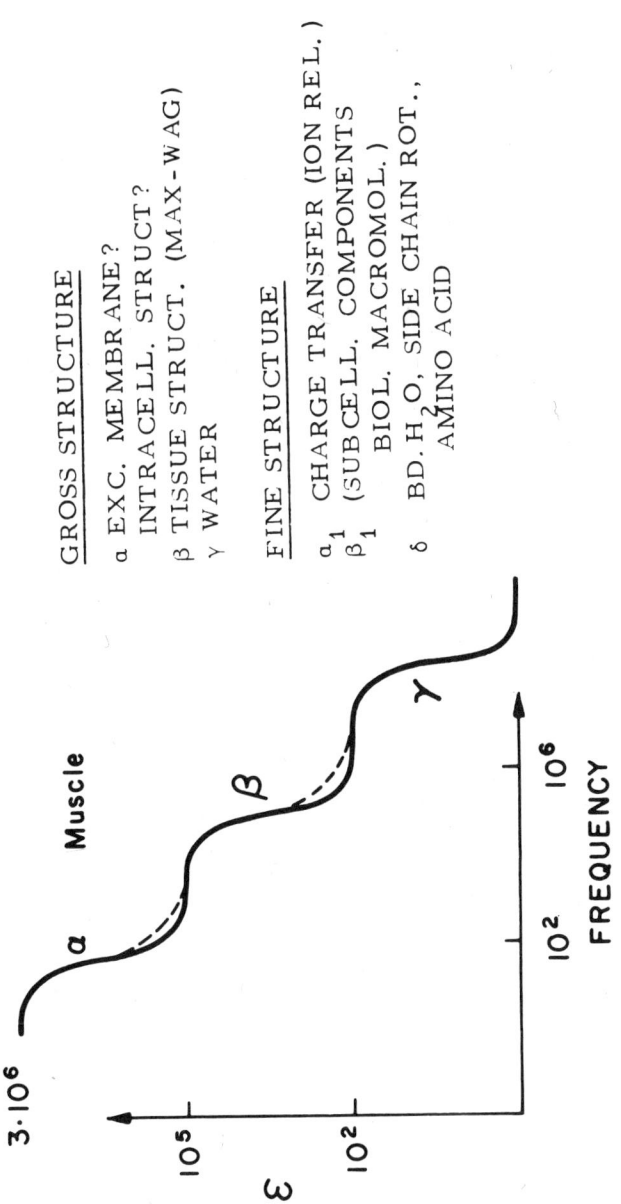

Figure 1

Gross and fine structural relaxation contributions to the dielectric constant of muscle tissue. Dashed lines indicate fine structural contributions. The data and various structural contributions are typical for all tissues of high water content.

smaller in size than the surrounding cell, their relaxation frequency is higher, but their total dielectric increment smaller. They, therefore, contribute another addition to the tail of the β-dispersion (β_1).

The γ-dispersion is solely due to water and its relaxational behavior near about 20 GHz. A minor additional relaxation (δ) between β and γ-dispersion is caused in part by rotation of amino acids, partial rotation of charged side groups of proteins and the relaxation of protein bound water which occurs somewhere between 300 and 2000 MHz.

The α-dispersion is presently least clarified. Intracellular structures such as the tubular apparatus in muscle cells, which connect with the outer cell membranes could be responsible in all such tissues which contain such cell structures. Relaxation of counter ions about the charged cellular surface is another mechanism suggested by us. Last but not least, relaxational behavior of membranes per se such as reported recently for the giant squid axon membrane can be account for it.[1] The relative contribution of the various mechanism varies no doubt from one case to another and needs further elaboration.

Electrophoretic relaxation is the counter part of the relaxation due to counter ion movement. It results from the oscillatory movement of charged particles with the alternating field. Its magnitude can be calculated. It is too small to noticeably contribute to the other relaxation mechanism indicated in Table 1.

No attempt is made to summarize conductivity data. Conductivity increases similarly in several major steps symmetrical to the changes of the dielectric constant. These changes are in accord with the theoretical demand that the ratio of capacitance and conductance changes for each relaxation mechanism is given by its time constant, or, in case of distributions of time constants, by an appropriate average time constant.

Figure 2 indicates the variability of the characteristic frequencies for the various mechanism α, β, γ and δ from one biological object to another. For example, blood cells display a weak α-dispersion centered at about 2 KHz, while muscle displays a very strong one near 0.1 KHz. The β-dispersion of blood is near 3 MHz, that of muscle tissue near 0.1 MHz. Clearly there is considerable variation depending on cellular size and other

---

[1]Takashima, S., H. P. Schwan, J. Memb. Biol., 17:51-68, 1974

DIELECTRIC PROPERTIES OF BIOLOGICAL MATERIALS

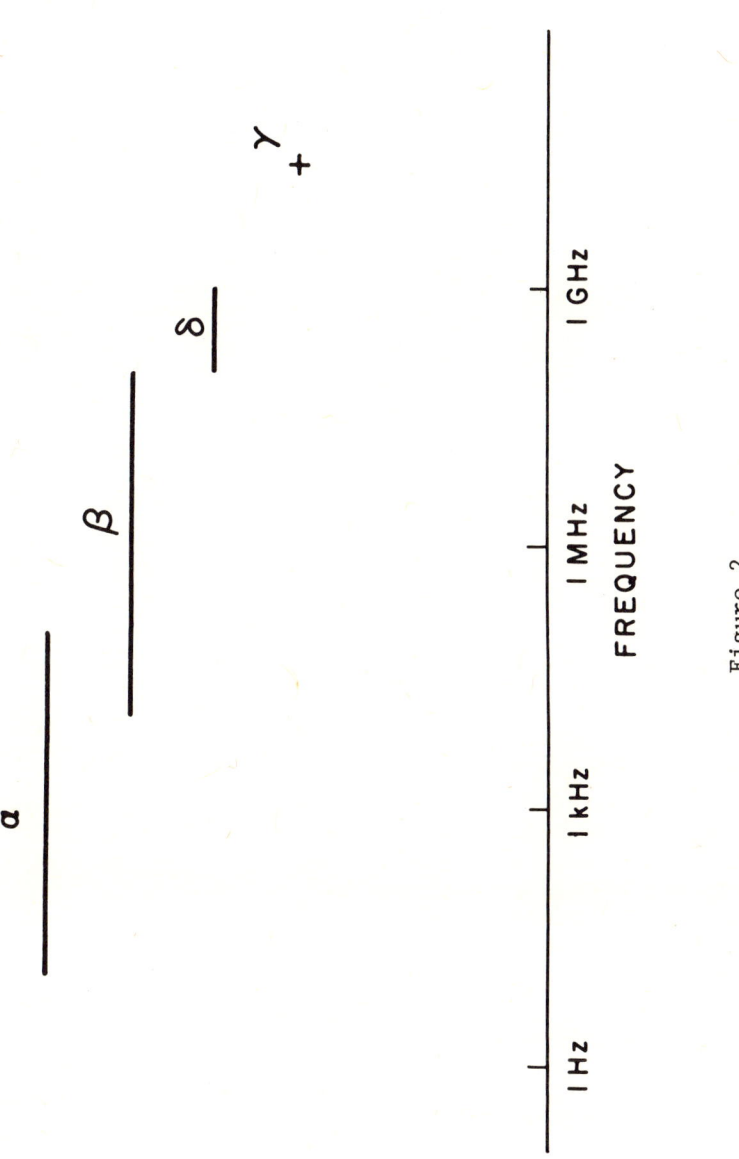

Figure 2

Ranges of characteristic frequencies for various biological systems.

factors. There may not be as strong a variation in the δ-case as there is for the α- and β-dispersion frequencies. The γ-dispersion, however, is always sharply defined at the same frequency range.

Table 2 attempts to summarize at what level of biological complexity the various mechanism occur. Electrolytes display only the γ-dispersion characteristic of water. Biological macromolecules in water add to the water's γ-dispersion a δ-dispersion. It is caused by bound water and rotating side groups in the case of proteins and caused by rotation of the total molecule in the case of the amino acids. And proteins and nucleic acids in particular add further dispersions in the β and α-range as indicated. Suspensions of cells free of protein would display a Maxwell-Wagner β-dispersion and the γ-dispersion of water. If they contain protein an additional comparatively weak β-dispersion due to the polarity of protein is added and a δ-dispersion. If the cells carry a net charge an α-mechanism due to counter ion relaxation is added and if their membranes on their own relax as some excitable membranes do, an additional α-mechanism may appear.

Table 2

Electrolytes	γ
Biol. Macromolecules	
Amino Acids	δ+γ
Proteins	β+δ+γ
Nucleic Acids	α+β+δ+γ
Cells, free of protein	β+γ
Charged	α+β+γ
Exc. membranes	α+β+γ

SOME CONCLUSIONS

Obviously, much is known about the mechanism responsible for the electrical properties of biological materials. In all cases where mathematical tools can be applied to fairly simple shapes such as in the case of spherical cells and even erythrocytes, the theoretical prediction is precisely in agreement with the experimental data.

In cases where the complexity of biological structure exceeds mathematical power, at least semiquantitative agreement with approximating theories has been achieved. In any case, our present understanding is sufficient to warrant confidence in the various models chosen.

A variety of experiments have been conducted to check the validity of the model concepts indicated above. For example, if cell membranes are broken down, the β-dispersion disappears as it ought to if the polarization of membranes is responsible for it. Furthermore, the extracellular fluid's ionic strength can be changed for example with suspensions of erythrocytes. Then the dispersion shifts since the time required for the accumulation of charges on the membranes through the extracellular medium changes inversely with ionic strength, i.e., concentration of charge carriers. Counter ion relaxation on the other hand has been well demonstrated with colloidal suspensions containing particles which are electrically charged and consequently surrounded by a counter ion atmosphere.

All this by no means is meant to imply that there are no more unresolved problems. As a matter of fact the following attempts to summarize some of the work which has to be done in the future:

The precise origin of the δ-dispersion and the relative contribution of bound water and polar side chain rotation requires further clarification;

The theory of counter ion relaxation has been so far only developed under fairly restricting assumptions and further work is required to fully assess the contribution of counter ions to the low frequency data;

There are still more data to be gathered for amino acids, proteins and nucleic acids particularly as function of concentration and field strength. Macromolecular interactions may affect dielectric data at typical biological concentrations. There is need to better understand the various relaxations typical of nucleic acids. There is a particular need to gather data as function of field strength in order to find out where macromolecular dielectric saturation occurs. This is needed to assess the possibility that high field strength levels as may be applied with pulsed fields may go beyond saturation levels and perhaps cause denaturation.

Last, but not least, the origin of the α-dispersion needs further elaboration. What is the relative contribution of counter ion effects, of dielectric effects in excitable membranes caused by time dependent sodium and potassium fluxes, and of dielectric effects due to intracellular membrane structures?

It is quite possible that such work may provide further insight into the various ways that electrical fields of whatever frequency and magnitude may reversibly and irreversibly affect biological structures.

In the meantime, the past work and the internal consistency achieved can be summarized in some fairly simple conclusions. The electrical data are entirely consistent with biological matter consisting of membranes surrounding and being surrounded by intra- and extracellular fluids which contain macromolecules. There appears little doubt that the large part of the water contained is "free" since both dielectric data and conductivities at microwave frequencies are essentially those of electrolytes containing biopolymers and the characteristic frequencies in the γ-range are identical for tissues and free water. From this well supported picture emerge the following conclusions of interest here:

1. At frequencies above those characteristic of the α-dispersion all biological membranes have a fairly frequency independent capacity of about 1 $\mu F/cm^2$, with the precise value varying between 0.5 and 1.5 $\mu F/cm^2$. This capacitance range which is presently unchallenged corresponds, say at 3 GHz, to an impedance of about 50 microohms!! Let us now consider a current density of about 3 mA/cm^2, i.e., a current density which corresponds to a flux of 10 mW/cm^2 in tissues of a typical resistivity value of 100 to 200 Ohm-cm at 3 GHz. An alternating membrane potential of only 0.15 μ Volt is induced by such a current density. This value is easily 1000-fold, smaller than potentials which are reported to cause effects on excitable membranes. This potential varies far more rapidly than those potentials which are known to cause excitation. It has been known for many decades that the ability to excite membranes declines rapidly as the frequency increases above a few hundred Hertz.

2. All the properties reported above are "linear" ones in the range of potentials at which they were studied. I.e., the reported dielectric constants and conductivities are independent of field strength at field strength levels of interest here. Membrane data change at first at field strength levels which almost compare with those corresponding to a typical resting potential of -70 mV, i.e., at imposed field levels of the order of 100 KV/cm or more. Biological macromolecules, judging from the restricted body of data available, respond to fields in a linear manner up to field levels of many KV/cm. This means that the energies which are imparted by thermal collisions are greater than those which can be applied with fields of the

order of some V/cm. It is therefore difficult to see how the
field strength values of interest here could irreversibly
affect macromolecules while the comparatively much stronger
forces imposed by thermal collisions at normal temperatures
obviously do not! The present knowledge therefore does not
indicate a possibility of denaturation by the field strength
levels of interest here. Moreover, the characteristic frequencies quoted are smaller than those of microwaves, making it
thereby unlikely for biopolymers and membranes to respond to
microwave frequencies.

3. Some people have speculated about the possibility that
"resonances" may occur. However, no one has indicated why such
a resonance effect should do irreversible harm. Moreover, there
are reasons to believe that a resonant type of response is
unlikely at frequencies up to many Gigahertz since the viscosity
of the water surrounding biopolymers in vivo is so high. Indeed,
such work has been done on the response of biopolymers to
alternating fields. The changes which have been observed so
far are relaxation effects, i.e., degenerated resonance effects
which occur when resonant behavior is prevented by high viscosity
of the medium in which the biopolymer is suspended. These
relaxational responses are not very frequency selective and not
likely to do irreversible harm as discussed above.

FORCES CAUSED BY EM-FIELDS

Above discussed electrical properties provide much
insight into the mode of interaction of alternating electrical
fields with biological systems. However, they do not reveal
readily whatever mechanical effects may be imparted. It is
known for quite some time that not only DC fields but also
alternating ones can evoke forces which may or may not be
significant. The study of some of the manifestations of these
forces has been of considerable interest over a number of years
in our laboratory. The well-known pearl chain formation effect
and the orientation of non-spherical particles are some of the
effects which have been studied (Figure 3).

In general, it may be said that such forces become significant in comparison with random thermal ones if the ratio of
the electrical potential energy of the system considered or its
change as the observed effect takes place is larger than the
thermal energy. This ratio is proportional to the expression
given in Figure 4, and the proportionality constant depends on
the particular phenomena considered. For example, in the case
of the movement of a particle in an inhomogeneous field, termed
"Dielectrophoresis" by Pohl, the proportionality factor
characterizes the inhomogeneity of the field and the force

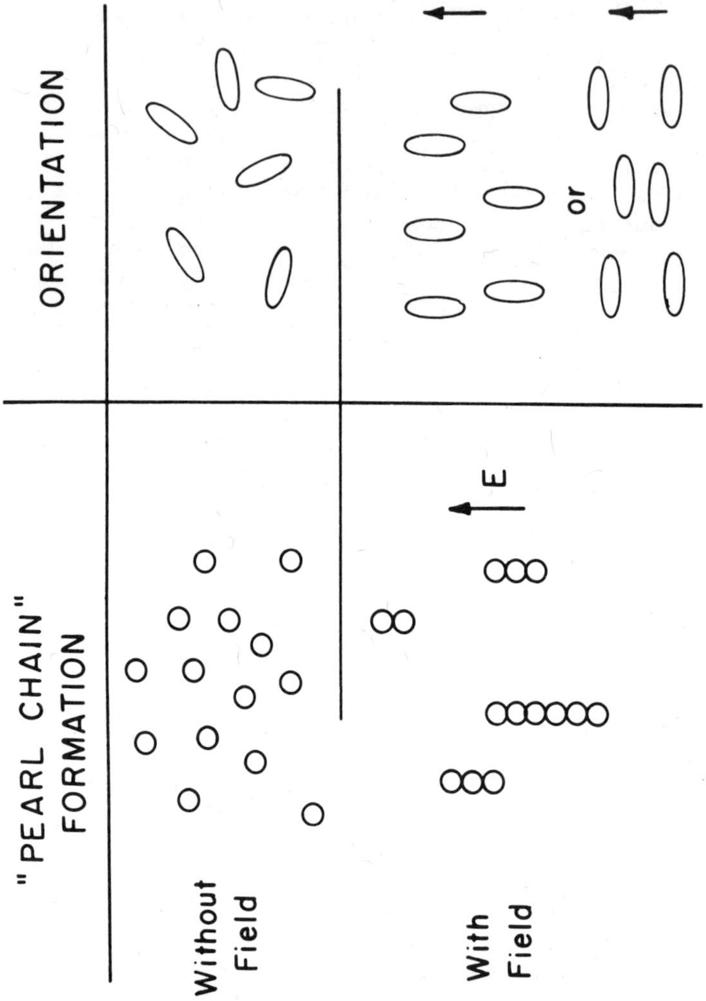

Figure 3

Schematic presentation of some effects of alternating electrical fields on particle and cellular arrangements.

DIELECTRIC PROPERTIES OF BIOLOGICAL MATERIALS

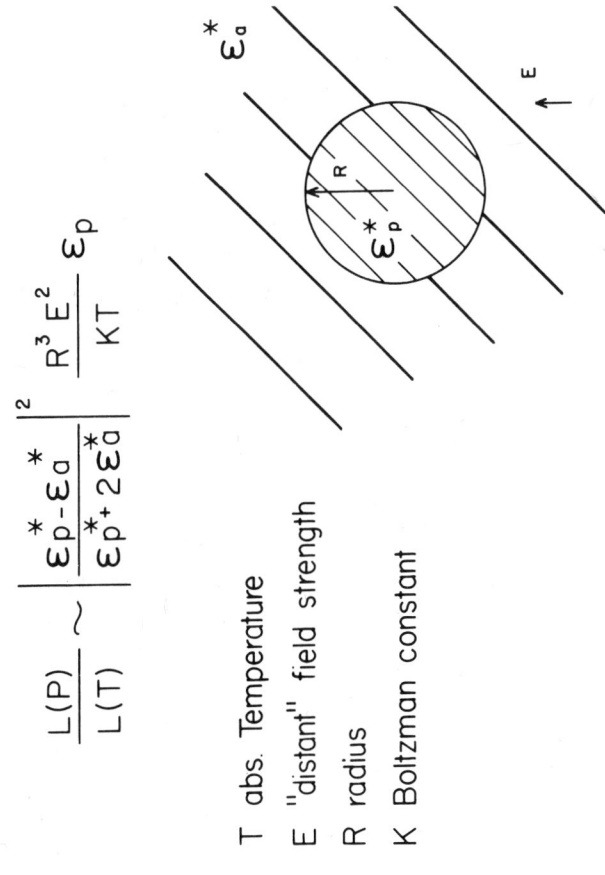

$$\frac{L(P)}{L(T)} \sim \left|\frac{\varepsilon_p^* - \varepsilon_a^*}{\varepsilon_p^* + 2\varepsilon_a^*}\right|^2 \frac{R^3 E^2}{KT} \varepsilon_p$$

T abs. Temperature
E "distant" field strength
R radius
K Boltzman constant

Figure 4. Ratio of electrical potential energy L(P) to thermal energy L(T) of a particle of a complex dielectric constant ε_p^* and radius R in a medium of complex dielectric constant ε_a^* and exposed to a field E.

generated is given by[2]

$$F = 2\pi R^3 R_e \left[\frac{\varepsilon_p - \varepsilon_a}{\varepsilon_p + 2\varepsilon_a} \varepsilon_a^* \right] \nabla |E_a|^2$$

Here, subscripts p and a refer to suspended particle and suspending medium; ε, complex dielectric constants $\varepsilon = \varepsilon' - j\varepsilon''$; ε^* denotes the complex conjugate; E_a external field; and R, radius of particle.

It is apparent from the expressions given that the threshold field strength to evoke such effects as mentioned changes inversely with the square root of the volume of the particle exposed. Indeed, typical cellular dimensions of the order of some μ require field values of several hundred volt/cm, i.e., values which have been reported in the literature for erythrocyte pearl chain formation, E coli orientation, etc. On the other hand large unicellular organisms have been recently subject of investigation in our laboratory and respond dramatically if the field strength is only of the order of 1 V/cm. It is therefore indicated that forces which are caused by alternating fields are not likely to be significant at the molecular and microscopic cellular level unless huge field strength values are assumed, values which are so large that accompanying thermal effects would be overwhelming. However, on a macroscopic level these effects may well be significant. We have pointed out before at several occasions that the phenomena of hearing pulsed microwave fields may well be explained by the forces rythmically applied to the middle ear structures as the field is turned on and off. Since only the strength of the applied field dictates the force generated, the observation by several investigators that the peak power is the important parameter is readily understood. This observation is in no contradiction to our statement that for "steady state" effects such as pearl chain formation or orientation, the effect of a pulsed field can at most equal that of a continuous field of the same average power.

SUMMARY AND ADDITIONAL CONCLUSIONS

The electrical properties of biological systems are well known and understood. This understanding does not indicate so far the existence of nonthermal effects at the molecular or cellular level. However, forces which can be generated by the

---

[2]Sher, L. D., Nature, 220(5168):695-696, Nov. 1968.

application of alternating fields (field evoked forces, dielectrophoresis, electromechanical effects as various authors have termed them) deserve consideration and may well be responsible for the phenomena of "hearing" pulsed field and, hence, some of the behavioral effects reported in the 1 to 10 mW/cm^2 range.*

The foregoing discussions aid us in putting the standards of safe exposure in both the Western and Eastern countries in proper perspective. Table 3 helps in this attempt. The medical profession routinely applies for therapeutic purposes short wave and microwaves to parts of the human body with intensities up to 1000 mW/cm^2 without apparent ill effects. However, in animal experiments ill effects have been noticed with levels of somewhere above 100 mW/cm^2 in the case of total body exposure. The most comprehensive work in this regard has been conducted by Michaelson[4,5]. These reported effects are accompanied by significant temperature elevations. Near and above the American standard of long time exposure of 10 mW/cm^2 behavioral responses have been reported and the sensation of hearing pulsed microwave fields occurs. These effects are not necessarily nonthermal in nature nor representative of a direct effect on the central nervous system. The threshold of microwave-induced thermal perception occurs for man in the same energy range and may well be lower for some animals. The auditory phenomena studied by Frey[6,7] may be a field force or other "macroscopic" effect and need not necessarily indicate a direct influence on the central nervous system. The Department of Health,

---

*Foster and Finch[3] have just recently suggested another macroscopic explanation of the hearing sensation of pulsed fields. They demonstrated convincingly that the minute and instant temperature elevation which results from the absorption of a pulse gives rise to a pressure wave of sufficient magnitude to evoke the hearing sensation.

[3] Foster, K.R., E.D. Finch, Science, 185: 256-258, July, 1974.

[4] Michaelson, S.M., J.W. Howland, W. Diechman, Ind. Med. Surg., 40:18, 1971.

[5] Michaelson, S.M., R.A.E. Thompson, J.W. Howland, Tech. Rept. No. RADC-TR-61-461, Rome Air Dev. Center, Air Force Systems Command, Griffis Air Force Base, New York, Sept., 1967.

[6] Frey, A. H. Aero. Med., 32:1140-1142, 1961.

[7] Frey, A. H. J. Appl. Physiology, 17: 689-692, 1962.

Table 3

Flux	Description	Effect type
1000 mW/cm²	Therapeutic application by medical profession	Thermal effects
100	Threshold of observed hazardous effects	
10	U.S.A. standard Audible effects Behavioral responses Threshold of thermal sensation	Subtle effects
1	HEW oven standard	
0.1		
0.01	Russian standard TV, radio	Nonthermal (?)

Education and Welfare standard for microwave ovens is set at the 1 mW/cm² level and the Russian long time exposure standard 100-fold lower than this level. There is no doubt that significant temperature elevations can happen above the 100 mW/cm² level. More subtle thermal effects[8,9] and the threshold of sensation due to microwave-induced thermal gradients[9-11] occur in the range between 1 and 100 mW/cm². There appears to be no basis for anticipating thermal effects below approximately 1 mW/cm². Hence, any effects evoked by fluxes below 1 mW/cm² appear to be of the "nonthermal" type.

It should be stressed that the subtle responses indicated above have not been demonstrated to establish a health hazard. Quite obviously a small sensation which may be evoked either by a small thermal stimulus or a field induced force can be effectively used to let a properly trained animal respond to

---

[8] Tolgskaya, M.S., Z.V. Gordon, Consultant's Bureau, New York, 1973.

[9] Hendler, E., J.D. Hardy, D. Murgatroyd, in *Temperature, Its Measurement and Control in Science and Industry*, Part 3, Ed. J. D. Hardy, Reinhold, New York, 1963.

[10] Hendler, E., in *Thermal Problems in Aerospace Medicine*, Ed. J. D. Hardy, Sirca, New York, 1968.

[11] Vendrik, A.J.H., J.J. Vos, J. Appl. Physiol., 13:435, 1958.

the field. The demonstrated perception of a weak field by an animal need not indicate that the field is dangerous to man.

At the recent WHO International Symposium on Biologic Effects and Health Hazards of Microwave Radiation, Warsaw (1973), Z.V. Gordon and H. P. Schwan accepted the responsibility to draft a resolution which, after appropriate discussion by the assembly, recommends to WHO the recognition of three ranges of flux levels as far as their biological effectiveness is concerned. It states:

"Microwave intensities may be divided into three categories:

1. High intensities at which distinct thermal effects occur; in many instances, such effects may be hazardous. These intensities range from 10-100 mW/cm^2 upwards (the region of thermal effects).

2. A range of subtle effects from about 1 to 20 mW/cm^2. In this range exist in part weak thermal, but noticeable effects, direct field effects as for example, the phenomena of hearing pulsed fields, and, perhaps, a group of other effects of a microscopic or macroscopic nature whose details are presently unclarified (region of subtle effects).

3. The region at intensities below 1 mW/cm^2. In this region thermal effects are improbable (the region of nonthermal effects).

The border limits between these regions are approximate and may be different for various species of animals and may also depend on a variety of parameters such as frequency, modulation, etc."

Quite obviously the standards of the Western countries are chosen below the range of potential danger, while the standards of the Eastern Countries attempt to include the total range of subtle effects. However, it must be recognized that a subtle transient response to a pulse of energy, if repeated steadily every once in awhile, can be made to correspond to average intensities which are far below even the Eastern standards. It remains to be seen if mankind can afford to set standards so conservatively that man is protected from any subtle effect, dangerous or not.

REFERENCES

It is not possible to adequately mention the many hundred references which establish the body of facts on dielectric data of biosystems summarized in the preceding paper. However, the following books, reviews, and book chapters may serve the interested reader further.

Electrical properties of cells and tissues:

(a) Cole, K.S.: Membrane Capacity. In: Ions, Impulses and Membranes. Ed.: Tobias, C.A. University of California Press. Berkeley and Los Angeles. 1968. p.12.

(b) Schwan, H.P.: Electrical Properties of Tissue and Cell Suspensions. In: Advances in Biological and Medical Physics. Eds.: Lawrence, J.H. and Tobias, C.A. Academic Press. New York. 1957. p. 147.

(c) Schwan, H.P.: J. Cell. and Comp. Physiol. 66, 5, 1965.

(d) Schwan, H.P.: Proc. IRE.' 47, 1841, 1959.

Electrical properties of macromolecules:

(a) Boettcher, C.J.F.: Theory of Electrical Polarization. Elsevier Publishing Co. Amsterdam, 1952.

(b) Hill, N.E., Vaughan, W.E., Price, A.H., Davies, M.: Dielectric Properties and Molecular Behavior. Van Nostrand. London, 1969.

Field evoked forces:

(a) Pohl, H.A.: J. of Biol. Phys. 1, 1, 1973.

(b) Schwan, H.P., Sher, L.D.: J. of Electrochem. Soc. 116, 170, 1969.

(c) Schwan, H.P., Sher, L.D.: Electrostatic Field-induced Forces and their Biological Implications. In: Dielectrophoretic and Electrophoretic Deposition. Eds.: Pohl, H.A. and Pickard, W.F. Electrochemical Society, Inc. New York. 1969. p. 107.

-DISCUSSION-

BEISCHER - Twenty years have passed since Doctors Schwan and Piersol (The absorption of electromagnetic energy in body tissues - A review and critical analysis. Part I Biophysical Aspects. Am. J. Phys. Med. 33: 371-404, 1954. Part II Physiological and Clinical Aspects. Am. J. Phys. Med. 34: 425-448, 1955) published a paper on a similar subject. This paper has been quoted numerous times and it is expected that the present paper will find the same interest.

NYBORG - I might mention, since this conference deals with two subjects, microwaves and ultrasonics, that the phenomenon of pearl-chain formation, which occurs in electromagnetic fields, also appears in ultrasonic fields. There is quite a different physical explanation for it, however.

VOGELMAN - I would like to make one comment concerning peak power. Even if you postulate that is is purely thermal, the expansion and contraction of the membrane in the ear is just adequate to allow you to hear the modulation (which is really the pulse repetition) without any concern about pearl-chain formation or anything of that sort. It could be explained as a pure thermal phenomenon, because it occurs only at repetition rates where the ear would normally hear.

CZERSKI - May I add also, as Dr. Beischer pointed out, this approach presented in this paper is only a part of the story. During the New York Academy of Sciences conference, Adey's group demonstrated that you can get disturbances in bioelectric functions and efflux from the brain if you use 147 MegaHertz amplitude modulated between 8 and 16 Hertz (Ann. N. Y. Sci. February, 1975. See also Bawin et al. Brain Res. 58: 365-384, 1973). If you use exactly the same conditions, but the amplitude modulation is below 8 or above 16 Hertz, the effect is not observed.

It seems that the approach presented in this paper is limited, and the limitation is predicted on the theoretical concept underlying this approach. It seems that by using quantum mechanical models perhaps better understanding and better prediction of expected biological phenomena can be had.

ACOUSTIC PROPERTIES OF BIOLOGICAL MATERIALS

F. Dunn

Bioacoustics Research Laboratory

University of Illinois, Urbana, Illinois 61801

ABSTRACT

The propagation relations for compressional waves in isotropic, elastic media are described and discussed with reference to ultrasonic propagation in biological media. A selected review of experimentally obtained results from tissues, organs and aqueous solutions of biologically pertinent macromolecules is presented.

INTRODUCTION

The acoustic spectrum can be divided into three major regions, as shown in Fig. 1. The audio region, most easily defined as extending approximately from 20 Hz to 20 kHz, is flanked at its low and high frequency extremities by the infrasonic and ultrasonic regions, respectively. The ultrasonic region can, of course, be subdivided further, however, for the purposes of toxicity the region from 1 MHz to 10 MHz may be considered sufficient to include most clinical exposures. As the speed of sound in soft tissues, largely excluding lung, is approximately 1.5×10^5 cm/sec, the wavelengths being dealt with range from millimeters to tenths of millimeters (mineralized tissues exhibit considerably greater speeds of sound). Extension at both ends of this portion of the ultrasonic spectrum may, at times, be essential for full treatment of associated phenomena and application, for example, in treating subharmonic cavitation emissions which requires observations to 500 kHz for the frequency range identified above (6), the full spectra of pulse mode signals which can involve a bandwidth of approximately one decade to describe completely a short pulse (34), and higher frequency modes of operation such as in acoustic microscopy where current technology can perform at 220 MHz (11).

By "ultrasonic properties of biological materials" is generally meant knowledge of the behavior of those measurable acoustic parameters, as functions of state and acoustic variables, which characterize the fate of acoustic signals propagating within the biological environment. The speed with which the signal propagates in the biological medium, the attenuation and/or absorption of the wave energy by the medium, and the features of the medium responsible for reflection and scattering phenomena, generally embodied in the impedance concept, constitute the ultrasonic propagation properties. The present discussion seeks to treat briefly these properties, at several biological levels of structure, to allow for later applications to problems of toxicity and dosimetry. A more detailed account has been prepared by Dunn et al (9).

PROPAGATION RELATIONS

Presentation is made first of pertinent relations describing acoustic phenomena in ideal fluid media. A brief discussion then follows on their alterations necessary to describe events occurring in media which extract energy from the propagating wave process.

The wave equation describing propagation of mechanical disturbances in dissapationless, isotropic, elastic media is

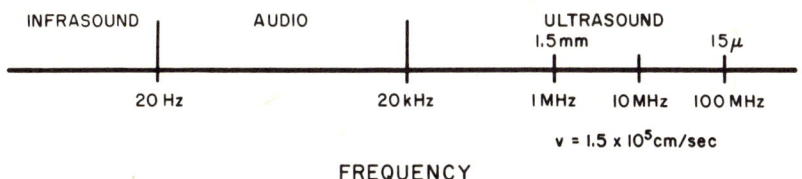

Figure 1. The acoustic spectrum.

$$\frac{\partial^2 \xi}{\partial t^2} = \frac{K + \frac{4}{3}G}{\rho_o} \nabla\nabla \cdot \xi - G \nabla \times \nabla \times \xi \qquad (1)$$

where ξ is the instantaneous displacement of an element of the medium from its equilibrium position, t is time, K is the adiabatic bulk modulus, G is the modulus of shear rigidity, and ρ_o is the mean density of the medium. The specialization for lossless fluid media is obtained by setting G equal to zero since these are characterized by an inability to support an elastic shear strain. The one-dimensional elastic wave equation, as it applies to an ideal, linear, homogeneous, perfectly elastic (dissipationless), liquid medium is, therefore,

$$\frac{\partial^2 \xi}{\partial t^2} = \frac{1}{\rho_o \beta_s} \frac{\partial^2 \xi}{\partial x^2} \qquad (2).$$

where β_s, the adiabatic compressibility, is the reciprocal of the adiabatic elastic bulk modulus. These relations are approximations of the more general hydrodynamic equations valid under the conditions that the particle velocity amplitude, $(\partial \xi/\partial t)_{max}$ is small compared with the speed of sound in the medium and that the adiabatic compressibility is not significantly dependent upon pressure over the range of pressure variations occurring in the sound field. Under such conditions the free-field sound propagation for compressional waves is

$$v = \frac{1}{\sqrt{\rho_0 \beta_s}} \tag{3}$$

The solution to eq. 2 is

$$\xi(x,t) = A e^{j(\omega t - kx)} + B e^{j(\omega t + kx)} \tag{4}$$

which includes waves propagating along both directions of x. k is the wave number ω/v and A and B are the amplitudes of the two wave components. The acoustic pressure at any point in the field is defined as

$$p = P - P_0 \tag{5}$$

where P is the instantaneous pressure at the point and P_0 is the constant equilibrium pressure in the medium. The acoustic pressure and the particle displacement are related as

$$p = -\rho_0 v^2 \frac{\partial \xi}{\partial x} \tag{6}$$

The quantity $\rho_0 v$, the product of the medium density and the sound velocity, is the characteristic impedance of the medium, i.e.,

$$Z_0 = \rho_0 v = \sqrt{\frac{\rho_0}{\beta_s}} \tag{7}$$

For plane traveling waves this quantity is numerically equal to the specific acoustic impedance which is defined as the ratio of the sound pressure p to the particle velocity $\dot{\xi}$ at any point in the field. For other field configurations, e.g., standing waves, the specific acoustic impedance differs numerically from $\rho_0 v$ and is, in general, a function of position. Further, the characteristic acoustic impedance is dependent upon the type of wave propagating since the velocity of, for example, shear waves is different from that of compressional waves.

The intensity I of the acoustic wave is defined as the time average of the rate of transport of energy through unit area normal to the direction of propagation. It is related to the sound pressure amplitude P_a and the particle velocity amplitude $\dot{D}_a$ as follows for plane traveling waves

$$I = \frac{P_a^2}{2Z_0} = \frac{P_a \dot{D}_a}{2} = \frac{Z_0 \dot{D}_a^2}{2} \tag{8}$$

The energy density E_o of the wave motion at a specific position in the field is the sum of the kinetic energy per unit volume of the moving element and the potential energy per unit volume of compression (or expansion) of the element. For plane traveling waves this becomes equal to the ratio of the intensity to the wave velocity, i.e.,

$$E_o = \frac{\rho_o \dot{D}_a^2}{2} = \frac{P_a^2}{2\rho_o v^2} = \frac{I}{v} \tag{9}$$

The assumptions made above leading to the linearization of the hydrodynamical equations can now be expressed symbolically as

$$\dot{D}_a/v \ll 1 \tag{10}$$

and

$$\frac{(\beta_s)_{P_o+P} - (\beta_s)_{P_o-P}}{(\beta_s)_{P_o}} \ll 1 \tag{11}$$

Nonlinear or second-order effects may be of importance for values of $\dot{D}_a/v$ smaller than 0.01, but the linearized equations constitute a good first approximation for calculating values of the physical parameters when this numerical limit is applied. When an ultrasonic wave propagates in a real fluid, wave energy is irreversibly absorbed by the medium, by any of a variety of physical mechanisms. The occurrence of this phenomena modifies the phenomenological description of lossless plane-wave propagation by the introduction of an absorption factor, i.e., eq. 4 becomes, for unidirectional propagation in the positive x-direction

$$\xi(x,t) = Ae^{-\alpha x} e^{j(\omega t - kx)} \tag{12}$$

where α is the amplitude absorption coefficient per unit path length. The Stokes-Kirchhoff classical absorption coefficient is

$$\alpha = \frac{\omega^2}{2\rho_o v^3} \left(\frac{4}{3}\eta_s + \eta_B + \frac{\gamma-1}{C_p}K\right) \tag{13}$$

where ω is the angular frequency, η_s is the shear viscosity coefficient, η_B is the bulk viscosity coefficient, γ is the ratio of specific heats, C_p is the heat capacity at constant pressure, and K is the thermal conductivity. This relation expresses the view that energy transported by the wave process is converted to heat by virtue of its possessing finite viscosity and thermal conductivity. The former results from the fact that, though a fluid cannot support a static shear, it can support a dynamic one in the form of a viscous drag. The latter considers that during passage of an ultrasonic wave, temperatures will be greater in the regions of higher acoustic pressure than in the neighboring regions of lower pressure, and heat will flow in response to such gradients. The thermal conductivity term is, for biological media, very much less than the viscosity terms and can be ignored. Equation 13 indicates that the absorption coefficient should vary with the square of the frequency and that the temperature dependence should follow that of the viscosity. However, liquids in general and biological media in particular, do not often exhibit these dependencies in so simple a fashion and it is necessary to consider that other relaxation processes are involved. Such processes reflect the presence of mechanisms of energy transfer which require time (15). In the following equation the term A represents the classical absorption terms and the second term describes a single relaxation process with f_r being the relaxation frequency:

$$\frac{\alpha}{f^2} = A + \frac{B}{1+(f/f_r)^2} \qquad (14)$$

With regard to the second term, it is seen that at very low frequencies such that the period of the wave is long compared to the time required for energy transfer, it is constant and has the value B, while at very high frequencies, such that the period of the wave is short compared to the time required for energy transfer, energy is not extracted from the wave process and this term vanishes leaving only the classical term. Again, it does not often happen that a single relaxation process is present and in order to describe observed absorption spectra, it is necessary to consider a number of such processes occurring and to sum on all of them (15), and the following equation symbolizes this:

$$\frac{\alpha}{f^2} = A + \sum_i \frac{B_i}{1+(f/f_i)^2} \qquad (15)$$

VELOCITY AND ABSORPTION OF ULTRASOUND IN BIOLOGICAL MEDIA

Goldman and Hueter (12) have compiled a detailed list of ultrasonic velocity and absorption data in animal tissues. Therein it is seen that the velocity in soft tissues, excluding lung, is very nearly that of dilute salt solutions and varies only slightly among the various tissues, with the exception being that fatty tissue has a velocity about 10% less than that for non-fatty tissues.

ACOUSTIC PROPERTIES OF BIOLOGICAL MATERIALS 27

The reflection properties, which are of greatest importance in clinical diagnostic procedures, can be discussed in terms of elementary considerations for unbounded lossless, homogeneous and isotropic media for which the magnitude of the reflection coefficient R is, in terms of the characteristic impedances of the two contiguous media (See eq. 7),

$$R = \frac{Z_1 - Z_2}{Z_1 + Z_2} = \frac{(\rho_1/\beta_{s1})^{1/2} - (\rho_2/\beta_{s2})^{1/2}}{(\rho_1/\beta_{s1})^{1/2} + (\rho_2/\beta_{s2})^{1/2}} \qquad (16)$$

As the inertial property exhibits minimal variation among tissues, viz., variations less than about 1% in density, $\rho_1 \approx \rho_2$ and eq. 16 can be written as

$$R \approx \frac{(1/\beta_{s1})^{1/2} - (1/\beta_{s2})^{1/2}}{(1/\beta_{s1})^{1/2} + (1/\beta_{s2})^{1/2}} \qquad (17)$$

It is apparent that the reflection characteristics of these biological media are dependent, for the most part, upon the elastic properties of their components (10). The speed of sound in lung tissue increases nearly linearily with frequency, in the low megahertz frequency region, but has a value considerably less than other soft tissues (7) and depends upon inflation (2).

Hueter has measured the speed of sound in specimens of human skull and found it to be approximately twice that of the soft tissues. As the density of bone is considerably greater than that of soft tissues, its impedance thus becomes nearly four times greater (see eq. 7).

The following is a selected review of the experimentally observed ultrasonic absorption characteristics of biological media. Figure 2 shows the measured absorption of several biological materials vs frequency on a log-log plot wherein the slope of the curve is the exponent on frequency upon which the absorption coefficient depends. This figure shows materials of increasing biological complexity and illustrates correspondingly more complicated absorption behaviour (17). For example, the 10-molar urea solution exhibits a slope of 2 indicative of classical viscous absorption. Homogenized milk, a suspension of fat molecules and casein complexes, exhibits a slope of nearly unity, a behaviour which cannot be explained in terms of simple classical absorption. Egg albumin, brain tissue, liver and striated muscle exhibit slopes between 1 and 2 in the neighborhood of 1 MHz and approach slope 2 at higher frequencies. Figure 3 again shows the complicated dependence of the ultrasonic absorption coefficient on frequency (12).

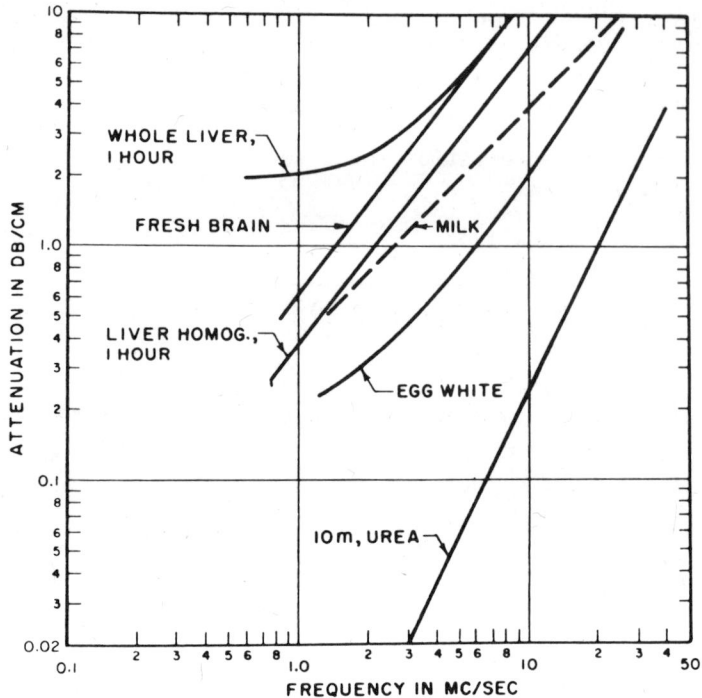

Figure 2. Acoustic absorption in biological media of increasing complexity vs frequency (after Hueter, 1958).

Here, the absorption per cycle is plotted against the log of frequency, for which classical theory would predict a linear dependence with positive slope. The broad shaded regions of Fig. 3 are partially accounted for by the inclusion of all available data (at time of preparation), i.e., the figure includes measurements by numerous investigators employing different experimental techniques under a variety of physical and biological conditions. Nevertheless, it is possible to discern several relatively simple relationships, e.g., the absorption per cycle is generally constant over the frequency range considered. Further, this quantity increases slightly from 1 to 10 MHz for fat, and striated muscle and liver appear to exhibit minima in the neighborhood of 2 MHz. The important dependence of the absorption coefficient upon temperature has been established only recently (8a,b). Figure 4, which is the most complete set of data to date, includes observations on mammalian central nervous tissue of neonatal mice (essential poikilotherms) and adult cats (homoeotherms).

Several tissues are known to exhibit unique properties. Fresh skull bone in the frequency range 0.6 to 3.5 MHz (25 to 35°C) exhibits a quadratic dependence of the absorption coefficient upon

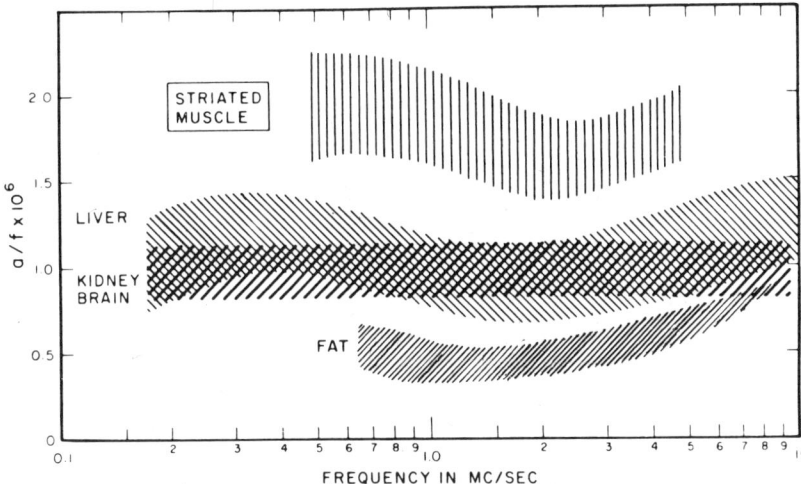

Figure 3. Acoustic absorption per wavelength (in dB/cm) for several mammalian tissues vs frequency (after Goldman and Hueter, 1956).

frequency with a transition to a lower power dependence at higher frequencies (16). An average value for the acoustic amplitude absorption coefficient per unit path length in skull bone, in the neighborhood of 1 MHz, is of the order of 1 cm^{-1}, i.e., approximately an order of magnitude greater than that of soft tissue at the same temperature. The acoustic amplitude absorption coefficient in freshly excised dog lung at a fraction of residual inflation has recently been shown to increase with frequency from 4.5 cm^{-1} at 1 MHz to 10.3 cm^{-1} at 5 MHz (7). This frequency dependence has also been found for formalin-fixed preparations which has provided some information on the dependence of absorption on inflation(2).

In an attempt to determine the details of ultrasonic absorption in tissues, investigators have studied blood and solutions of its components. Blood provides a material for study which possesses some cellular features of more highly organized tissues, but is also sufficiently homogenous to allow a relatively high degree of accuracy in the measurements. Carstensen et al discovered that the acoustic properties of blood are determined largely by the protein content, and that the absorption coefficient is directly proportional to the protein concentration whether in solution or contained within the cell (5). Several years later Carstensen and Schwan showed that a small fraction of the absorption was due to the cellular organization of the blood, i.e., it was not of molecular origin (4).

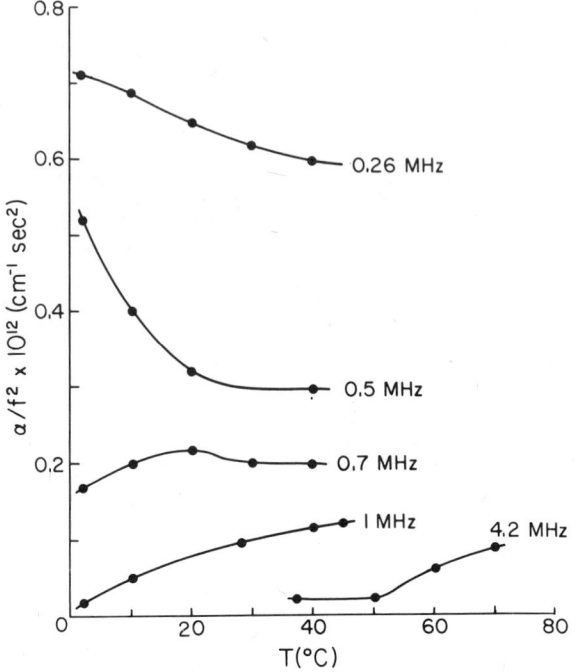

Figure 4. Frequency and temperature dependence of ultrasonic absorption in tissue (after Dunn and Brady, 1974).

A second study, also performed in the 1950's but only recently published, showed that about two-thirds of the ultrasonic absorption in liver tissue occurs at the macromolecular level with the remaining one-third due to the structural features of tissue (31). Thus, in order to obtain detailed information of the ultrasonic absorption properties of tissue, the absorption processes which occur in aqueous solutions of the molecular constituents of biological cells were examined.

When biological macromolecules, such as the globular protein hemoglobin, are in aqueous solution, a certain amount of the solvent becomes an inherent part of the molecule since the polymer possesses ionic and polar groups which associate with water molecules. In addition, proteins contain a number of nonpolar side chains such that within the vicinity of the macromolecule, some water structuring occurs. Thus, it is possible that the structure of liquid water, the hydration layer, increases in the neighborhood of the biological macromolecule. It is considered, therefore, that as an acoustic wave propagates through an aqueous solution of biopolymers, it perturbs the hydration layer manifesting absorption of energy by a structural relaxation process. The roles of molecular conformation of the biopolymer and pH of the solution have also been considered as the origin of the observed ultrasonic absorption. Figure 5 shows the excess frequency-free absorption per unit concentration for several biomacromolecules and supports the view that structuring contributes to the ultrasonic absorption spectra. Both dextran (14), a carbohydrate, and polyethylene glycol (22), a synthetic polymer, assume random coil configurations in aqueous solution and exhibit absorption magnitudes similar to that of gelatin. Hemoglobin has a quaternary structure, bovine serum albumin and ovalbumin have tertiary structure, polyglutamic acid (3), a synthetic polyamino acid, has secondary structure, and the double helix DNA has a rigid rod conformation. However, hemoglobin in 5 molar aqueous guanidine hydrochloride solution exists as a random coil (20) and yet exhibits ultrasonic absorption spectra similar to that of hemoglobin in aqueous solutions (27). The importance of molecular spatial arrangement is thus an unsettled question, though the polypeptide structure appears to be of considerable significance.

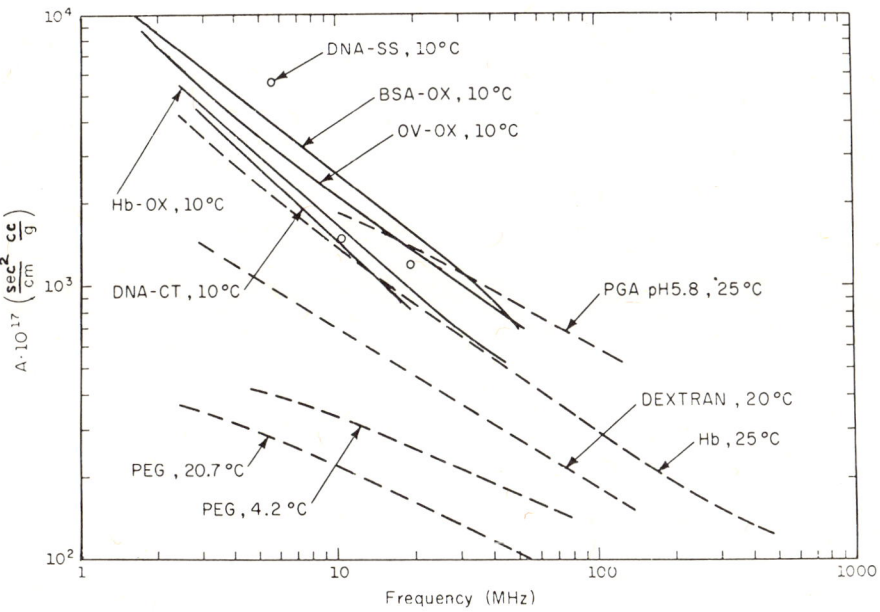

Figure 5. Excess frequency-free ultrasonic absorption per unit concentration (after O'Brien and Dunn, 1971).

Within Figure 5, bovine serum albumin and hemoglobin have a molecular weight of 68,000 while that of ovalbumin is 46,000. β-lactoglobin (25) with a molecular weight of 35,000, falls in the range determined by hemoglobin, bovine serum albumin and ovalbumin while lysozyme (25) with a molecular weight of 14,600 appears between hemoglobin at 25°C and dextran. Two random coil polymers, dextran and polyethylene glycol, have been studied as a function of molecular weight. For dextran solutions, the frequency-free absorption per unit concentration increases with increasing molecular weight to a molecular weight of about 10,000, which corresponds to approximately 100 monomer units, beyond which it is independent of molecular weight (13). Aqueous polyethylene glycol solutions show similar absorption behavior in that beyond a molecular weight of about 4500, which also corresponds to a chain length of about 100 monomer units, the absorption is independent of molecular weight (22). Thus, ultrasonic absorption depends to some degree upon molecular weight. Possibly beyond about 100 monomer units the macromolecule assumes random coil characteristics.

Two synthetic polyamino acids have been examined extensively under varying environmental conditions with respect to their ultrasonic absorption. The primary mechanism proposed to explain the excess ultrasonic absorption in poly-L-glutamic acid solutions by Burke et al. (3) is solvent-solute interaction whereas Schwarz (33) attributes it to the helix-coil transition. The examination by Wada et al. (35) of poly-L-glutamic acid revealed that, at 50 kHz, the absorption mechanism is the helix-coil transition while at 3 MHz the absorption is attributed to side chain dissociation. Parker et al. (30) concluded that the observed ultrasonic absorption behavior in aqueous poly-L-lysine solution can be attributed to the helix-coil transition.

The pH of solutions has been varied to study the ultrasonic absorption processes and a number of aqueous solutions of amino acids have been investigated, viz., serine and threonine (36), glycine (1,18), glutamic acid, aspartic acid and alanine (1) and arginine and lysine (19). Absorption maxima have been observed within the pH ranges 2 to 4 and 11 to 13, with such peaking being described quantitatively by assuming that the proton transfer reaction dominates the absorption process. When the pH of the aqueous biopolymer solutions is varied alteration of the molecular charge distribution occurs thus affecting the conformation of the biopolymer. Here, buried amino acid side chains of globular proteins become exposed to the environment of the solvent and thereby affect the acoustical properties of the solution. The excess frequency-free absorption per unit concentration for bovine serum albumin versus pH (21) shows maxima around pH 2 to 4 and a steep increase in absorption beyond pH 10. Similarly, hemoglobin exhibits, in addition to the peak in the acid region, one in the alkaline region around pH 11 to 13 (28). These peaks have been correlated with the proton transfer reaction in which it is postulated that the pressure variations of the propagating acoustic wave perturb the proton from the solvent (water) to the solute, an amino acid side chain, and vice versa. The pH values of the titration peaks have been correlated with the appropriate conditions under which the proton can be transferred at both the alkaline and acidic pH regions. Ovalbumin shows similar absorption behavior as a function of pH (26) in addition to β-lactoglobulin and lysozyme (25). Previously the absorption peaks in the bovine serum albumin spectrogram were discussed in terms of conformal changes (21) although it now appears that a more acceptable cause is the proton-transfer reaction. However, the role of conformation cannot be completely discounted as the evidence for each mechanism is less convincing than one would like and does not exclude the other from being present to some degree. Other mechanisms such as solvation equilibria (3) and keto-enol equilibria (25) may also affect the absorption behavior.

Deoxyribonucleic acid (DNA), a rod-like molecule possessing a double helix structure, also exhibits absorption maxima in the pH range 2 to 3 (29) and 11 to 13 (24). Acidic and basic denaturation of DNA result in collapse of the secondary structure and the splitting of the two hydrogen-bonded strands yielding two single-stranded randomly coiled nucleotide chains, and this occurs in the same pH ranges as do the absorption peaks. In this denaturation process, the base-base hydrogen bonds are replaced by base-solvent bonds. It has been suggested that this transfer of hydrogen bonds from base-base to base-solvent is responsible for the absorption peaks in both the acidic and alkaline regions (29).

As the studies of solvent-solute interactions do not appear capable of yielding explanations for the absorption behavior, at the molecular level, at physiological pH, Kremkau and Carstensen (23) have suggested that solute-solute interactions may be a more profitable area for inquiry.

The above discussion was not intended to be a comprehensive treatment of knowledge of this field, but rather a selected review of the pertinent literature. The comprehensive treatment by Dunn et al. (9) and the definitive Workshop Proceedings of the Seattle Conference (32) should be consulted for detailed information. It is apparent that a great deal must be accomplished before a complete understanding of ultrasonic velocity and absorption by biological media is obtained. From the previous work it is possible to determine types of dependencies of the absorption as a function of some of the important field and state variables and it is possible to classify tissues into rather general categories as regards these dependencies. However, the present state of experimental data does not allow for the desired understanding of the mechanisms for any tissue structure or component relative to even a single physical variable.

REFERENCES

1. APPLEGATE, K. L., SLUTSKY, L. J., and PARKER, R. C.: Kinetics of proton-transfer reactions of amino acid and simple polypeptides. J. Am. Chem. Soc. 90 (1968) 6909-6913.

2. BAULD, T. J.: Ultrasonic study of excised canine lung tissue. Ph.D. thesis, University of Pennsylvania (1973).

3. BURKE, J. J., HAMMES, G. G., and LEWIS, T. B.: Ultrasonic attenuation measurements in poly-L-glutamic acid solutions. J. Phys. Chem. 42 (1965) 3520-3525.

4. CARSTENSEN, E. L., and SCHWAN, H. P.: Absorption of sound arising from the presence of intact cells in blood. J. Acoust. Soc. Am. 31 (1959) 305-311.

5. CARSTENSEN, E. L., LI, K., and SCHWAN, H. R.: Determination of the acoustic properties of blood and its components. J. Acoust. Soc. Am. 25 (1953) 286-289.

6. COAKLEY, W. T.: Acoustical detection of single cavitation events in a focused field in water. J. Acoust. Soc. Am. 49 (1971) 792-801.

7. DUNN, F.: Attenuation and speed of ultrasound in lung. To be published.

8a. DUNN, F., and BRADY, J. K.: Pogloshchenie Ul'trazvyeka V Biologicheskikh Sredakh. Biofizika 18 (1973) 1063-1066.

8b. DUNN, F., and BRADY, J. K.: Temperature and frequency dependence of ultrasonic absorption in tissue. Proc. 8th Int'l. Congress on Acoustics (1974).

9. DUNN, F., EDMONDS, P. O., and FRY, W. J.: Absorption and dispersion of ultrasound in biological media, Ch. 3, Biological Engineering (Schwan, H. P. Ed.). McGraw-Hill. New York (1969). pp 205-332.

10. FIELDS, S., and DUNN, F.: Correlation of echographic visualizability of tissue with biological composition and physiological state. J. Acoust. Soc. Am. 54 (1973) 809-812.

11. FIELDS, S. I., KESSLER, L. W., and DUNN, F.: Microstructure in mammalian kidney as seen with the acoustic microscope. J. Acoust. Soc. Am. 55, Supplement (1974) 85.

12. GOLDMAN, D. E., and HUETER, T. F.: Tabular data of the velocity and absorption of high-frequency sound in mammalian tissues. J. Acoust. Soc. Am. 28 (1956) 35-37

13. HAWLEY, S. A., and DUNN, F.: Ultrasonic absorption in aqueous solutions of dextran. J. Chem. Phys. 50 (1969) 3523-3526.

14. HAWLEY S. A., KESSLER, L. W. and DUNN, F.: Ultrasonic absorption in aqueous solutions of high-molecular-weight polysaccharides. J. Acoust. Soc. Am. 38 (1965) 521-523.

15. HERZFELD, K. F., and LITOVITZ, T. A.: Absorption and Dispersion of Ultrasonic Waves. Academic Press. New York (1959).

16. HUETER, T. F.: Messung der Ultraschall absorption im menschlichen Schadelknocken und ihre Abhangegkeit von der Frequenz. Naturwissenschaften 39 (1952) 21-22.

17. HUETER, T. F.: WADC Tech. Rept. 57-706 (1958).

18. HUSSEY, M., and EDMONDS, P. D.: Ultrasonic examination of proton-transfer reactions in aqueous solutions of glycine. J. Acoust. Soc. Am. 49 (1971a) 1309-1316.

19. HUSSEY, M., and EDMONDS, P. D.: Ultrasonic examination of proton-transfer reactions at the α-amino and side-chain groups of arginine and lysine in aqueous solution. J. Acoust. Soc. Am. 49 (1971b) 1907-1908.

20. KAWAHARA, K., KIRSCHER, A. G., and TANFORD, C.: Dissociation of human CO-hemoglobin by urea, guanidine hydrochloride and other reagents. Biochemistry 4 (1965) 1203-1213.

21. KESSLER, L. W., and DUNN, F.: Ultrasonic investigation of the conformal changes of bovine serum albumin in aqueous solution. J. Phys. Chem. 73 (1969) 4256-4263.

22. KESSLER, L. W., O'BRIEN, W. D., Jr., and DUNN, F.: Ultrasonic absorption in aqueous solutions of polyethylene glycol. J. Phys. Chem. 74 (1970) 4096-4102.

23. KREMKAU, F. W., and CARSTENSEN, E. L.: Molecular interaction in sound absorption, Interaction of Ultrasound and Biological Tissues (J. M. Reid and M. R. Sikov, Eds.) DHEW, Rockville (1972), pp. 37-42.

24. LANG, J., and CERF, R.: Absorption ultrasore dans des solutions d'acide desoxyribonucleique: étude de la denaturation alcaline. J. Chem. Phys. 66 (1969) 81-87.

25. LANG, J., TONDRE, G., and ZANA, R.: Effects of urea and other organic substances on the ultrasonic absorption of protein solutions. J. Phys. Chem. 75 (1971) 374-379.

26. O'BRIEN, W. D., Jr.: The absorption of ultrasound in aqueous solutions of biological polymers. Ph.D. thesis, University of Illinois (1970).

27. O'BRIEN, W. D. Jr., and DUNN, F.: Ultrasonic examination of hemoglobin dissociation process in aqueous solutions of guanidine hydrochloride. J. Acoust. Soc. Am. 50 (1971) 1213-1215.

28. O'BRIEN, W. D., Jr., and DUNN, F.: Ultrasonic absorption mechanisms in aqueous solutions of bovine hemoglobin. J. Phys. Chem. 76 (1972) 528-533.

29. O'BRIEN, W. D., Jr., CHRISTMAN, C. L., and DUNN, F.: Ultrasonic investigation of aqueous solutions of deoxyribose nucleic acid. J. Acoust. Soc. Am. 52 (1972) 1251-1255.

30. PARKER, R. C., SLUTSKY, L. J., and APPLEGATE, K. R.: Ultrasonic absorption and the kinetics of conformational change in poly-L-lysine. J. Phys. Chem. 72 (1968) 3177-3186.

31. PAULY, H., and SCHWAN, H. P.: Mechanism of absorption of ultrasound in liver tissue. J. Acoust. Soc. Am. 50 (1971) 692-699.

32. REID, J. M., and SIKOV, M. R., Eds.: *Interaction of Ultrasound and Biological Tissues*. DHEW, Rockville (1972).

33. SCHWARZ, G.: On the kinetics of the helix-coil transition of polypeptides in solution. J. Mol. Biol. 11 (1964) 64-77.

34. SOKOLLU, A.: private communication.

35. WADA, Y., SASABE, H., and TOMONO, M.: Viscoelastic relaxation in solutions of poly-(glutamic acid) and gelatin at ultrasonic frequencies. Biopolymers 5 (1967) 887-897.

36. WHITE, R. D., SLUTSKY, L. J., and PATTISON, S.: Kinetics of the proton-transfer reactions of serine and threonine. J. Phys. Chem. 75 (1971) 161-163.

ACOUSTIC PROPERTIES OF BIOLOGICAL MATERIALS

-DISCUSSION-

CARSTENSEN - In Schwan's paper (this conference) he emphasized the fact that these biological media are linear from the dielectric point-of-view. Can you comment on that from the acoustic point-of-view?

DUNN - From the acoustic point-of-view, non-linearities can arise from several aspects. The equation of state is non linear, that is, the medium, itself, is non linear. A wave propagating in such a medium, energy, can be transferred from an initially monochromatic wave to harmonics. This generally requires a rather appreciable propagation distance, i.e., the distance the wave propagates must be sufficient for an appreciable transfer of energy from the monochromatic wave to its harmonics. Generally such conditions do not exist experimentally. However, they should not be ruled out completely as the hydrodynamic equations are non linear, and one would expect that non linear effects could be quite prevalent under appropriate conditions. I think, though, that we rarely see them because of the way experiments and measurements are carried out. It should, however, be possible to produce them. There is a fundamental difference in that Maxwell's equations are linear, and the hydrodynamic equations are non linear.

LEHMANN - Dr. Dunn, could you comment on the contribution of shear waves to the absorption in a structured medium.

DUNN - I don't believe there is much that is known. The shear wave should be attenuated within about one skin depth, such that if the propagating wave has any shear component, then the absorption should be very drastic. But, if you are propagating the wave through a liquid medium that does not propagate the shear wave, and this is the usual laboratory situation, the opportunity to introduce the shear wave at a boundary, can only occur from some form of mode conversion.

LEHMANN - At our laboratory we theoretically studied the development of shear wave, depending on the angle of incidence at interfaces of differing acoustic impedance. Shear waves contribute significantly to the heating at the interface because they are very quickly attenuated.

DUNN - Another thing that has to be remarked is that researchers who make absorption measurements using different schemes and different fundamental relations for making absorption measurements, seem to arrive at the same values for attenuation and absorption. Thus, if there is very much mode conversion, say, in one of these schemes that is not present in another scheme,

DUNN - one would expect to get a different value for the absorption coefficient, and this is not observed.

MILLER - Quite some time ago El'Piner (Ultrasound; Physical, Chemical and Biological Effects. Consultants Bureau, New York, 1964, page 335) found when he measured striated muscle he obtained quite a different absorption if he exposed it along the direction of the fibers versus transducer fiber direction.

DUNN - Yes, but that difference wouldn't have to be entirely associated with shear phenomena. That could be simply associated with difference in structure. In propagating along the bands as against the bands, one is observing a completely different structure, i.e., the medium is not isotropic.

TAYLOR - I should like to quote some clinical evidence to support Dr. Dunn's concept that echo amplitude is dependent upon the difference in the bulk moduli of the media forming the interfaces. It means virtually that what we are looking at in all our diagnostic echo techniques are the structures with high bulk moduli, such as collagen and elastin, which form the fibrous skeleton. Now, this is tremendously important and has been extremely useful to me as a clinician because if one finds an increased amplitude of echoes emanating from a liver or spleen, then you know that this is due to an increased fibrous content in it. Thus, increased echo amplitudes will be found in chronic inflamatory spleens rather than neoplastic ones. High level echoes emanating from liver will be due to a chronic inflamatory process or cirrhosis but cannot be due to a neoplastic process. In the same way, if the normal fibrous skeleton of the liver is replaced by a tumour this will be seen as a definite defect in the liver structure. In practice this is exactly what we see in clinical practice and adds strong support to the hypothesis proposed by Dr. Dunn.

JOHNSON - What is the state of the art more or less in scattering measurements?

DUNN - I don't want to cast this off, but I believe Dr. Hill, at the present time, is conducting a study. Perhaps he should remark.

HILL - I can say that the state of knowledge on scattering by tissues is pretty abysmal. It is very surprising that it should be so, because the whole basis of ultrasonic diagnosis is basically a scattering phenomenon. But, we don't really know why it occurs, how it occurs, and how the information that we do get out of the patient is related to structure in the formation of echoes.

HILL - I think the other point that perhaps ought to be made in connection with this, in the interest of this meeting, is that the data you (Dr. Dunn) were presenting on absorption in tissue, in fact, are usually measured as attenuation in tissue. The ultrasonic attenuation coefficient, strictly speaking, is made up of true absorption plus scattering, plus probably some diffraction correction, unless this has been properly looked after.

DUNN - That's partially true. In the two slides by Hueter and by Hueter and Goldman, that's probably very true. But, for the figure that showed the temperature variation, it is not true. The method used there, both for the mice by us and on the adult cat by Dr. Lele, is one that gives true absorption.

HILL - I didn't want to put holes in your data, but basically I think it is an important point that a lot of the data reported are based on measurements made fifteen to twenty years ago. Attenuation measurements are difficult to do, and I think we ought to do quite a bit more to really look to see what is going on and to separate the different mechanisms that contribute to overall attenuation.

BEISCHER - Yesterday evening in looking at the presentation by Waag and Gramiak ("Development in Cardiac Imaging") I had the impression that there was little gradation in the pictures. Can this be improved?

VOGELMAN - Has anybody taken muscle out of a living organism and exposed it to ultrasound, and varied the tension and seen how much that would modify the absorption characteristic? I have a feeling the reaction of the living organism affects what you are measuring in absorption.

DUNN - I don't know that anyone has performed the experiments that you have suggested. I suspect that no one has. The kinds of things that have been done have been to observe an evoked potential or propagating electrical response through a muscle while irradiating with sound and observing the changes in that electrical signal (Welkowitz and Fry, J. Cell and Comp. Physiol. $\underline{48}$: 435, 1956).

VOGELMAN - In the loudspeaker design business the density of the medium that exists behind the speaker makes a marked difference in how much is absorbed and how much is reflected. So, I can see where in a living organism if you tense the muscle you could get a different response.

LEHMANN - The answer is: no one has done it. I think what you propose is very logical.

LELE - Well, we have done it in a way, trying to measure absorption in a beating heart, in connection with the myocardial infarction program. Firstly, it seems that the absorption characteristics do not change even after death at normal physiological conditions, provided one maintains body temperature up to 3 to 4 hours at normal mean temperatures of 24° or 37°. Secondly, in a beating heart, there is no difference observable whether the heart is in diastole or in systole. Most of these experiments are done by arresting the heart with electrical shock or potassium. We have been unable to measure any differences with the contracture of the muscle tissue itself.

HILL - This result of Dr. Lele's might not be very surprising because, as far as I know, the effect you suggest would only involve the scattering component of the attenuation coefficient. In the frequency range at which one is working in ultrasonics, in other words 1-10 MHz, the scattering contribution to attenuation is around 10 percent of the total. It may be rather higher in muscle, but usually it is in that range. So that, unless you had very sensitive experimental techniques, you might find it difficult to see a small change in a 10 percent contribution to a total attenuation.

SYNTHESIS OF FREQUENCY RESPONSE OF ELECTRIC FIELD PROBES

Tadeusz M. Babij

Technical University of Wrocław
Wrocław, Poland

ABSTRACT

It is known that the behavior of electromagnetic field probes is limited by the averaging of the measured fields over the finite size of the probe. This is the major restriction on the size of dipoles used in a very small array as an electric field probe. The two major requirements of an EM-field strength meter are:

1. The type of measured quantity
2. Accuracy of measurements

The first case should be determined by biologists and medical experts carrying out studies in the biological effects of hazardous EM fields. The exact specifications of this case have not yet been determined. This has led to the necessity of constructing a general purpose electromagnetic flux (mW/cm^2) instrument. The second is an engineering problem; the accuracy should be maintained during measurement of any field irrespective of its configuration and in the presence of primary and secondary sources of radiation.

Introduction

Measurements of electromagnetic fields are quite often made in the neighborhood of the Fresnel zone. The problem of measuring in the Fresnel zone is quite complicated (1, 4, 7, 9). Individuals are often exposed to fields in situations where a high level of multipath interference exists. Moreover, it is sometimes necessary to take the measurements is the presence of several signals of different frequencies. In addition, electromagnetic field intensity meters designed for measurements in plane-wave fields cannot be used for measurements in the Fresnel zone (1, 7).

The literature provides examples of experimental results on the basis of which conditions to be fulfilled by the antennas used for measurement are formulated (1, 7, 9). The sensor for the electric field component should have:

a. isotropic antenna pattern

b. small dimensions of the antenna, usually $h \leq 0.1\lambda$; h = the physical half length of the antenna

c. high resistance leads transparent to the RF field which convey the detected signal to readout electronics.

Fulfillment of all these requirements at the same time leads to the construction of the ideal probe. It seems that today's probes with the sensor consisting of three orthogonal dipoles and transparent line fulfill the above-mentioned requirements (2, 6).

The type and range of the quantities measured are determined by biologists. These parameters have not been agreed upon unequivocally because of disparity in understanding as to what criteria should be applied to the estimation of the electromagnetic field's biological noxiousness (3, 5, 8).

In the remainder of this paper, analysis of the radiation characteristics of an antenna consisting of the three orthogonal dipoles is presented. The analysis is performed for different kinds of detectors (square law and linear characteristics). It shows analytically why short dipoles should be used for electric field measurements. The analytical conclusions are in agreement with previously obtained experimental results (2, 7).

Field Pattern of Three Orthogonal Dipoles

A sensor composed of three orthogonal thin short dipoles ($\beta_o h \ll 1$) is shown in Figure 1. In the figure the dipoles are marked with a thick line. In the case of a dipole, the open circuit voltage U induced in the antenna is related to the magnitude of the incident field E by a factor termed the effective height h of the antenna. For the case $\beta_o h < 1$, the open circuit voltage on the input terminals of each dipole equals

$$U = h\vec{E} \cdot \vec{s} \tag{1}$$

where $\vec{s}$ - unitary vector orienting dipole /Fig.1a/

For three orthogonal dipoles /Fig.1b/ we have

$$\vec{s}_1 [1, 0, 0], \vec{s}_2 [0, 1, 0], \vec{s}_3 [0, 0, 1] \tag{2}$$

$$E = \{E \cos\phi \sin\theta, E \sin\phi \sin\theta, E \cos\theta\} \tag{3}$$

$$E = |\vec{E}|$$

SYNTHESIS OF FREQUENCY RESPONSE

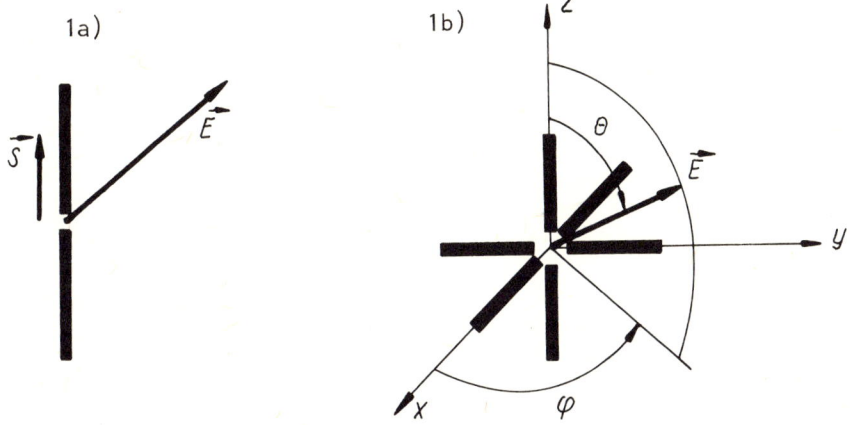

Figure 1a, b. Coordinate system for three orthogonal dipoles

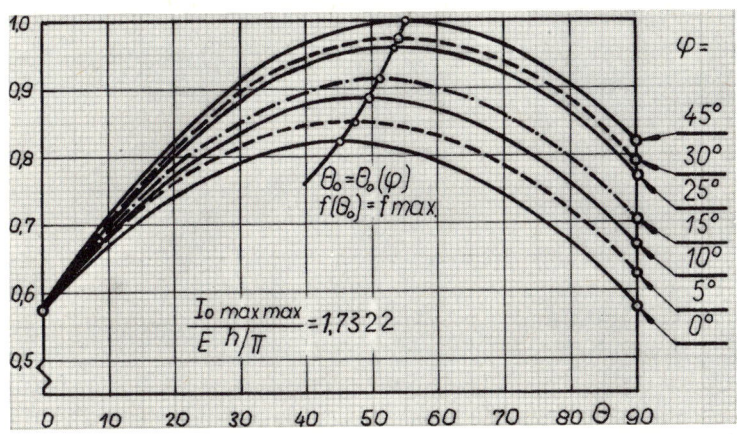

Figure 2. Runs $\dfrac{A_o}{A_{omax}} = f(\theta)$

Therefore,

$$U_1 = hE\cos\phi\sin\theta \qquad (4a)$$

$$U_2 = hE\sin\phi\sin\theta \qquad (4b)$$

$$U_3 = hE\cos\theta \qquad (4c)$$

Applying: a/ diode detectors with linear characteristics connected between the arms of the dipoles,
b/ addition of currents,

we get

$$I_o = \sum_{k=1}^{3} I_{ok} = \frac{Eh}{\pi R}\{(\sin\phi + \cos\phi)\sin\theta + \cos\theta\} \qquad (5)$$

where

$$I_{o1} = \frac{1}{\pi}\frac{Eh}{R}(\cos\phi\sin\theta)$$

$$I_{o2} = \frac{1}{\pi}\frac{Eh}{R}(\sin\phi\sin\theta)$$

$$I_{o3} = \frac{1}{\pi}\frac{Eh}{R}(\cos\theta)$$

R - a load resistance terminating the dipole arms

It is sufficient to examine the function

$$I_o = I_o\{Eh, \phi, \theta\}$$

for the range $0 \le \phi \le \frac{\pi}{4}$; $0 \le \theta \le \frac{\pi}{2}$

Then $\quad A_o = \frac{R\pi I_o}{Eh} = \{(\sin\phi + \cos\phi)\sin\theta + \cos\theta\} \qquad (6)$

The plot of $\frac{A_o}{A_{omax}} = f(\theta,\phi)$ is shown in Figure 2.

The sensor with linear diode detectors reacts to the maximal quantities of the rotating resultant E vector's projections on the axis of the set and its indication depends on the sensor orientation to the field.

Applying:
a/ diode detectors with square law characteristics, or thermocouples connected between the arms of the dipoles,
b/ addition of currents

we get

$$\frac{A_o}{A_{omax}} = a_2 \frac{E^2 h^2 R^2}{\pi^2} \qquad (7)$$

where, a_2 - proportionality factor.

SYNTHESIS OF FREQUENCY RESPONSE

The sensor with diode detectors of square law characteristics indicates the quantity equal to the maximal value of the square resultant modulus of the field and its indication does not depend on the sensor orientation to the field.

Irregularities of the Radiation Pattern of Three Orthogonal Dipoles Versus $\beta_o h$

It can be assumed that the element of the sensor is loaded with a detector of square law characteristics, e.g. thermocouple. The flow of current in each of these thermocouples is proportional to EMF squared induced in each dipole of the sensor. Furthermore, the EMF is proportional to the function $f(\alpha)$ of the dipole orientation given by the equation

$$f(\alpha) = \frac{\cos(2\pi\xi)\sin\alpha - \cos(2\pi\xi)}{\cos\alpha} \tag{8}$$

where: α - angle between the direction of the E vector and the direction of the dipole, $\xi - h/\lambda$. The vector E can have an arbitrary direction in cartesian coordinates oxyz:

$$\vec{E} = E(\cos\phi\sin\theta, \sin\phi\cos\theta, \cos\theta), \quad E = |\vec{E}| \tag{9}$$

then α_1, α_2, and α_3 denote the angles made by the vector E with the axis ox, oy and oz respectively, along which three dipoles of the sensor are situated. The dependences between α_i in relation to ϕ and θ are as follows:

$$\cos\alpha_1 = \cos\phi\sin\theta; \quad \sin\alpha_1 = \sqrt{1 - \cos^2\phi\sin^2\theta} \tag{10a}$$

$$\cos\alpha_2 = \sin\phi\sin\theta; \quad \sin\alpha_2 = \sqrt{1 - \sin^2\alpha\sin^2\theta} \tag{10b}$$

$$\cos\alpha_3 = \cos\theta; \quad \sin\alpha_3 = \sin\theta \tag{10c}$$

The radiation pattern of the three orthogonal dipoles under these conditions can finally be represented by the equation:

$$F(\xi,\phi,\theta) = [f(\alpha_1)]^2 + [f(\alpha_2)]^2 + [f(\alpha_3)]^2 =$$

$$= \frac{[\cos(2\pi\xi\sqrt{1 - \cos^2\phi\sin^2\theta}) - \cos(2\pi\xi)]^2}{\cos^2\phi\sin^2\theta} +$$

$$+ \frac{[\cos(2\pi\xi\sqrt{1 - \sin^2\phi\sin^2\theta}) - \cos(2\pi\xi)]^2}{\sin^2\phi\sin^2\theta} +$$

$$+ \frac{[\cos(2\pi\xi\sin\theta) - \cos(2\pi\xi)]^2}{\cos^2\theta} \tag{11}$$

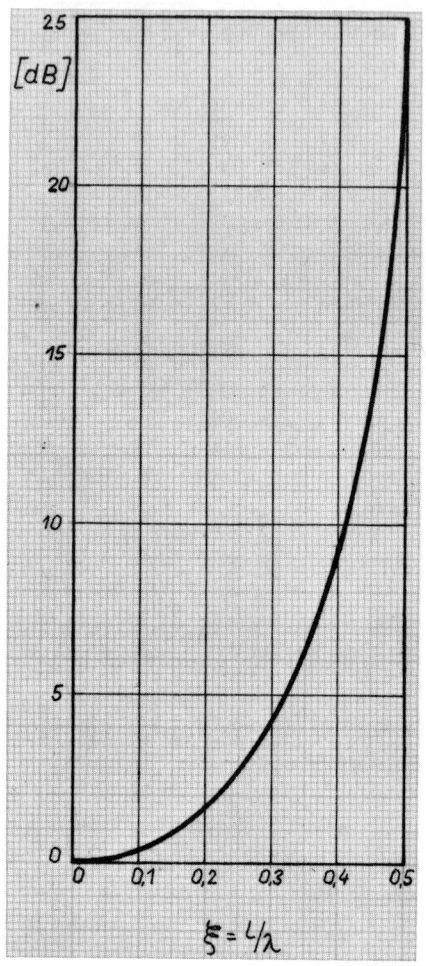

Figure 3. Irregularity B as a function of ξ

SYNTHESIS OF FREQUENCY RESPONSE

The irregularity B of the radiation pattern of the sensor can be defined using the equation

$$B = B(\xi) = 20 \log \frac{F_1(\xi,\phi_1,\theta_1)}{F_2(\xi,\phi_2,\theta_2)} \qquad (12)$$

where: $F_1(\xi,\phi_1,\theta_1)$ and $F_2(\xi,\phi_2,\theta_2)$ are maximal and minimal values of the formula $F(\xi,\phi,\theta)$ according to the values of the arguments ϕ and θ within the intervals

$0 \leq \phi < 2\pi$

$0 \leq \theta < \pi$

Thus the irregularity B is for fixed ϕ and θ, the relation in dB of the maximal value of the sum of EMF's obtained from each dipole of the sensor to the minimal value of the sum, within all possible positions of the sensor towards the field. The calculations of the value of the function B has been carried out for the values ξ within the interval

$0 \leq \xi \leq 0.5$

The graph of the function B is shown in Figure 3. With the determined length h of the arms of the dipole as well as with the assumed irregularity B, it is possible to calculate the upper frequency of the sensor using the curve shown in Figure 3. For example: when h = 0.1 m and B = 2 dB (minus); ξ = 0.22 and the required value of the minimum wavelength λ_{min} = 0.454 m. In the same way it is possible, with the assumed maximal frequency and irregularity B_{max}, to calculate the maximal length h_{max} of the dipole arms, e.g. for f_{max} = 1 GHz and B_{max} = 1 dB; ξ = 0.155 and the maximum length of the dipole arm h_{max} = 4.6 cm.

Conclusions

The regularity of the radiation pattern of three orthogonal dipoles is greater for shorter dipoles. Figure 3 shows irregularities of the radiation pattern with regard to long dipoles. Therefore, Figure 3 is useful for sensor designing.

References

1. BABIJ, T.M., TRZASKA, H.: The Problem of Measurements of Electromagnetic Field Strength of Nonionizing Radiation. XVII Conf. Gen. URSI, Warsaw, Poland, 1972.

2. BOWMAN, R.R., LARSON, E.B., BELSHER, D.R.: Semiannual Progress Report: Electromagnetic Hazards Project, NBS Technical Note 9761, p. 31, 1969.

3. GUY, A.W.: Analysis of Electromagnetic Fields Induced in Biological Tissues by Thermographic Studies on Equivalent Phantom Models. IEEE Trans. vol. MTT-19, No. 2, 1971.

4. KUCIA, H.R.: Accuracy Limitation in Measurements of HF Field Intensities for Protection Against Radiation Hazards. IEEE Trans., vol. IM-21, 1972.

5. MICHAELSON, S.M.: Human Exposure to Nonionizing Radiant Energy - Potential Hazards and Safety Standards. Proc. IEEE, vol. 60, No. 2, 1972.

6. PUNTENNEY, D.G., VETTER, R.J., WEEKS, W.L., ZIEMER, P., BORN, G.S.: Microwave Dosimetry Using Electrochemical Effects. The Journal of Microwave Power, vol. 9, 1974.

7. RUDGE, A.W., KNOX, R.M.: Near Field Instrumentation. Bureau of Radiological Health. BRH/DEP 70-16, 1970.

8. SCHWAN, H.P.: Microwave Radiation: Biophysical Considerations and Standards Criteria. IEEE Trans., vol. BME-19, May 1972.

9. WACKER, P.F., BOWMAN, R.R.: Quantifying Hazardous Electromagnetic Fields: Scientific Basis and Practical Considerations. IEEE Trans., vol. MTT-19, 1972.

SYNTHESIS OF FREQUENCY RESPONSE

–DISCUSSION–

BEISCHER - The three orthogonal dipole probe is at present calibrated up to 5 GHz. Can the probe be used at higher frequencies and what is the accuracy then?

BOWMAN - Even though you can calibrate the probe at higher frequencies, you lose field resolution. In other words, once you enter the resonance region of the dipole, you begin to lose field resolution. So even though it can be used, the use is more restricted than for the flat response region of the probe. We have concentrated more on trying to make the probe much smaller than on trying to calibrate it in the resonance region.

VOGELMAN - What is the resonant wave length of this probe?

BOWMAN - The dipoles used in the probe have a self-resonance of about 20 GHz. But the dipole-diode resonates at about 6 GHz because of the inductance of the diode package. For various technical reasons, we could not then get beam-lead diodes with the right characteristics for that probe. Now, the beam-lead package diode has become available for the type of diode we need; a high-breakdown-voltage Schottky diode. Using these diodes, the same probe that Dr. Beischer has mentioned would work flat up to about 5 GHz.

VOGELMAN - When you have loaded the probe with diodes, the resonance frequency is in the order of 10 GHz; so it is really operating at half of resonance.

BOWMAN - When loaded with the excess inductance, the resonance is more like 6 GHz.

The diode package used has about 2 manohenry's of inductance, whereas the beam-lead package has about 2/10's of a nanohenry.

BEISCHER - Has somebody used other probes recently?

GLASER, Z. - We have used some of the commercial (Narda) probes for shipboard measurements (both CW and pulsed). Provided that exposure to fields in excess of their design maximum did not occur, the probes were quite satisfactory, and in good agreement with the Bowman probe.

ASLAN - The probe you're discussing is the thin film probe; its capability is in the order of 300 watts per centimeter square, peak power, which I think is quite agreeable with any diode-type probe.

BOWMAN - I don't know that the diode-type probe can burn out except at very high field levels. They may fail to give the right answer in a high intensity field, but there is a question as to whether they would be damaged or not. They may recover. To destructively test this type of probe requires fields in excess of kilowatts per square centimeter. I don't have a generator that powerful. When you exceed the inverse breakdown rating on a diode, the stored charge in the electronic circuitry suddenly dumps through the diode. Since the diode is isolated by a high-resistance line, it is unlikely that this charge dumping will damage the diode.

Many people would like a three-orthogonal H-field probe which would measure an H-field in the same simple way as the E-field probe. Unfortunately, there are some technical problems with loop antennas because they respond to both electric and magnetic field components, whereas a dipole responds only to the electric field component. That is, you can express a dipole's output in terms of the unperturbed electric field. I think the Bureau of Radiological Health is interested in developing a three-orthogonal-loop probe. At the Bureau of Standards, we are progressing a little more modestly. We are currently building single loop probes that have the loop placed at the end of the wand in such a way that by rotating the wand to three 120° positions you can measure three orthogonal components of the field. Frank Greene at the Bureau of Standards is developing such probes that will work up to several hundred megahertz; perhaps 500 MHz. This development, the "skewed loop" probe, is for NIOSH (National Institute for Occupational Safety and Health). The next step is to find some way of combining three loops in such a way that they do not mutually couple, but this is a tricky problem because you have both magnetic coupling and E-field coupling to worry about.

MITCHELL - What are the requirements for the H-field measurements in addition to the E-field, for all biological work, in terms of everything from hazards to the work in the laboratory? I wonder whether these requirements are coming to NBS, in terms of measuring E^2.

BOWMAN - The work that we are doing on an H-field measuring probe is an external field probe. This work is prompted by the need to measure magnetic fields around some diathermy machines that operate below 50 MHz. As we all know from the work of Dr. Guy and others, the unperturbed H-field represents the greater hazard at these lower frequencies than does the unperturbed E-field. That is: the eventual heating in the tissue can be more closely related to the unperturbed H-field than to the unperturbed E-field. This is due to the induced "Eddy currents" in the tissue. The

SYNTHESIS OF FREQUENCY RESPONSE

tissue heating is still due to the internal E-field; so the need to measure internal H-fields, as far as I know, is non-existent. The need to measure external H-fields is most important at frequencies below 20 or 30 MHz. Fortunately, at the lower frequencies single loop antennas are fairly easy to make. The only issue in question is whether or not one can make a three orthogonal loop probe that will provide direct measurements of complicated H-fields.

VOGELMAN - Why don't you let the probe spin and just make measurements at the right places, doing the arithmetic with electronic logic?

BOWMAN - There's no basic reason why you couldn't. However, the wand may be fairly long and spinning it could be a mechanical problem. You might want the wand to be 10 meters in length. I suppose if it's only a meter or less, you could probably spin it and sample at any three 120° spaced positions. All you have to do is sample it three times per revolution symmetrically and add the results.

LIN - I think your idea of trying to measure the E-field inside is a very good one, but I sort of detect that you have some reservations about the sensitivity. I'd like to make reference (C. C. Johnson & A. W. Guy. IEEE Proc. 60:692-718, 1972; J. C. Lin, A. W. Guy & C. C. Johnson. J. Microwave Power 21:791-797, 1973) to some studies we did using both theoretical and phantom models in the range of 900 MHz to 22,000 MHz which show, for instance, at 918 MHz for a model of the human head that one could have a considerably enhanced power deposition inside the head. Therefore, your sensitivity problem there probably would not be as severe as you have indicated. The same situation holds for the animal case as 2450 MHz, since most of us use animals for our experimentation. Again, the internal absorption could be higher in the center of the brain than close to the surface.

BOWMAN - Yes, I agree.

LIN - Secondly, you alluded to the use of the scattered field for internal field measurements at very high frequencies. The numbers you quoted were 2 to 3 GHz.

BOWMAN - That's a questionable lower limit. The technique may not be useful at those frequencies.

LIN - That's the point. I think the limit is too low. You have to get quite a bit higher in order for that technique to be useful.

BOWMAN - I believe that scattering techniques to determine internal fields would be difficult at 2 to 3 GHz.

LIN - At 2000 MHz, there may be considerable absorption in the center which may be difficult to detect with this technique. I just sort of like to make that point.

BOWMAN - May I return to your first question? If you take a cut through a subject's head, you may get fairly high heating in the fat and also around the center at some frequencies.

LIN - Right.

BOWMAN - However, the reason that there is significant heating anywhere is because of the very lossy nature of the material. The internal field is quite small compared to the external field. There is quite a sensitivity problem, but having a peak in the middle takes care of the attenuation problem. That is, the sensitivity problem is, for some frequencies, not particularly worse at the center of the head than it is near the surface. But there is still a sensitivity problem because the field inside the material drops due to the large dielectric constant of tissue.

LIN - No. I don't think that is exactly the case. What I am referring to is, suppose you start using resonance effects of, say, around a couple of gigahertz for a small animal head, not a human head. There are, of course, larger dimensions in the body than the head. So, if you're worried about whether or not the field can penetrate to the heart or something like this, you may get an answer at even a few gigahertz by using the boundary value technique. But I think the accuracy of the answer will be marginal; at, say 10 GHz, it should work out fairly well, and be much more applicable.

OSEPCHUK - A question of terminology. I hear professional radiation people speak authoritatively about the need to use the term densitometry, as opposed to dosimetry, and you've intermixed all of this. I am somewhat unhappy about all this confusion on terms.

BOWMAN - I don't like either densitometry or dosimetry. I just don't think they are meaningful. If the biological effects have a threshold, then the ideal of dosimetry really cannot be applied. It's just nonsense. Though I don't exactly like the word, I sometimes use the term "causimetry" in my own thinking. That is, the "measurement" of some kind of cause of the effect. "Causimetry" may sound strange, but it is semantically correct.

OSEPCHUK - Somehow or other, we have to use a common language.

BOWMAN - Yes. The best thing to do would be to throw dosimetry out, and speak about what one is measuring; thermometry, or whatever.

JOHNSON - It seems to me that, as you indicated, frequencies well above 2 to 3 GHz are needed to accomplish what we normally think of as imaging. Yet, at these frequencies tissue absorption is so great that I really question whether this is a viable approach. Are you proposing imaging in a normal context, or observations of superficial layers? How deep did you think one could obtain useful signals?

BOWMAN - I was not thinking about imaging. The possibility of microwave imaging is quite interesting, but that's not what I was referring to. At very high frequencies the absorption does take place in a thin layer, but that is the area of interest. If you're interested in energy absorption, then that's the only place you're interested in.

JOHNSON - But for deep internal anatomical imaging, one would be interested in the heart or the lungs. The medical applications you have in mind are for very superficial observations.

BOWMAN - Right. Boundary measurements are useful for determining what's happening at certain layers just inside. You may want to know, for instance, what the E-field is in the fatty layer under the skin or at the surface of the muscle under the skin. At high enough frequencies, only surface absorption is of interest. I think you can determine this absorption by a boundary field measurement. The question is: how else would you determine it?

SUESS - I just thought I would comment on what you have suggested, as it is a very important point. Being involved in international work, I feel it is very important to establish a clear language among scientists, not only among those speaking the same mother tongue, but also among those who speak different languages. There is now an intense activity to establish something that could become equivalent to committees in the ionizing radiation field, such as the ICRU (The International Commission for Radiological Units and Measurements) and the ICRP. I think that such a body would then probably be established on an international basis and should be able to resolve this question and clarify the meaning of "dosimetry." In the meantime, however, when a term is being used, one should explain what he means by using it.

HILL - The question of the use of the term "dosimetry" comes up in ultrasound. I dislike it in ultrasound as much as you obviously dislike it in microwaves. But I don't really know what the answer is. Maybe it's exposimetry or some horrible sort of word like this, because what one is trying to do is to measure some physical exposure of an individual to a form of radiation or what have you.

I am not even sure about causimetry, because that seems to imply that the radiation you're measuring causes something or other. In ultrasound, we're not even sure whether it causes something.

I am interested in hearing Dr. Suess talk about ideas for establishing something analogous to ICRU. I don't know whether this is the right moment, but sometime I would be interested to hear him perhaps say a bit more about this. I don't know how familiar people here are with this situation, but in the ionizing radiation field, there are these rather remarkable organizations, the International Commission on Radiological Units and Measurements and the International Commission on Radiological Protection, which have succeeded in setting themselves up as organizations of the scientific community, rather than as agencies of government.

JANES - As a point of information, the U.S. National Council on Radiation Protection has formed a scientific committee to develop definitions of dose and units for use in the field of non-ionizing radiation. Dr. George Wilkening, of Bell Laboratories, is its current chairman.

BEISCHER - Our interest concentrates on the energy absorbed by man as he is exposed to a given microwave field. We have made measurements which demonstrate that man reflects about 50% of the incident radiation in the wavelength range of 1-10 GHz. This reflection is as Schwan showed previously in a theoretical deduction fairly independent of polarizations of the radiation and slightly dependent on the wavelength.

The reflected beam interacts with the incident beam and standing waves are formed on the exposed side of man. Direct measurements of energy reflected, diffracted and transmitted by an object, can provide information concerning the absorption properties of that object.

LEHMAN - As a biologist looking at hazards and/or desired biological effects in relation to "dosimetry", I think we can relate the biological effect to measurements of the field into which we place the individual or the part of the individual. However, this prediction is a difficult one and may not hold all the time because of what Dr. Lin said before, i.e., the possibility of focusing and the concentration of energy. We have to know how the energy is distributed, where it is absorbed within the organism. Now, if we know the sites of absorption, and we know what the field is there, then we still cannot predict unless we know what the mechanism is by which this electromegnetic field produces the biologic reaction. We cannot conclude by extrapolation from one single phenomenon to the next. I think one of the real problems in the whole hazard area is that whatever phenomena have been found, we have only very limited information as to how they are brought about.

This holds to a large degree also from the therapeutic point of view, where we obviously apply much higher energy levels because we want to produce a reaction. If you consider the so-called non-thermal effects, we are in the same dilemma. Again, we know they exist. We know what happens with some of them. But for most of them we do not know how they are brought about exactly and therefore we cannot generalize. Because of this lack of information, we still don't know whether or not this effect would occur under therapeutic conditions. Therefore, in either case, if the effect may be beneficial, we don't know whether to avoid it.

CZERSKI - I think the same question exists in ionizing radiation biology, and I think the approach used there could also be useful. You can distinguish, and this is a classic distinction in ionizing radiation biology, between what is called the primary interaction in the biophysical terms of the radiation with the living system; and then you have direct biologic effects. Analyzing what happens further, usually a chain of events, you may have immediate secondary biologic effects. Also, distant effects in other parts of a body that involve radiation exposure, and delayed effects. This may very well be taken on a strictly thermal basis. Let's say you have an increase in the kinetic energy of a group of molecules and an increase in temperature; this induces a thermal compensatory reaction in tissue in terms of increased blood flow and a changed metabolic rate. But, an increased blood flow influences also the metabolism of this region. So, you have a certain sort of sub-chains. If such a thing happens, let's say in a secretory cell, it will affect its function, and may affect the level of, let's say, an endocrine secretion in the blood, and then you will have specific effects in other organs responsive to the effects of the given hormone. In the non-ionizing radiation hypothesis this will certainly reflect on the function of other endocrine organs, and I feel that the thermographic approach is very important in that it could show various temperature gradients especially within cranial structures and effects such as thermal stimulation around the hypothalamus. I would imagine that this would induce several effects, but to complicate the matter slightly more, I remember that about seven years ago there appeared a paper on blood vessel reaction in muscles exposed to diathermy. The authors demonstrated that the blood flow might increase or decrease, so there is no specifically determined reaction in terms of cooling.

I am impressed by a certain sort of disproportion between the really imposing trend of advances and measurements on the side of the technical part, by physicists; and a lack of physiological examination using these techniques, and such readily available techniques as radio-isotope determinations of blood flow or even simple metabolic investigation. I would like to stress especially that concerning the whole system, you should speak about the interaction, and remember that a living system unlike a physical model will react with certain mechanisms of adaptive compensation. Certainly the approach to the analysis of all these effects should be much more dynamic.

JOHNSON - I think we all agree that thermography is extremely valuable for obtaining rapid two-dimensional heating information from phantom models or killed and dissected animals. There are a few limitations with thermography that should be recognized. Firstly, as Mr. Bowman explained quite well, you must have the animal severed in a plane parallel to tissue current flow. On any real animal there really is no such plane, particularly with asymmetrical irradiation.

Secondly, we found in tests on phantom models in Dr. Guy's laboratory, using the fiberoptic liquid crystal temperature probe, that as soon as the microwave power is turned off the temperature changes are very rapid in the first two or three seconds. By the time the model is split and presented to the thermographic camera the temperature was considerably reduced from its peak. If the objective is good quantitative dosimetry, or densitometry, if you prefer, there will be some problems with the thermography. Thermography isn't the answer to all our dosimetry problems but nonetheless it does give quick and valuable physical indications of power absorption in the animal.

LIN - Dr. Johnson has just brought out a point saying thermography is a poor quantitative method of doing things. In order to render some justice to the thermographic method, I would like to start with a spherical model, where mathematically one can get an exact solution. In fact, the result can be obtained to any degree of accuracy desired. If we take this result and compare it with the result one can obtain with thermography using the split model of the same size and the biological properties, the result compares very favorably. In fact, this is what gave us the confidence in applying this technique. There's no question that in animal and human bodies it is hard to find a plane of symmetry. However, this working method has given us quite a bit of information which I don't think could be disputed very easily.

Another point I would like to bring up in conjunction with dosimetry. Up to this point, we have used animals to get at dosimetry and used theoretic models of different shapes to obtain some information in terms of average absorption or peak absorption as a function of distance. However, the model shapes are often different from an animal or human. Recently we have carried out some studies using scale models of man. We have developed a homogeneous

SYNTHESIS OF FREQUENCY RESPONSE

dielectric material which simulates muscle. Using this model and the thermography method, we have shown that the power absorption is quite a bit different from what one would expect with spheres. A paper was published some time ago where, in using a sphere exposed to an H-field, we showed enhanced absorption at the periphery of the body due to magnetically induced electric field. This result has been corroborated with this true-to-shape model. The circulating current and enhanced absorption are evident in the larger cross sections of the body (upper torso). However, at the narrow parts of the upper and lower extremities, you get a different kind of picture. Instead of predominantly magnetically induced, the electrically induced absorption becomes gradually important. What this indicates is that one really has to take both electric and magnetic fields into consideration.

ULTRASONIC MEASURING TECHNIQUES

Harold F. Stewart, Ph.D.

Bureau of Radiological Health, FDA

Rockville, Maryland 20852

INTRODUCTION

Within the past few years significant advances have been made in the application of ultrasound in medical practice, particularly in the diagnostic area. In fact, at the present time diagnostic ultrasound is one of the most rapidly growing areas in medical electronics. With the increasing use of this modality comes the responsibility and need for the correct assessment of ultrasonic field parameters in order to relate exposure levels to the likelihood or extent of biological effects in living tissue. This is particularly important in studies attempting to understand interaction mechanisms and their subsequent application to the establishment of exposure guidelines. Such studies can only be accomplished with the aid of accurate and reliable measurement techniques. In the establishment of an acceptable standard for the measurement of ultrasonic field parameters, particularly power from ultrasonic medical devices, comparisons are needed among measuring techniques currently used. These techniques for the measurement of ultrasonic power levels may be grouped into four general categories: (1) radiation force techniques; (2) thermal detectors, including calorimeters, thermocouples and thermistors; (3) optical methods; and (4) microphone receiver techniques, including piezoelectric, magnetostrictive transducers, and capacitance transducers. Each of the preceding measurement techniques can be used to measure total acoustic power and intensity patterns. This paper discusses some of these techniques and the intercomparison of total power measurements made in the Acoustics Branch at the Bureau of Radiological Health.

Contribution of the Bureau of Radiological Health, U.S. Public Health Service.

Radiation Force Techniques

When an object is placed in an ultrasonic field it experiences a steady force due to the alternating pressure variations in the propagating medium. This force is usually called radiation pressure and is related to the acoustic power or intensity. This force is independent of frequency and for plane waves, the radiation force (F) is given by

$$F = PD/c \qquad \text{Equation 1}$$

where P is the acoustic power, c the propagation velocity of the wave, and D a factor which is determined by the type of interface an obstacle presents to the ultrasonic field (10). In the case where only part of the total beam is intercepted an intensity averaged over that area is detected and thus a force per unit area is measured. Values for D are shown in Table I. Thus, by knowing the type of interface a target presents to the ultrasonic wave and by measuring the magnitude of the radiation force, the total power or spatial averaged intensity of the acoustic field can be computed. The relationship in Equation 1 also applies for a pulsed ultrasonic field, provided P is taken as a time averaged value. Thus the experimental problem to be solved for determining the power or intensity of an ultrasonic field is the design of a suitable technique for measuring the force produced by the ultrasonic field on a target placed in it. Various techniques have been used to measure this force. Experimental arrangements have included the use of various types of balance arrangements, the use of buoyant force, and observations of the deflections of pendulum type targets.

LABORATORY MEASUREMENT METHODS

Radiation Force Float Method

One radiation force method which has been used in laboratories around the world is the radiation force float system (17, 12). Figure 1 shows a schematic drawing of a float system built by the Bureau of Radiological Health and patterned after the one developed by Oberst and Rieckmann (17). It consists of a thin inverted conically shaped reflecting target that forms a cover of a hollow, air-filled cylinder with a coaxial stem below. The float is designed so that it is slightly heavier than water. When placed in a vessel of water which contains a beaker of carbon tetrachloride the float sinks, immersing more and more of its stem into the beaker of carbon tetrachloride, until a condition of neutral buoyancy is obtained. When exposed to a radiation force from a vertically incident ultrasonic beam, the float is forced down further into the carbon tetrachloride until that radiation force and the increase in buoyant force cancel. The amount of displacement of the float

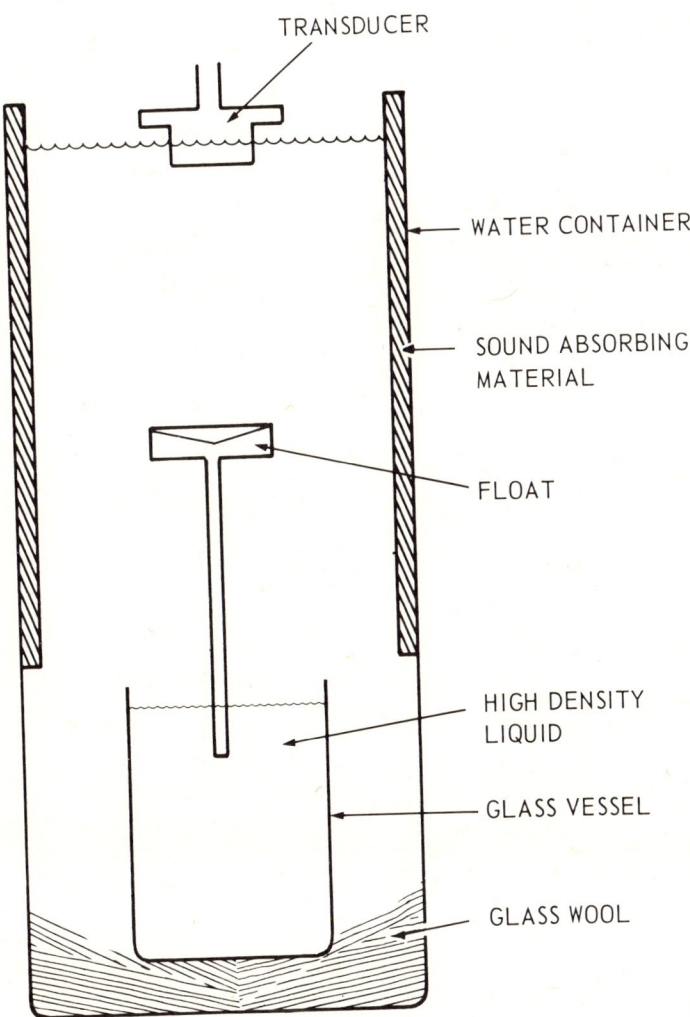

Fig. 1. Radiation Float System

which occurs is proportional to the incident ultrasonic power. The relation between the force (F) in newtons exerted on the float and the incident power (P) in watts is given by the following equation

$$F = P(2\cos^2\theta/c)$$

where c is the velocity of sound in water (c = 1.5 x 10^3 m/s at 30°C) and θ is the angle between the normal to the reflecting surface and the incident acoustic beam axis. The float is calibrated by determining the displacement corresponding to the force of known mass (m), corrected for buoyancy, placed on the float, i.e. F = mg, where g is the gravitational acceleration, 9.8 m/s^2. The sensitivity of the system is dependent upon the diameter of the coaxial stem. In our system, stem diameters of 8 mm. and 3 mm. are used for measuring powers from 1 to 30 watts and 100 mw to 3 watts respectively. This system which utilizes a reflecting target has the advantage of being self-centering because of the horizontal component of the radiation force which tends to center the float in the field. The included angle of the float is 130° which provides an angle to give a horizontal centering force while at the same time avoiding a situation where the sound reflected from one side of the float will impinge on the other side. If such a system used an absorbing target, it would not have the self-centering feature because there is no horizontal component to the radiation force as demonstrated experimentally by Herrey (8).

When operating at power levels requiring the use of degassed water it is necessary to replace the water in the vessels daily with fresh degassed water. This float system provides a simple inexpensive technique for the determination of total acoustic power output using radiation force.

Analytical Balance Radiation Force Techniques

The use of sensitive analytical balances using either absorbing or reflecting targets has been reported by many users (21, 29, 13, 27). Such a system, currently in use in our laboratories, is shown schematically in Figure 2. It consists of a Mettler precision microbalance model no. H20T with a sensitivity of 0.01 milligrams and a target suspended in the water bath below the balance. Measurements are made by mounting the ultrasonic transducer whose output power is to be measured directly over the target. When the transducer is energized, the ultrasonic energy from the transducer strikes the reflector and the radiation force is measured using the precision balance. It is important to these measurements that the size of the target be larger than that of the ultrasonic beam to insure that the entire ultrasonic field is intercepted. On the other hand, the volume of the target should be kept to a minimum to

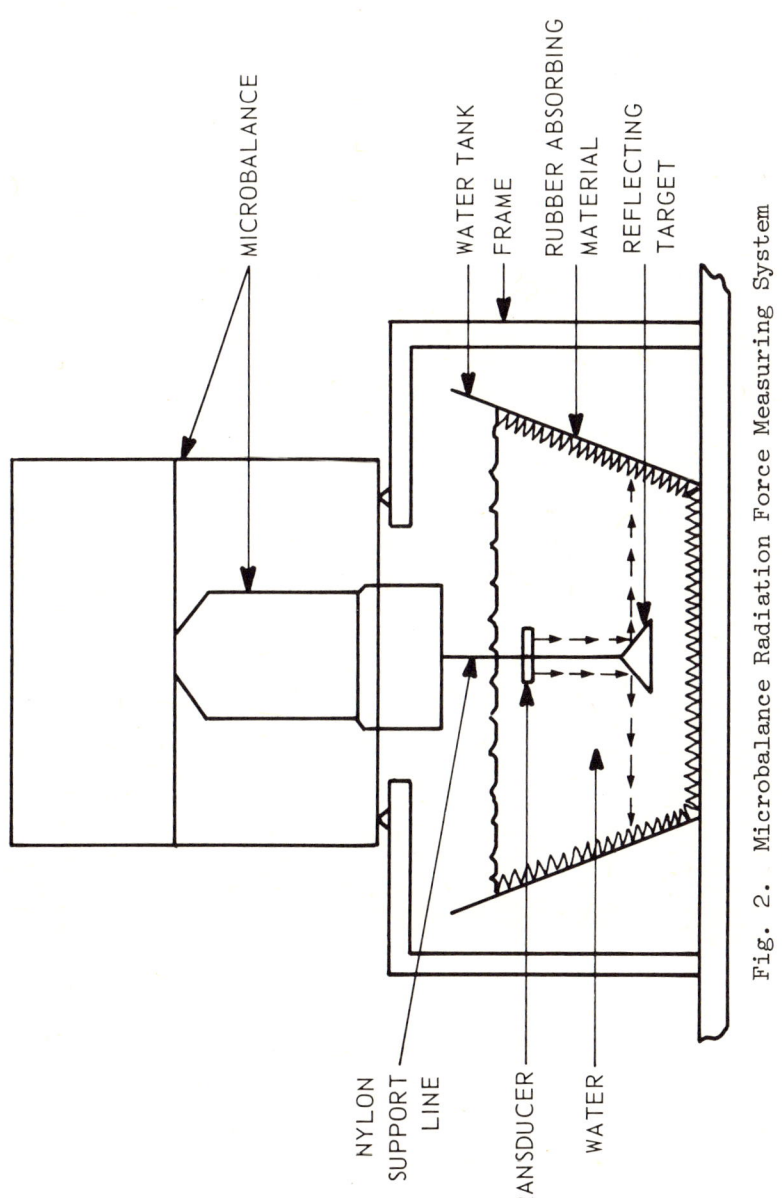

Fig. 2. Microbalance Radiation Force Measuring System

Table I

Value of constant D for various physical situations (10)

Physical situation	D
Perfect absorber, normal incidence, $r = 1$*	1
Perfect reflector, normal incidence $r = 0$ or ∞*	2
Perfect reflector, incident at angle θ to sound beam, *$r = 0$ or ∞	$2\cos^2\theta$
Nonreflecting interface, normal incidence *$r = 1$ $c_1 \neq c_2$	$1 - c_1/c_2$ For $c_1 < c_2$ force in direction of propagation $c_1 > c_2$ force opposite to direction of propagation
Partially reflecting interface, normal incidence $Z_2 \neq Z_1$, $c_1 \neq c_2$	$2[(r-1)^2/(r+1)^2]$

*$r = Z_2/Z_1$ the impedance ratio at an interface
where $Z = \rho c$

eliminate excess noise in the system. To eliminate the small drifts due to the effects of air movement the sides of the balance system are enclosed. The microbalance system using the Mettler microbalance, as reported by Kossoff, Ziedonis, have sensitivities in the milliwatt range (29, 13). A more sensitive system, using a Cahn electrobalance, has been described by Rooney (21). Both absorbing and reflecting targets have been employed. SOAB, a rubber from B.F.Goodrich, has been shown to be a good material for use as an absorbing target (26). An absorbing target has the advantage that external absorbers are not required since the problem of reflected beams is not present. A disadvantage of the absorbing target method is the difficulty of constructing a target which is a perfect absorber. Such mismatches are not easily measured and as can be seen in Table I the value of the constant D for a partial reflecting interface requires knowledge of the impedance in both the transmitting medium and target material, as well as the value of the velocity of propagation in each medium. Perfect reflecting targets have the advantage that they are more easily constructed and since the value of the constant D can then be known, they allow for higher accuracy in the measurements provided the establishment of standing waves can be avoided. Targets are generally mounted at an angle of $45°$ with respect to the beam axis. As can be seen from the value of the constant D, ($D = 2\cos^2\theta$), in Table I, at this angle the target experiences the same force along the direction of propagation of the beam as an absorbing target at normal incidence.

The balance can be calibrated by noting the deflection resulting when calibrated masses are placed on the balance. For example, an ultrasonic beam of 1 watt power produces a radiation force of 0.067 gram weight on a totalling absorbing target at normal incidence, or a totally reflecting target at $45°$. Because of the sensitivity requirements of these systems it is necessary to insure vibration isolation.

Portable Radiation Force Systems

Portable radiation force devices have long been used for the measurement and calibration of the output from ultrasonic therapy equipment. Such portable measurement devices, using both absorption and reflecting targets, are discussed with illustrated drawings by Hueter and Bolt (pp. 45-50)(10). These devices are no longer commercially available. However, a device similar to the Siemens sonotest is currently marketed by the Birtcher Corporation and a new prototype instrument with direct electronic readout was recently demonstrated by Burdick Corporation. Other investigators have reported various methods for designing and constructing portable radiation force meters (25, 20). In our laboratories we have investigated the use of strain gauges, magnetostrictive transducers, and magnetic forces produced by current passing through coils. Figure 3

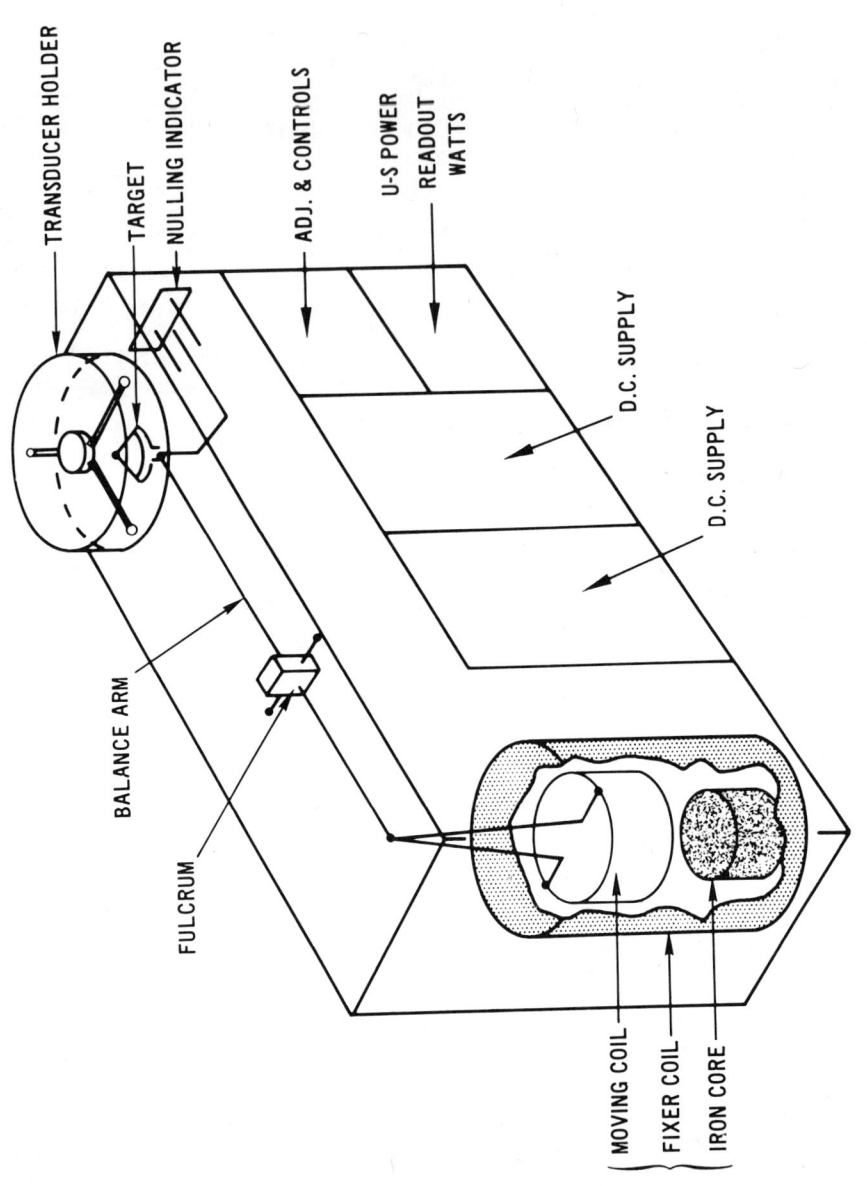

Fig. 3. Portable Ultrasonic Therapy Radiometer

is a schematic diagram of a portable ultrasonic radiometer developed in our laboratory. This radiometer incorporates a solonoid balance arm nulling technique. When a transducer is placed over the target, and energized, the radiation force drives the target downward from its null position. A restoring force or torque is applied to the ultrasonic target by means of a balanced arm on which a moving coil is attached. This coil moves inside another fixed coil. The force of attraction established between the two helical coils when current is passed through them provides a restoring force to offset that radiation force produced by the ultrasound so that the target is returned to its original null position. Since the restoring voltage across the moving coil is an electrical analog of the ultrasonic power, the instrument can be calibrated with masses to read out directly in watts of ultrasonic power.

Intensity Distribution Measurements

A technique which has been used for the measurement of radiation pressure involves the measurement of the deflection of a target suspended in a sound field in a pendulum arrangement. Intensity distributions can be made using a small target provided the dimensions of the target are greater than several wavelengths. This provides a measurement averaged over the area of the ultrasonic field the target intercepts. Because only a fraction of the toal beam is intercepted, the sensitivity of this measurement method is correspondingly reduced.

A second technique, commonly used for radiation force intensity distribution measurements, involves the measurement of the deflection of a small metal ball suspended in pendulum fashion in a nonviscous fluid in a travelling plane wave ultrasonic field. Ideally, the deflection of this pendulum arrangement from the vertical should be small so that the tan θ can be approximated by the deflection divided by the length of suspension to compute the force components. For such small deflections, the force (F) produced is given by the following relation:

$$F = y\theta r^2 I/c$$

In terms of intensity it can be expressed as follows where F is mgx:

$$I = (mgxc/\gamma l r^2 y) 10^{-7}$$

where m is the weight of the sphere in water in grams, g the acceleration due to gravity, x the displacement of the ball in centimeters, c the velocity of sound in water, l the length of the supporting string in centimeters, r the radius of the ball in centimeters, and y is the acoustic radiation force function which depends on the material and dimensions of the ball (4, 22).

Thermal Methods

The absorption of ultrasonic energy in materials and the resulting temperature rise can be used as a dosimetric method for measuring ultrasonic energy. There are numerous variations using this basic idea. This is a standard measurement method for ionizing radiation. The calorimetric evaluation of ultrasound has been used since the the early part of this century (19).

Calorimetry Systems

Various designs for ultrasonic calorimeters have been reported in the literature (25, 15, 23). The calorimetric method developed at the Bureau of Radiological Health employs a constant temperature environment calorimeter shown schematically in Figure 4. This method provides a unique means for assessment of the energy in an ultrasonic field since it is least affected by beam shape and pulse duration, and is not dependent on plane wave assumptions. It consists of a conically shaped cup mounted in an outer brass cylinder container. The calorimeter is placed in a water bath which provides both a constant temperature environment and an effective medium for the propagation of ultrasound. The calorimeter cup is filled with carbon tetrachloride which provides a good impedance match with water, is an effective absorption medium, and has a low viscosity. The low viscosity enables the carbon tetrachloride to be stirred by acoustic streaming, which equilibrates the temperature in the absorbing medium. The conically shaped absorbing cup permits a long path length in the absorbing medium (for maximum absorption of the ultrasonic energy) to be seen by the entire ultrasonic wavefront, while minimizing the volume. The temperature difference between the absorbing fluid and the outer brass cylinder is measured with a bank of chromel-constantan thermocouples. This temperature difference is a measure of the acoustic power. Calibration of the system is accomplished with a constant resistance heating coil immersed in the carbon tetrachloride absorbing fluid. This system was used in making comparative measurements with the radiation force float and acousto-optic method, using National Bureau of Standards sources.

A second method currently under investigation for total beam measurements using thermal methods involves the use of parabolic cones with a small thermocouple or thermistor imbedded in a sound absorbing medium at the focal plane. This method involves the same principles as those employing thermocouple probes used for the measurement of spatial intensity distributions of ultrasonic fields (5, 2). A parabolic reflector has the property that a plane ultrasonic wave, travelling parallel to the reflector's axis of symmetry, will be focused, upon reflection, at the geometric focus

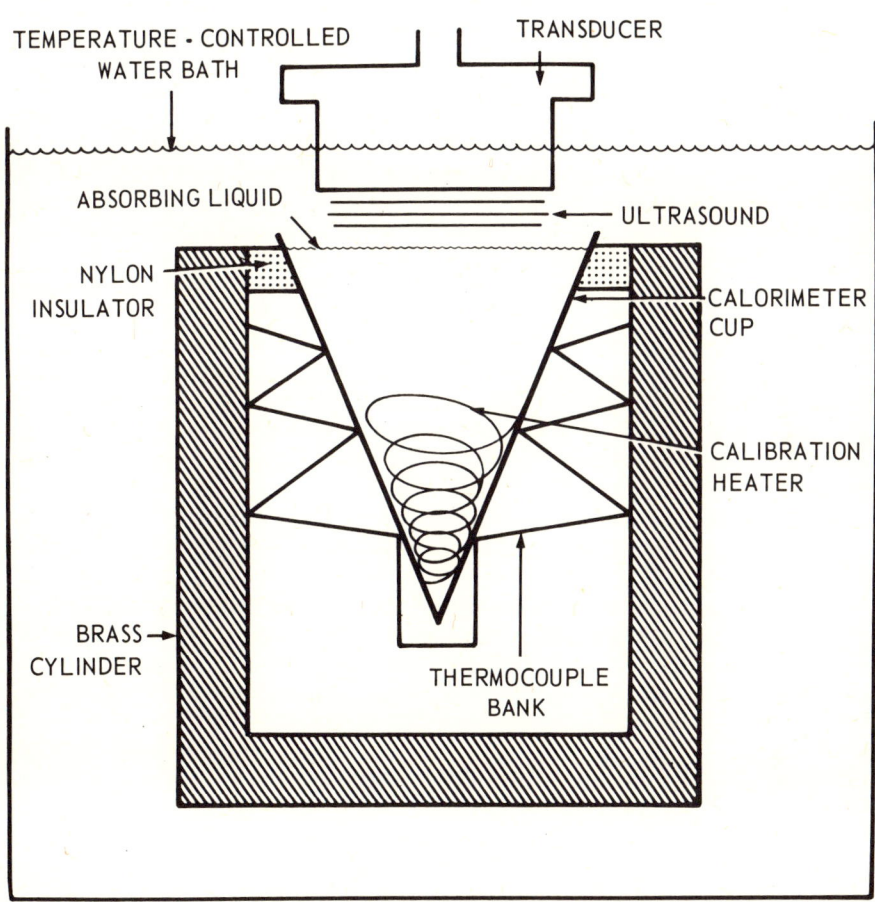

Fig. 4. Ultrasonic Calorimeter

of the parabola. This property may be employed to determine the total ultrasonic power output from a transducer.

The operation of the parabolic thermal probe is based on the measurement of the approximately linear initial rise in the thermoelectric signal produced by the small temperature rise as a function of time. This is caused by absorption of sound in the absorbing material surrounding the thermal sensor. Thus when the parabolic cone is filled with a transmitting medium such as distilled degassed water and an ultrasonic beam directed into the parabolic cup the rate of the temperature rise as a function of time for the first few seconds is a measure of the total ultrasonic power in the beam. The advantage of this measurement method is the increased sensitivity by concentrating all the energy in the ultrasonic field into a small area. Although this is not an absolute measurement method it has promise as a measurement system which could be made portable and calibrated for use in determining the output from diagnostic and therapeutic ultrasonic medical devices.

The calibration of such a device to determine ultrasonic power from diagnostic devices presents some difficulty however. Since the absorption properties of most materials are a strong function of frequency, the energy deposited in the material by a pulsed ultrasonic beam will be determined by the frequency spectrum found in the beam. This means that one could measure different temperature rises when determining the power in radiated ultrasonic beams having the same power but different frequency components.

The calibration of the instrument for the measurement of single frequency ultrasound, such as is found usually in therapeutic units and doppler diagnostic devices, should present no major difficulties.

ACOUSTO-OPTIC METHODS

Introduction

Most acousto-optic phenomena depend upon diffraction of the light by the ultrasound. Usually in the low megahertz region (below 5 MHz) the light is normally incident to the sound field. In the high megahertz region (greater than 10 MHz) the light should be incident at the bragg angle (11). Methods employing this phenomenom can be divided into two general categories: visualization and intensity measurement. A newly developed area of acousto-optic interaction depends upon the interferometric measurement of the displacement of a thin film placed across the acoustic beam.

Schlieren Visualization of Ultrasonic Fields

The schlieren technique relies on the interaction between the acoustical energy and the medium of transmission. Water is the transmitting medium commonly used. As the acoustic wave propagates the medium is subjected to increases and decreases in pressure which cause localized volume change, and hence density variations. These density variations caused by the compression and rarefraction phases of the ultrasonic wave in the optically transparent medium result in corresponding variations in the index of refraction. A schematic diagram of such an experimental setup is illustrated in Figure 5. In this arrangement, all the light passing through the medium is focused on a stop when no ultrasonic energy is present. When ultrasound is present these variations in the refractive index cause the light to be deviated from the stop and projected on to a screen behind the stop. These diffraction orders projected on the screen produce a light pattern which provides a visualization of the ultrasonic beam pattern. Schlieren visualization systems provide a good method for qualitative analysis of the detection of beam patterns such as divergence, focusing, detection of the presence of side lobes and so forth. In recent evaluations of ultrasonic therapy applicators in our laboratory, the schlieren photographs taken from a multi-crystal applicator containing 4 individual crystals illustrated that these 4 crystals produce 4 discreet ultrasonic beams as would be expected, rather than a uniform single beam as claimed. Another schlieren photograph of an ultrasonic beam produced by a large single crystal illustrated that only the center portion of the crystal was generating ultrasound because of the electrode arrangement.

Optical Measurement Methods

The study of ultrasonic fields by means of optical methods has the advantage of providing means of measurement while avoiding disturbance of the field. In addition, optical methods may be used for absolute measurements. Systems employing the basic Raman-Nath theory assume the ultrasonic field presents a phased grating to a normally incident beam of plane of monochromatic light. This is a method used by a number of investigators to determine ultrasonic beam intensity based on pressure amplitude measurements (9, 3). The diagram of such an optical system used in our laboratory is shown in Figure 6.

A lens collects the diffracted light and produces a pattern of dots (Fraunhofer diffraction pattern) at the focus. Without sound there is no diffraction and only a central spot is present. The ultrasonic light diffraction responds to peak acoustic pressure amplitude. The relative intensities of light (I_n) in each diffraction order (n) provides a measure of acoustic intensity as

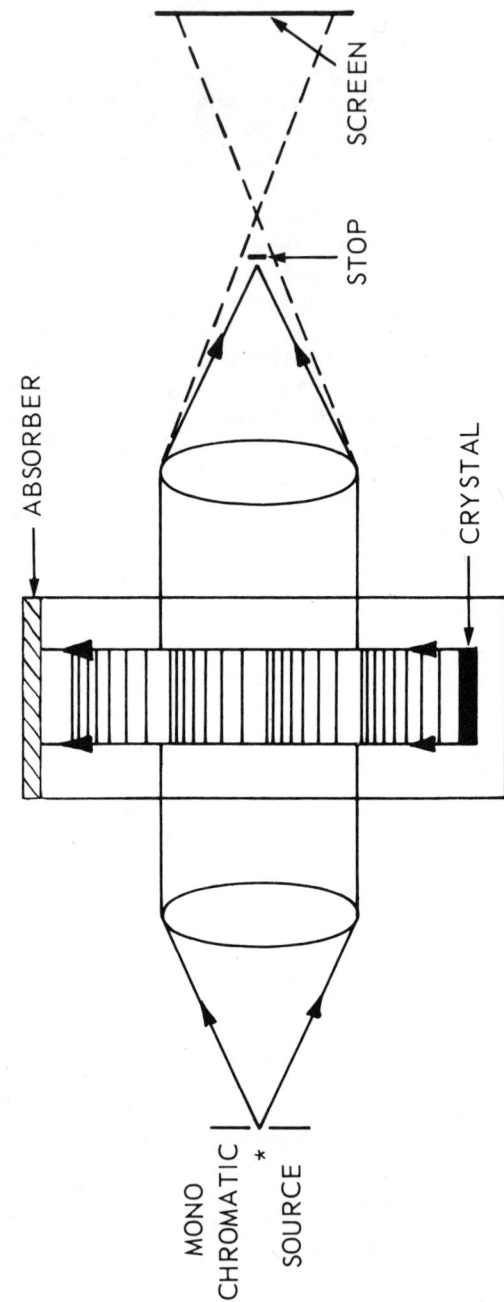

Fig. 5. Basic Schlieren Apparatus

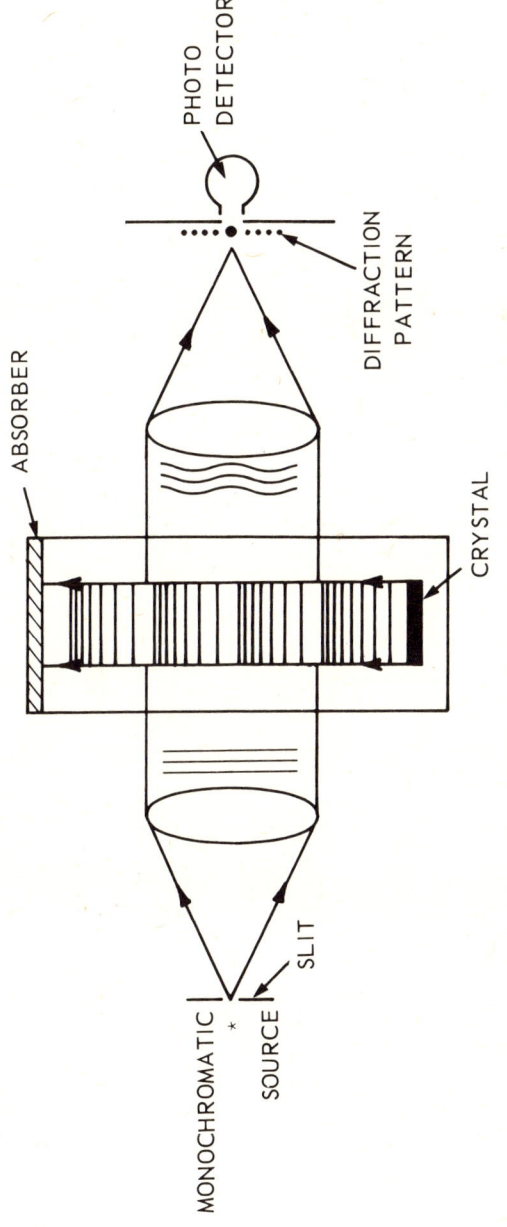

Fig. 6. Acousto-Optical Measurement Apparatus

given in the following relationship

$$I = [J_n(v)]^2$$

where J_n is the nth order Bessel Function and

$$V = \frac{2\gamma kpL}{\lambda}$$

where k is the piezo-optic coefficient and p is the peak acoustic pressure, L is the length of the ultrasonic field through which the light beam passes and λ is the wavelength of the light.

The theory for this technique was proposed by Raman and Nath and developed into a practical method by Hiedeman (18, 9).

In a recent set of measurements, using a "plane piston" source of known size, a correction was made for the axial distance from the face of the transducer to a point sampled by the light. This correction, based on an additional theory (6), eliminates the need to know the pathlength (L). This optical technique is the one used for the comparison measurements reported later.

Interferometric Measurement and Visualization Method

An optical system which can provide both qualitative and quantitative information about intensity distribution of acoustic wavefronts was presented at the 87th Annual Meeting of the Acoustical Society of America in New York, April 24, 1974 (14). This system, which was developed by the RCA Laboratories, is uniquely suitable for measuring ultrasonic wavefronts for medical diagnostic ultrasonic equipment. The system is capable of measuring acoustic wavefronts in water at frequencies from 0.5 to 5 megahertz and in intensities as low as 250 nanowatts per square centimeter with a linear response up to several watts per square centimeter. The arrangement of this system is basically that of a Michelson interferometer as illustrated in Figure 7.

The Michelson interferometer arrangement consists of one mirror which is external to the acoustic field and a second mirror which is the pellicle suspended in the water bath normal to the path of the ultrasonic wave.

The pellicle is thin enough (8 microns) so that it is essentially transparent to the ultrasonic field. Since the pellicle is transparent to the ultrasound the motion of this flexible pellicle membrane corresponds to that of the transmitting medium. Therefore its displacement is equal to the particle displacement amplitude of the ultrasonic wave.

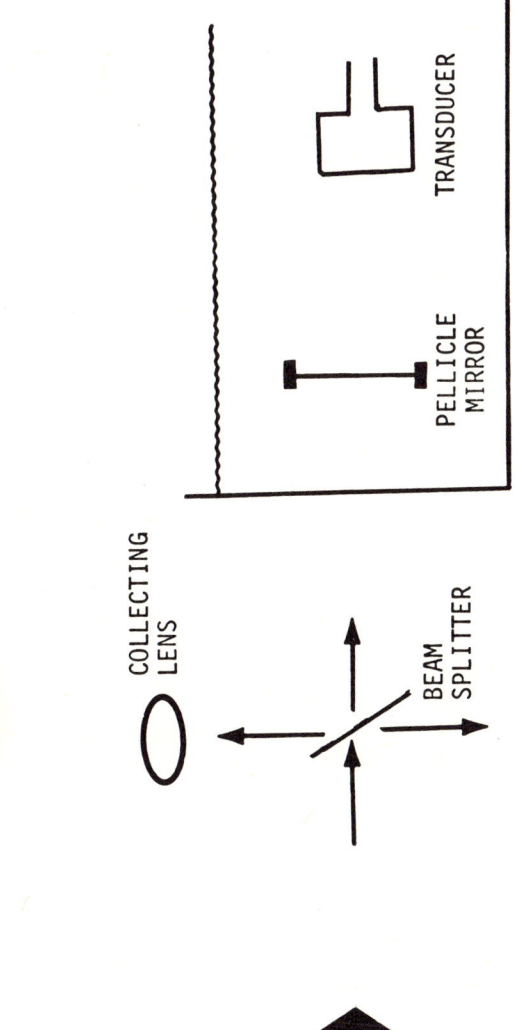

Fig. 7. Basic Interferometric Arrangement for Measurement and Visualization of Ultrasonic Fields

The displacement of the wave is related to the acoustic intensity by the following relationship

$$I = 1/2 \, Zw^2 \Delta^2$$

where Δ is the particle displacement amplitude, w the angular frequency of the ultrasound and Z the acoustic impedance. Thus the measurement of pellicle motion provides a detailed measurement of the particle displacement amplitude and intensity. A laser beam is raster scanned across the pellicle. The current from the photodiode provides a measure of the pellicle displacement which oscillates with the frequency of the acoustic wave. This photocurrent may be measured directly or used to brightness-modulate a cathode ray oscilloscope and thus produce an image of the sound field. The features of the system include: high sensitivity, large dynamic range, wide angular response, good resolution and broad frequency range which encompasses those predominantly used in medical diagnostic ultrasound. With this system, ultrasonic wavefronts can be examined either directly from the transducer or after being acted upon by other acoustic elements and provide point acoustic intensity data as well as the distribution of that intensity. The use of this system to measure the spatial distribution of intensity in wavefronts from diagnostic ultrasonic transducers should provide a convenient and rapid method for evaluation of ultrasonic diagnostic transducers.

Hydrophones

There are many instances when a fast, small, sensitive detector is needed, such as in the measurement of peak pulse intensities from diagnostic systems. Piezoelectric probes presently are the most practical means for making such measurements, their sensitivities ranging down to approximately 10^{-9} watts/cm^2 (16,1). Some of the most serious problems associated with these detectors are directional sensitivity, lack of wideband frequency response, and the need for calibration.

One of the hydrophones used in our laboratory to measure beam profiles and pulse characteristics from ultrasonic therapy fields consists of a lead zirconate titanate hollow cylinder (PZT-5H Vernitron Corporation), 1.6 mm. in diameter, 1.6 mm. in length and with a 0.25 mm. wall thickness. It is mounted on the end of a length of 1.6 mm. outside diameter aluminum tubing. This hydrophone is used as part of a scanning system for beam profile measurements from ultrasonic therapy transducers in the 1 megahertz frequency range. Because of the directional sensitivity of this hydrophone, it is mounted in a mechanical positioning system so that the probe axis remains at a right angle to the ultrasonic beam during the scanning process.

A commercially available hydrophone of small size and which reportedly can be used as an omnidirectional receiver is marketed by Helix Ultrasonics. This ultrasonic hydrophone transducer consists of a 460 micron diameter, 10 megahertz piezoelectric ceramic disc built into the end of a modified number 21 gauge hypodermic needle.

The calibration of such hydrophones can be accomplished by substitution techniques using suspended spheres or thermal probes in an ultrasonic field. The second technique for calibration is that of the graphical integration of a beam profile which provides a value proportional to the total output power of the transducer. Then total power measurement techniques, such as the float technique, can be utilized in calibrating hydrophones. A comparison of the calibration constant obtained using these two techniques as part of a study on the measurement of beam profiles from ultrasonic therapy transducers has recently been reported by Herman et al. (7).

Intercomparison Measurements of Various Techniques

Recently the results from three of the independent measurement methods discussed previously were intercompared in cooperation with the National Bureau of Standards (24). For this set of measurements, the National Bureau of Standards provided two transducers whose characteristics had been carefully determined. Based on measurements of the impedance of the generating ultrasonic transducers and applied voltage predicted output values were compared against values measured by three of the Bureau of Radiological Health measurement systems. These systems include the radiation force float, the Raman-Nath acousto-optic, and the calorimetry systems. Tables II and III present the data collected for measurements on the two National Bureau of Standards air-back quartz crystal transducers, each with a resonant frequency of 2 MHz. The quartz crystals were circular discs with active surface diameters of 1.905 cm (3/4 inch) and 1.270 cm (1/2 inch). The majority of the measurements made in this preliminary intercomparison by the National Bureau of Standards and Bureau of Radiological Health were within 5% of the mean value at each power level. This data was recently presented at a workshop on ultrasonic therapy equipment conducted by the Bureau of Radiological Health (24). In this particular set of measurements, the large variations in the float readings in Table II at an applied voltage of 203 volts have not been explained. In another recent intercomparison of measurements made in our laboratory between the radiation force float method and the acousto-optic method, variations in the values measured using the floats were much smaller (6). This experiment involved 20 measurements of the float system and 20 measurements of the acousto-optic system at power settings from 70 milliwatts to

Table II.   Comparative Measurements of Acoustic Power from 3/4" Quartz Transducer

Power (milliwatts)
(See note b below)

Voltage[a] (volts)	Electrical Method (NBS)	Rad. Force Mehod (BRH)	Acousto-Optics Method (BRH)	Calorimetric Method (BRH)
232.6	603(-4.4)	658(+4.4)		
203.0	459(-1.0)	377(-18.7) 535(+15.4)	479(+3.3)	469(+1.1)
100.8	113(-0.5)	111(-2.3)	116.7(+2.8)	
50.54	28.5(-3.9)		30.8(+3.9)	
30.26	10.2(-3.8)		11.0(+3.8)	

a--rms voltage measured across the transducer

b--numbers in ( ) indicate percent deviation from the mean (measured-mean/mean). Where the mean is determined by averaging the values obtained for all techniques at one voltage setting.

Table III.   Comparative Measurements of Acoustic
              Power from 1/2" Quartz Transducer

Power (milliwatts)
(See note b below)

Voltage[a] (volts)	Electrical Method (NBS)	Rad.Force Method (BRH)	Acousto-Optics Method (BRH)	Calorimetric Method (BRH)
203.0	204(+0.4)	200(-1.6)	219(+7.7)	190(-6.5)
100.8	50.3(-1.9)		52.6(+2.5)	51 (-0.6)
30.26	4.53(-4)		4.9(+4)	

a--rms voltage measured across the transducer.

b--numbers in ( ) indicate percent deviation from the mean
   (measured-mean/mean). Where the mean is determined by
   averaging the values obtained for all techniques at one
   voltage setting.

2 watts. The experiment was performed using a 1-MHz 1-inch diameter lead zirconate titanate crystal. The slopes and variances of the slopes were calculated and compared. The square root of the variance is one standard deviation for a gaussian distribution; twice that value gives a 95% confidence level for repeatability. This 95% confidence level approximation for repeatability for optical measurements was 1.15% of the mean, over the range of the power investigated and for the floats it was 6.25% of the mean. These values refer to the reproducibility of each given technique itself, and thus indicate a much smaller variation for the radiation force float method than was seen in the intercomparison study conducted by the National Bureau of Standards and Bureau of Radiological Health.

REFERENCES

1. BRENDEL, K.: Hydrophones, D.H.E.W. Publication (FDA) 73-8008 (Sept. 1972).

2. DUNN, F., FRY, W.J.: Ultrasonic field measurement using the suspended ball radiometer in thermocouple probe. Interaction of Sound and Biological Tissues Workshop Proceedings, Ed. J.M.Reid and M.R. Sikov. D.H.E.W. Publication (FDA) 73-8008 (1972) 173-176.

3. ERIKSON, K.R.: Calibration of standard ultrasonic probe transducers using light diffraction. D.H.E.W. Publication (FDA) 73-8008 (Sept. 1972) 193-197.

4. FRY, W.J., DUNN, F.: Ultrasound: Analysis and experimental methods in biological research, Physical Techniques In Biological Research. (Nastuk, W.L. Ed). Academic Press. New York 4 (1962).

5. FRY, W.J., FRY, W.B.: Determination of absolute sound levels and acoustic absorption coefficients by thermocouple probes. J.Acous.Soc.Amer. 26 (May 1954) 294-310

6. HARAN, M.E., COOK, B.D., STEWART, H.F.: A comparison of an acousto-optic and radiation force method of measuring ultrasonic power, presented at 87th Meeting of the Acoustical Society of America (23-26 April 1974).

7. HERMAN, B. et al: Measurement of beam profiles of ultrasonic therapy transducers. D.H.E.W. Publication (FDA) 73-8029 (March 1973).

8. Herrey, E.M.J.: Experimental Studies on Acoustic Radiation Pressure. J. Acoust. Soc. Amer. 27 (1955) 891-896.

9. HIEDEMANN, A.A., COOK, B.D.: The measurement of acoustical pressure by optical techniques. A report to Office of Naval Research for Contract NORN-2587 Project No. 384-304. Technical Report No. 19 (June 1968).

10. HUETER, T.F., BOLT, R.H.: Sonics, Section 2.12, Radiation pressure. Wiley, New York (1955) 43-53.

11. KLEIN, W.R., COOK, B.D.: Unified approach to ultrasonic light diffraction. IEEE Transactions on Sonics and Ultrasonics, Su-14 (July 1967) 123-134.

12. KOSSOFF, G.: Calibration of ultrasonic therapeutic equipment. Acustica, 12 (1962) 84-90.

13 KOSSOFF, G.: Balance technique for measurement of very low ultrasonic powers. J. Acoust. Soc. Amer. 38 (1965) 880-881.

14 MEZRICH, R., ETZOLD, K.F., VILKOMERSON, D.: Ultrasonovision: An interferometric device for ultrasonic visualization and measurement. 87th Annual Meeting of the Acoustical Society of America (23-26 April 1974).

15 MIKHAILOV, I.G.: Methods of measuring the absolute intensity of ultrasonic waves in liquids. Ultrasound 2 (July 1964) 129-133.

16 NEWMAN, D.R.: Measurement of diagnostic level of ultrasound using small piezoelectric transducers. D.H.E.W. Publication (FDA) 73-8008 (Sept. 1972).

17 OBERST, H., RIECKMANN, E.: Methods of measurement used by the Federal Institute of Physical Technology for Evaluation and Certification of Medical Ultrasonic Instruments. Amtsblatt Der Physikalisch-Technischen Bundesanstalt, No. 2 (1953) 35-46.

18 RAMAN, C.F., NATH, N.S.N.: The diffraction of light by high frequency sound waves. Pro. Indian Acad. Sci. 2 (1935) 406; 2 (1935) 413; 3 (1936) 75; 3 (1936) 119; 33 (1936) 459.

19 RICHARDS, W.T.: Heating of liquids by absorption of sound, and its relation to the energy of intense high frequency sound waves. Proc. Nat. Acad. Sci. Wash. 17 (1931) 611.

20 ROBINSON, R.A., STEWART, H.F.: A portable ultrasonic radiometer. D.H.E.W. Publication (FDA) 73-8029 (1973) 534-541.

21 ROONEY, J.A.: Acoustic radiation pressure and its use in power and intensity determinations. Presentation at 84th Meeting of the Acoustical Society of America. 28 Nov. - 1 Dec. (1972).

22 ROONEY, J.A.: Determination of acoustic power outputs in the microwatt-milliwatt range. Ultrasound in Med. and Biol. 1 (1973) 1-4.

23 SOKOLLU, A.: Absolute measurement of total irradiated power of ultrasonic transducer for biomedical use. Bulletin du Laboratoire d'Electroacoustique Universite de Liege. 9 (1966) 23-27.

24 STEWART, H.F., Editor: Proceedings of Workshop on Ultrasonic Therapy Equipment. In preparation as BRH report (1974).

25 VanDenENDE, H.: A radiation power calorimeter and radiation force meter for small ultrasonic beams. Med. and Biol. Engng. $\underline{7}$ (1969) 411-417.

26 WALLACE, H.C.: High underwater attenuation measure in Ivory Bar Soap. J. Acous. Soc. Amer. (1973) 1187.

27 WEMLEN, A.: A milliwatt ultrasonic servo-controlled balance. Med. and Biol. Engng. $\underline{6}$ (1968) 159-165.

28 YOSIOKA, K., HASEGAWA, T., OMURA, A.: Comparison of ultrasonic intensity from the radiation force on steel spheres with that on liquid spheres. Acustica $\underline{22}$, 3 (1969/70) 145-152.

29 ZIEDONIS, J.G.: Pressure balance design for measuring ultrasonic energy. 84th Meeting of the Acoustical Society of America, 28 Nov. (1972).

-DISCUSSION-

JOHNSON - What is that pellicle material made of, and how thick is it?

STEWART - 8 micron gold coated mylar.

DUNN - This system you described is a very important one for several reasons; we always talk (or speak) in terms of measuring power or intensity or other second order quantities. But really, what we should be doing is trying to measure first order quantities, which is what your system allows us to do. Presumably, it measures displacement.

STEWART - Right. You can measure displacement amplitude and also particle velocity.

DUNN - Right. If you then have another way of getting the pressure and the phase relationship between those two, then you can say that you truly know everything you want to know about the field. But simply saying that you measure intensity or power, is really not what is ultimately desired; though many of us have been doing this only because that is the best we have been able to do.

Secondly, the thermoelectric method that you mention, in which you had the parabolic cone, can be an absolute method. At least the method is absolute the way we use it, and I presume that yours can be too.

STEWART - We have not tried to use it as an absolute measurement method.

HILL - What is the sensitivity of your thermal probe?

STEWART - It will be able to measure in the milliwatt/cm^2 range.

DUNN - Could the pellicle be used for measuring pulse?

STEWART - Yes.

NYBORG - The pellicle method also seems promising, I believe, for measuring pressure. The acceleration amplitude at a point (which you get immediately from the displacement amplitude and the frequency), according to the law of dynamics of fluids, is equal to the pressure gradient divided by the density.

JOHNSON - Is the pellicle really flexible enough to measure the medium displacement? If it has a sufficient degree of stiffness, it will not accurately measure displacement and velocity.

STEWART - We haven't done that work in our laboratory ourselves, but RCA has. And, at the present time, the answer appears to be "yes."

KAUFMAN - What powers do you use for diagnostic work?

STEWART - Average power levels are generally in the low milliwatt/cm^2 range. However, they are pulsed units, and so, the peak powers are on the order of Watts/cm^2. So, it depends upon whether you are talking about peak powers or average powers.

KREMKAU - In the thermal probe, where you have the parabolic mirror, you are not dealing with a plane wave situation at the probe. Have you measured the total power by another method to know whether or not the theory that's been worked out permits using this as an absolute measuring device?

STEWART - We have felt that because of the fact that we would be dependent upon getting it focused into an area that it would probably not be a very good absolute method. At least it would not have the accuracy that you might expect from some of the other methods. The problem is first getting it all focused into one area.

KREMKAU - Have you made any measurements using the approach that you would use if you wanted to use it for an absolute measuring device?

STEWART - No.

DUNN - One of the reasons that it doesn't lend itself too well as an absolute measuring method is that you have to know the absorption coefficient, the specific heat, and the density of the material. These are not so easily gotten. Thus, if you know the absorption coefficient within 10 percent, you are not going to have an absolute value better than 10 percent.

HILL - In the field of ultrasound dosimetry there is not too much of a problem in finding absolute methods. I think it is necessary to set up some absolute methods against which one can calibrate the devices which are appropriate for measuring the detailed parameters, which are going to be relevant to what you want. So, I don't know if it very much matters whether this particular device is an absolute device or not. It can perfectly well be calibrated, I would guess, and there are plenty of ways of doing that.

STEWART - We are working, in cooperation with the National Bureau of Standards, toward developing some type of calibration capability. The National Bureau of Standards has a contract

STEWART - from BRH to develop a calorimeter that can be used for calibration purposes. The calorimeter, hopefully, will be completed sometime in 1975. NBS has also done some work in looking at the characteristics of transducers in order to produce standard fields of calibration for devices.

NYBORG - With therapeutic devices and other applications, there is a need for knowing what the distribution is in the near field, where the acoustics are complicated. The pressure and the velocity are not necessarily in phase, and generally are not. And, the direction of the velocity varies over the face of the transducer. There doesn't seem to be very much in the way of techniques for scanning such fields, although the pellicle method might be one. Also, the thermocouple might be considered; there is a question though about how accurately a calibration could be relied on here. The theory for thermocouple techniques seems to depend on the assumption of a plane wave, whereas in the near field, one doesn't have a plane wave.

DUNN - One can make thermocouples very small, as small as a micron. In other words, it can be a very, very small fraction of the wave length. So, even though the wave is not a plane wave, it could give a good answer to what is there at that point in space. However, it only gives a second order parameter such as intensity. But the intensity, itself, is not a well-defined quantity in the near field.

DUNN - That's right. But that's all it gives you. It is an energy-integrating device. It can't give you anything else.

VOGELMAN - One of the problems in building ultrasonic filters is that they respond to acoustic waves, and you get a filter in which the resonant frequency is continuously changing. I am sure you are familiar with the Fresnel filter. Basically you have two grids that are spaced n wave lengths apart depending on how narrow you want the filter passband. If one of them is put into the medium, the other one sits somewhere else; you can get your resolution down to whatever you want by just changing the spacing. Put infrared through it and measure the relative intensity of what's received at some kind of photo cell.

STEWART - In this particular system this pellicle is not stationary, but moves and oscillates with the medium.

JOHNSON - Doesn't the pellicle have to be aligned with the field so that there is no phase change over the surface? Isn't that a pretty critical alignment?

STEWART - The beam axis needs to be lined up so that it is

STEWART - normal to the pellicle.

CARSTENSEN - Assuming that this device either doesn't work out or isn't universally available, what sort of accuracies would you expect in the intensity measurements by the various techniques that you can conceive?

STEWART - I think probably, at the present time, you would be doing good if you could get plus or minus 10 to 20 percent.

CARSTENSEN - What kind of probe would you use?

STEWART - Probably a piezoelectric or thermal probe, if you are looking at spatial intensities. And, of course, if you are looking at temporal intensities, then you are forced to go to something like a piezoelectric conductor.

CARSTENSEN - How would you get an absolute intensity, say, with a thermal probe?

STEWART - It would have to be calibrated against a radiation force device.

CARSTENSEN - You mean taking a beam pattern of the transducers?

STEWART - Graphical integration of a beam profile would provide a value that is proportional to the total power. You can also calibrate it using a substitution technique with metal spheres.

CARSTENSEN - You would use the metal spheres as the absolute standard?

STEWART - In that case, you would be forced to. Yes.

HILL - It seems to me that the hydrophone is basically a very important device to use because you can get very good spatial resolution, and temporal resolution. You can get band widths of the order of 20 MHz.

You can calibrate hydrophones against something like a radiation balance. But, I think, another point is that for most purposes, plus or minus 10 percent is terrific accuracy, because we are working in a region where we just don't know not only what the nature of the effects are, but usually whether there are any effects or not. So that, I think, we should not give a sort of false feeling of insecurity, we cannot do much better than 10 percent.

CARSTENSEN - I really question whether we can get that kind of accuracy in intensity measurement.

DUNN — I think under the best conditions, one can and does. I would seriously doubt that a physician in his office would get anything like 10 percent. But then, he probably doesn't need more than 100 percent. And that, I believe, he can get too.

CARSTENSEN — Is it reasonable then to agree that we are pretty lucky if we can get within a factor of 2 on intensity?

HILL — No, I think you can do better than that.

NYBORG — You are speaking of intensities in the far field now, I imagine? If so, one can get a more optimistic figure.

CARSTENSEN — Can you measure one watt per square centimeter with an accuracy of ± 1 or 2 db?

HILL — I would be pretty confident that provided you set out and wanted to do the job properly, you could do this within ± 20 percent.

CARSTENSEN — Using a thermocouple probe or a sphere?

HILL — I would use a calibrated hydrophone.

CARSTENSEN — Calibrated against what?

HILL — Calibrated against a standard beam which you have set up on to a radiation balance.

CARSTENSEN — You measure with the hydrophone the beam pattern of a transducer, measure its total power and then determine from that what the intensity is on the axis?

HILL — That's right. You set up, for example, a CW beam onto a radiation force balance and measure the total power. You then leave the beam on and replace the force balance with a hydrophone and scan across the region that you have intercepted with your force balance. And, you do the sums to connect the two.

DUNN — I think I would go a step further and say that the reliable manufacturer could actually produce these things (hydrophones) with the calibration sheets.

STEWART — We have seen good agreement in the calibration factors obtained using the integration technique and the substitution technique. This tends to give some confidence in the calibration. In one case you are calibrating the hydrophone using a steel

STEWART - sphere. In the other case, you are making a total power measurement and integrating.

Energy Absorption

TRANSIENT EFFECTS OF LOW-LEVEL MICROWAVE IRRADIATION ON
BIOELECTRIC MUSCLE CELL PROPERTIES AND ON WATER
PERMEABILITY AND ITS DISTRIBUTION*

Adolfo Portela, Osvaldo Llobera, Solomon M. Michaelson,
P.A. Stewart, Juan C. Perez, Ariel H. Guerrero,
Carlos A. Rodriguez, and Roberto J. Perez

Instituto de Investigaciones Biofisicas
Consejo Nacional de Investigaciones Cientificas y Tecnicas
Buenos Aires, Republica Argentina

Biomedical Sciences Division, Brown University
Providence, Rhode Island, U.S.A.

Comando General del Ejercito, Jefatura III, Operaciones
Departamento de Investigacion y Desarrollos
Buenos Aires, Republica Argentina

Department of Radiation Biology and Biophysics
The University of Rochester
Rochester, New York, U.S.A.

ABSTRACT

Microwave radiation effects on passive and dynamic electrical properties and on cell water parameters were studied in muscle cells from muscles of the South American Frog Leptodactilus ocellatus. Microwave exposure of 10 mW/cm^2 for a period of 120 minutes produced transient changes in specific membrane resistance R_m, the membrane capacitance C_m and the space constant λ. Those

---

*This work was supported by the Office of Naval Research, U.S. Department of the Navy; the Comando General de la Armada Argentina and the Comando General del Ejercito, Direccion General de Investigacion y Desarrollo, Ministerio de Defensa, Republica Argentina.

electrical parameters related to the excitation and propagation of the action potential, i.e., the rate constants k_r and k_K, the maximum rate of rise $\dot{V}+$ and fall $\dot{V}-$ of the action potential, the limiting membrane conductances (gNa and gK), the peaks of sodium inward and potassium outward ionic currents, the net ionic charge accumulation per action potential and the propagation velocity of the action potential, were all transiently altered. The water membrane permeability and the fraction of osmotically available cell volume were also transiently altered.

The analysis of these parameters has shown that the transient changes evoked by microwave radiation are larger in muscle cells from "winter frogs" than from "summer frogs." Seasonal differences in the observed transient microwave radiation effects were analyzed. It was concluded that microwave exposure to $10mW/cm^2$ did not produce permanent effects on electrical and cell water parameters.

INTRODUCTION

Microwave radiation at sufficiently high power levels is known to cause biological effects due mainly to the generation of heat in the organism. Effects observed at high levels include observable lesions, as a result of hyperthermia. Some workers have suggested that the nervous system is transiently affected by microwaves, even at low power density exposure. The extent and importance of more subtle changes which may occur at lower power levels, particularly with continued or long-term exposure, are not adequately known (Gordon 1962, 1970; Kamenskiy 1964, 1968; Lobanova 1962; McLees and Finch 1973; Michaelson 1971; Presman 1962, 1963, 1970). These reported low-level effects include neurasthenic responses and suspected behavioral changes, neuroendocrine effects and cardiovascular changes to apparently low power density exposure.

Power absorbed by an object from a radiation field will be dissipated as heat. This is especially true for microwave fields and has led to considerable controversy as to whether neural and behavioral effects noted in such irradiation fields are due simply to heating (thermal effects) or to specific stimulation of the neural network by the electromagnetic radiation field (non-thermal effects). The problem of changes in neuronal functions in a microwave field has received considerable attention (Lobanova et al. 1962, 1971; Kamenskiy 1964, 1968; McAfee 1961, 1962, 1969). The investigations conducted by Presman et al. (1962, 1963) have led to the hypothesis that microwave radiation alters those mechanisms involved in the function of excitatory structures (Lobanova 1971). This theory was advanced by Kamenskiy (1964) incorporating parameters of stimulation. Other investigators have refused to accept the possibility of non-thermal neural stimulation and have considered an explanation based solely upon local heating (McAfee 1969).

Several experiments have been done to study the effects of microwave irradiation on the nervous system of homeotherm and poikilotherm animals (McAfee 1961, 1962, 1969). The experiments were designed to separate thermal from non-thermal neural effects. Refrigerated poikilotherm animals were irradiated with a power level of 45 mW/cm^2 (10,000 MHz) for various periods of time. In addition, experiments at the same wavelength and power density were conducted on isolated nerve preparations.

The results from the peripheral nerve experiments indicate that previous reports of neural effects may be explained as due to local heating of peripheral nerves rather than as excitation of the central nervous system (McAfee 1961, 1962). The principal effect observed was a temperature dependent increase of motor activity of the irradiated animal. These experiments provided evidence that the observed effects of microwaves on isolated nerve preparations were reproducible by equivalent non-electromagnetic heating.

In many metabolic experiments, periodic daily changes have been observed. More striking are seasonal changes, when there is a marked difference in summer and winter climate (Florkin and Schoffeniels, 1969). Seasonal differences in respiratory rates, osmoregulatory mechanisms, etc., in several species may be correlated with the animal's biochemical composition during the various seasons (Vernberg and Vernberg 1972). It is known that many physiological functions adapt themselves to environmental changes (Proser 1958).

Recent work by Portela et al. (1974) has indicated that muscle cells from winter frogs exposed 120 min to microwaves (3 GHz) at power densities between 0.5 and 10 mW/cm^2 show transient changes in several osmotic and electrical cell parameters. The present study, which is a continuation of this work, was designed to analyze transient effects evoked by microwave irradiation at a power density of 10 mW/cm^2, and to detect seasonal variations and thermal effects in electrical and osmotic parameters of frog muscle cells.

METHODS

Muscle Preparation for Measurement of Cell Electrical Properties

Muscle tissue from the South American frog, Leptodactilus ocellatus, were used. Sartorius muscles (fresh weight, 45-47 mg) were carefully dissected from each frog with minimal damage to the fibers, leaving the lower side of the muscle practically free of connective tissue. The nerves were severed 1 cm from the muscle surface. The geometrical shape of this muscle corresponds to a plane sheet with the following dimensions: length, 2 cm; width, 0.4 cm; thickness, 0.08 cm. The muscle density was estimated to be about 1.07 g/cm^3. Membrane surface area measurements were carried out by light microscopy. Diameters and lengths of fibers (i.e., intact cells) gave mean diameters of 78-80 microns. The average cell membrane surface value was 500 $\pm$ 50 cm^2 per gram of muscle (volume/surface, ratio: 0.002 cm). The average extracellular space for sartorius muscle was 26% of muscle wet weight (Portela et al. 1965).

The nerve-muscle preparations were used as described by Portela et al (1970; 1970a, b; 1974). End plate regions and end plate free regions toward the end of the cells were carefully identified for each muscle cell studied according to the techniques of Fatt and Katz (1951), Thesleff (1955) and Portela (1970a, b). By carefully mapping the microscope field, it is possible to identify a particular cell and to return the microelectrode to within 50 micra of the same cell. Bioelectrical responses evoked by means of intracellular stimulation were measured repeatedly by short-term micropippette electrode insertions. Consequently, the electrical parameters of the muscle cell membrane free of motor end plates were analyzed.

Membrane electrical constants were determined following the general methods of Fatt and Katz (1951) and Portela et al. (1970, 1974). The calculations were based on the cable analysis of Hodgkin et al. (1946), Katz (1948), modified for use with the intracellular microelectrodes developed by Fatt and Katz (1951), Portela et al. (1974), and Hubbard (1963), Frant et al. (1964) and Plonsey et al. (1969). The minimum strength of the rectangular current pulse required to initiate an action potential was determined according to Portela et al. (1974).

Dynamic electrical parameters of the active membrane, including the magnitude of the ionic current associated with the action potential were calculated following the techniques and analytical procedures described by Jenerick (1963, 1964), Minorsky (1947), Graham et al. (1961) and Portela et al. (1974), assuming the genesis of the action current in muscle cells as postulated in the Hodgkin-Huxley model of the nerve axon (1952).

The nerve-muscle preparations were bathed with Ringer solution (pH 7.2 - 7.4) of chemical composition given in Table I. The osmolarity of solutions was checked cryoscopically. All experiments were carried out at 25°C; temperature dependence of the passive and active electrical membrane parameters (as well as cell water parameters) were also studied in the temperature range from 20° to 30°C. Biological preparations were perfused with Ringer solutions for 30 minutes following dissection, before experiments started.

Irradiation Procedures

Two types of experiments were run simultaneously following these protocols for determining the final criteria for irradiation procedure:

1. One chamber with each pair of sartorius muscles from the same frog was irradiated. Control muscles occupied similar plastic chambers outside the irradiation room during irradiation exposure and were otherwise treated identically. Within 5 seconds after completion of irradiation, the electrical measurements were obtained from the identified muscle cells as described above. This procedure is described in detail in Portela et al. (1974).

2. The electrical measurements were performed in the same muscle before and 5 seconds after irradiation, in the identified muscle cells. The control cells from protocol (1) measurements thus provided non-irradiated control data for protocol (2) measurements. The striking feature of these data is the constancy of the electrical membrane parameters for both controls. There was no statistical difference between the electrical parameters for controls determined following these two protocols. The irradiation data from these two experimental procedures were also similar. Thus, the data presented in this work were obtained following the criteria established in protocol (2).

The nerve muscle preparation was irradiated in the perfusion chamber, and exposed to 10 mW/cm^2 for 120 minutes, in the far field region on the axis of the horn antenna.

Preliminary work was recently published, reporting data from

TABLE I

Composition of Bathing Solutions

Solution	NaCL mM/l	KCL mM/l	CaCL$_2$ mM/l	Na$_2$HPO$_4$ mM/l	NaH$_2$PO$_4$ mM/l	Osmolarity mOsm/l
Normal Ringer	111.2	2.5	1.89	2.5	0.5	222
Test Ringer	59.3	2.5	1.89	2.5	0.5	111

muscle cells corresponding to frogs collected in the winter (Portela et al. 1974). The present study compares these data with those obtained in frogs collected in the summer. Data presented here were obtained between July 1971 and March 1974.

Maintaining Constant Temperature in the Apparatus for Nerve-Muscle Preparations

Special attention has been given to temperature control of the perfusion apparatus to insure constant temperature. The nerve-muscle preparation is held in a cylindrical water-jacketed chamber (internal diameter, 1 cm; length, 10 cm) with a sintered-glass filter disc sealed onto the top portion. The inflowing normal Ringer solution is delivered by means of teflon tubing (3 millimeter internal diameter) and the outflowing fluid is recirculated by a roller pump from a reservoir of 200 cm^3 containing the Ringer solution at the desired temperature, controlled by the thermoregulated bath system (Figure 1). The biological preparation is perfused at a flow rate of 50 cm^3/min. The additional circuitry for temperature control of the nerve-muscle chamber assures the chosen temperature, by means of a high flow rate of KCl-dioxane-water solution of 5 liters/min.

The temperature is recorded in the water circuitry at convenient points as indicated in Figure 1. In addition, the temperature of the biological preparation perfusion normal Ringer solution is monitored.

The reservoir containing the perfusion solution is enclosed in the KCl-dioxane-water solution temperature regulated bath, assuring equal temperature in the biological preparation perfusion chamber and water-jacketing system. The Pyrex water-jacketing system enclosing the perfusion nerve-muscle chamber is spherical (radius = 10.7 cm) and has a capacity of 5 l. The KCl-dioxane-water solution is supplied from the thermostatically controlled KCl-dioxane waterbath and circulator and is recirculated at a flow rate of 5 l/m. This equipment permits maintenance of the temperature of both perfusion solution and KCl-dioxane-water solution within $\pm$ 0.02°C, in the range 0° to 50°C. The nerve-muscle preparations were mounted in the teflon holder device at 120 percent of their resting length, adjusted by means of a special teflon micrometric system. This holder is plugged in the water spherical jacketing system as indicated in Figure 1. The sealed perfusion chamber has a capacity of 10 cm^3.

Biological preparations mounted as described were exposed to a power density of 10 mW/cm^2 for 120 minutes, which is the time in which reversible effects on the studied parameters were observed.

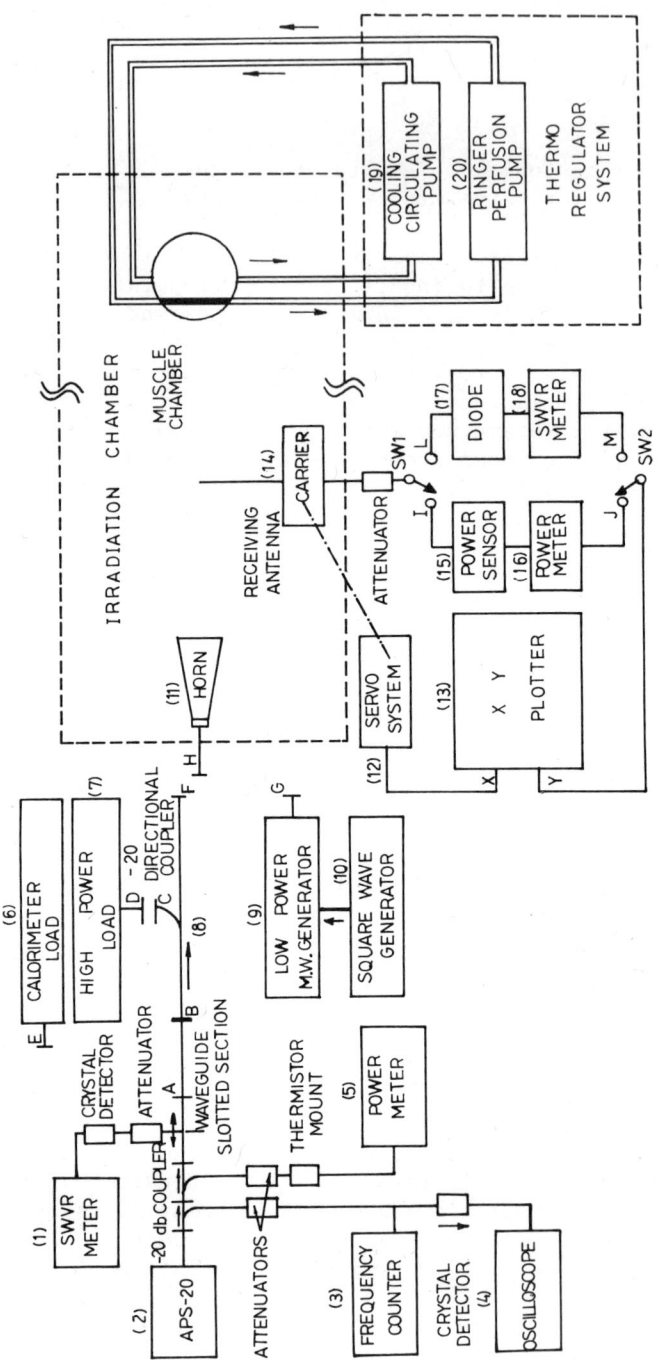

FIGURE 1. General block diagram of the Irradiation System.

The flux of 10 mW/cm^2 will increase the temperature of 1 g of muscle tissue (density 1.07 g/cm^3) by 1.5°C after 10 minutes of exposure, provided that this power is entirely absorbed. Temperature distribution in the biological preparation irradiated by the external microwave energy source was estimated by solving the thermal diffusion equation as described by Chan et al. (1973). The thermal conductivity and the specific heat coefficients were determined (Ponder 1962), giving the values 0.0012 cal cm/cm^2 sec °C and 0.085 cal/g °C, respectively. The Ringer solution at the chosen temperature was perfused into the nerve-muscle chamber directly cooling the preparation. The external thermal regulation device here described, permits a rapid heat transfer to the Ringer solution (Liang-Tseng Fan et al. 1971; A. K. Chan et al. 1973; Shitzer 1973).

Temperature Dependence of the Passive and Active Electrical, and Cell Water Parameters

To study the relationship of changes in the analyzed parameters to change in temperature, we used as a mathematical model, the Arrhenius equation. The temperature of the nerve-muscle preparation was monitored with thermistor probes placed in the vicinity of the recording microelectrodes. The "cable" as well as the active electrical parameters were determined as previously described. The values of several parameters of the action potentials depend on the magnitude of the membrane capacitance C_m (Nastuk and Hodgkin 1950). Since C_m has not been observed to be temperature dependent between 20° and 30°C, corrections for temperature are not required. Muscle cell water parameters were simultaneously determined following procedures described above.

Microwave Irradiation

The nerve-muscle preparations were exposed to microwave energy, corresponding to a power density of 10 mW/cm^2, by means of an APS20 Radar Transmitter System. The microwave generator has the following characteristics: frequency: 2.88 GHz; wavelength 10.41 cm; peak power 2 megawatts; pulse width 0.67 microseconds; pulse repetition rate 900 ppsec; duty cycle 6 X 10^{-4} and average power 1.2 kilowatts.

The calibration of the radar unit and power density measurements were done following the general procedures used by Schwan et al. (1961). The power delivered from the generator to the transmitting antenna was attenuated to obtain a power density in the indicated range. This attenuation was achieved by means of a 20 dB calibrated directional coupler, the main branch of which was connected to a higher power load termination, while the secondary branch was coupled to the horn antenna (15.62 dB at 2.88 GHz, similar to a Narda 644). The output power of the microwave generator was measured before and after exposure by means of a water calorimeter. The VSWR of the load was determined using a slotted line and a Standing Wave Meter. For all cases the VSWR magnitude was found to be 1.18, indicating that almost all the power delivered to the antenna was radiated. The equipment components are shown schematically in Figure 1.

The uniformity of the radiation pattern of the transmitting antenna was determined to ensure the absence of standing waves inside the irradiation room. The horn antenna was conducted to a low power microwave generator and a receiving system for measuring the relative field strength was used. The receiving system consists of a quarter wave dipole antenna connected to a detector diode. A VSWR meter was used to amplify and measure the signal from the receiving antenna. The receiving antenna was mounted on a carrier-servosystem which permitted the scanning of antenna position in an X-Y transverse plane relative to the radiation axis. Power density measurements were made using a quarter wave receiving dipole antenna connected to a Hewlett Packard (HP) 430 power meter with an HP 477 thermistor.

Muscle Preparation for Studying Muscle Cell Water Parameters

Using single compartment analysis techniques, it is possible to evaluate the osmotically available volume, W_{eff}, and the cell membrane water permeability, P_w, of a living cell from measurements of the transient changes in cell volume after a step change in the osmolarity of the bathing fluid. The procedure involves isolating a single muscle fiber in an appropriate bathing chamber, and measuring its diameter every 10 sec over a period of 5 min, following an abrupt change in osmolarity of the bathing solution.

The fiber is mounted so that length changes, which should not occur, can be detected. Effective fiber volume per unit length can be calculated from the diameter measurement, and volume changes with time can thus be computed. Frog Tibialis anticus muscles, from both legs, with fresh weights in the range of 45 to 47 mg, were dissected free, with as long a tendon as possible still attached to each end, and mounted in a special plastic perfusion chamber equipped with small plastic platforms to which the tendons could be secured by insect pins. Dissection was carried out under stereomicroscopic observation. With fine dissecting instruments, the tendon at one end was carefully sectioned longitudinally and the excess muscle cells teased away, until a single intact muscle fiber was left

attached to its tendons. Throughout this process, the muscle
was covered with normal Ringer solution maintained at 25°C.
The muscle and the supporting plastic platforms were arranged
so that the single fiber lay in the solution with a slight arc
downward, both to ensure that it was not under excessive tension
and to control the length of the fiber.

Chamber and Bathing Solution Changes

Since the cell volume changes occur over a period of about
two minutes after changing the bathing solution, it is essential
to be able to change the solutions in a few seconds. This was
accomplished by a specially made flushing valve, which connected
the chamber to the gravity-fed supply lines from the solution
reservoirs and to the vacuum-powered drain tube. When this
valve was turned on, the old solution was withdrawn from the
chamber at the same rate as the new solution entered, so that
the level of fluid in the chamber remained constant. Complete
flushing occurred within 3 seconds. Operation of the flushing
valve was carefully synchronized manually with the shutter of the
camera.

Bathing Solutions

Two bathing solutions were used in these experiments, normal
Ringer and half-osmolarity or "test" Ringer. Their compositions
are specified in Table 1. Normal Ringer has an osmolar concentration of 0.222 osmol/liter and produced no detectable changes
in muscle cell volume. Osmolarity of test Ringer is 0.111
osmol/l. All solutions used were bubbled with 95% O_2 - 5% CO_2,
had a pH of 7.2-7.4 and were maintained at 25°C.

Single Muscle Cell Radius Measurements and Volume Calculation

Once the single muscle fiber was prepared, the dissecting
microscope was replaced by a camera-microscope combination focused on the center of the fiber. Photomicrographs of the muscle
cell were taken every 10 seconds over the 5 minute period following each solution change, as indicated in Figure 2. After
development and magnification, the diameter could be measured
in this final image with a precision, about 2%. Cell volume per
unit length was calculated by assuming the cells to be uniform
cylinders.

In type A experiments, cells were perfused and photographed
with normal Ringer with test Ringer and with normal Ringer again,

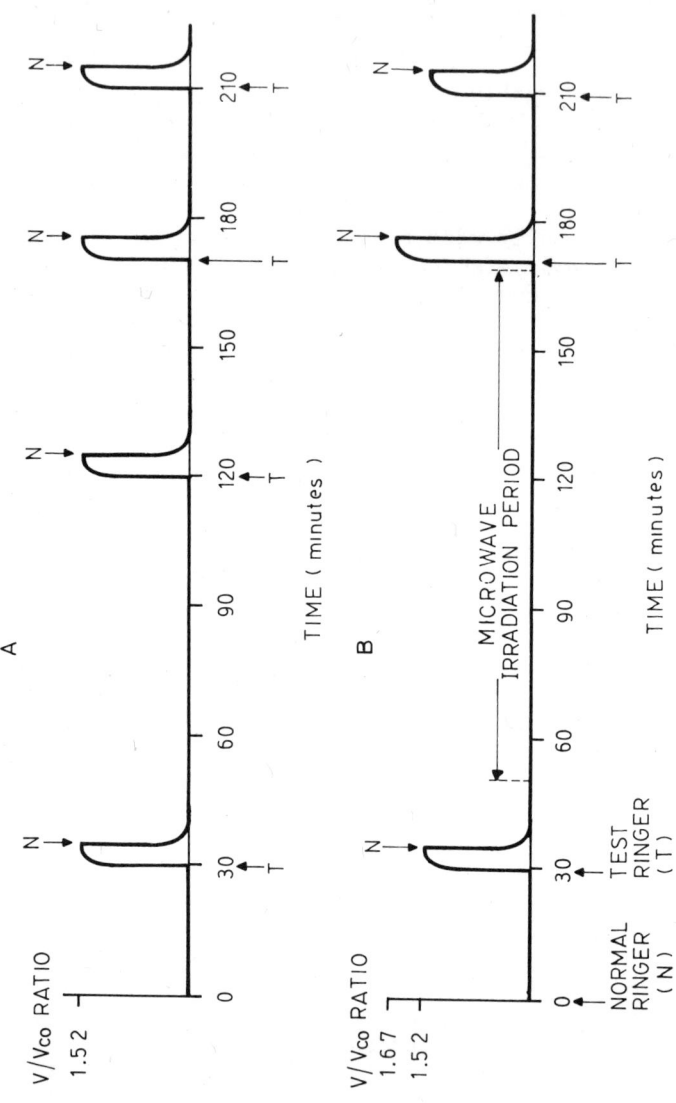

FIGURE 2. Relative muscle cell vol., V/V_{co} vs time in min (V_{co} – cell vol. in n Ringer sol.). Osmolarity of perfusing sol. change indicated by arrows. Upper curve A – average for 60 control muscle cells; lower curve B for 60 irradiated muscle cells as indicated.

each for five minutes. This entire cycle was subsequently repeated at 120m and 180 m as indicated in Figure 2.

In type B protocols, the second cycle of solution changes was replaced by a 120m irradiation period. The muscle cell was irradiated in the perfusion chamber. Within 60 s after completion of irradiation, the cycle of normal standard Ringer, half normal standard Ringer, and normal standard Ringer solution perfusions, was repeated. The type A measurements provided non-irradiated control data for the type B.

Determination of Osmotically Effective Volume Fraction, W_{eff} and Water Permeability, P_w

The derivation and details of the procedure are given in Portela et al. (1971, 1974). W_{eff}, of the test Ringer solution is given by:

$$W_{eff} = (V_c - V_{co}) / V_{co}$$

where V_{co} is the cell volume in normal Ringer and V_c in the test Ringer.

Curves for the determination of P_w were calculated by numerical solution of the non-linear differential equation describing cell volume V, as a function of time, incorporating the assumption of the previous paragraph. The equation is:

$$dV/dt = P_w \cdot 2\sqrt{\pi V} [C_o \cdot V_e / V - (V_{co} - V_e) - C_e]$$

where C_e is osmolarity of bathing solution, and the other symbols have already been defined.

RESULTS AND DISCUSSION

Seasonal differences in the observed transient microwave radiation effects on the passive and dynamic electrical properties and cell water parameters of muscle cells, from muscles of the South American frog (Leptodactilus ocellatus) are reported, indicating less effect on muscles from "summer" than from "winter" frogs, as a result of microwave exposures of 10 mW/cm^2 for 120m. Mean values and corresponding t-test for significance of muscle cell parameters determined before and immediately after irradiation, in experiments developed during winter and summer seasons for the period July 1971 - March 1974, are presented. Data for "winter" frogs were obtained from 60 cells from 20 different muscles and data corresponding to "summer" frogs were from 80 cells from 27 muscles.

Passive Electrical Membrane Parameters

An interesting feature of these data is the constancy of the passive and dynamic electrical membrane parameters of control muscle cells. Seasonal variations were not observed in these electrical membrane parameters. Immediately after microwave exposure, early transient effects have been observed in the specific membrane resistance R_m, the membrane capacitance C_m and the space constant λ of muscle cells. This microwave radiation effect was slightly greater in those muscle cells from winter frogs. Data are presented in Table II. The observed changes in the magnitude of the indicated electrical membrane parameters are transient, recovering their initial control values within time constants of approximately 3m for the observed value changes in muscle cells from summer frogs, and approximately 20m for muscle from winter frogs.

Action Potential Propagation

Immediately after microwave irradiation an early transient increase in the magnitude of the propagation velocity of the action potential θ, was observed. This effect was higher in muscle cell from "winter" frogs, returning to a control value with a time constant of approximately 20m. The induced effect on cells from "summer" frogs was characterized by a faster recovery of θ, with a time constant of approximately 3m. Data are presented in Table III.

Dynamic Electrical Parameters of the Active Membrane

The mean values of the phase plane trajectory (V, dV/dt) and membrane potential against time (V, t) are given in Tables III, IV and V. Data of muscle cell conditions corresponding to contral and immediately after irradiation are given with the corresponding t-test for significance. Microwave irradiation evoked a transient increase in the active membrane parameter $\dot{V}+$, $\dot{V}-$, V_{os}, V_{Na} and V_s.

The transient increase in the magnitude of V_{os}, V_{Na} and V_s as shown in Table III and Figure 3 were related to the increase in the maximum rate of rise of the propagated action potential. These parameters from the ionic current-membrane potential relations are given in Tables IV and V, and Figures 4, 5, 6 and 7. From the I-V relations it is clear that microwave energy absorption has enhanced the peak inward sodium current I_i. The inward ionic current from irradiated cells developed faster than that in control cells. Microwave irradiation, however, had

TABLE II Passive Membrane Electrical Parameters

Parameters Symbols and Units		Control (C)	Irradiated (I) Winter	Irradiated (I) Summer	Difference (I-C)/C, in % Winter	Difference (I-C)/C, in % Summer	"t" test Significance Level Winter	"t" test Significance Level Summer
V_r	mV	-87	-87	-87	0	0	none	none
$2a$	μ	78.6	75.1	76.6	-4.4	-2.5	.001	.05
λ	mm	1.18	1.06	1.12	-10	-5	.001	.001
ϕ	$K\Omega$	304	299	304	-1.6	0	none	none
R_m	$K\Omega \cdot cm^2$	1.77	1.49	1.64	-16	-7.3	.001	.001
G_m	mmho/cm^2	0.56	0.67	0.61	19.6	9	.001	.001
C_m	$\mu F/cm^2$	7.6	8.6	8.13	13	7	.001	.001
τ_m	msec	13.5	13	13.3	-3.7	-1.4	.001	none

In the last two columns are listed the calculated level of t-test significance of each difference.

TABLE III Potential Parameters from Phase Plane Trajectories and Conduction Velocity

Parameters Symbols and Units		Control (C)	Irradiated (I)		Difference (I-C)/C, in %		"t" test Significance Level	
			Winter	Summer	Winter	Summer	Winter	Summer
V_r	mV	-87	-87	-87	0	0	none	none
V_{os}	mV	27	34	30	26	11	.001	.001
V_{Na}	mV	32	39	35.5	22	11	.001	.001
V_{on}	mV	-73	-72	-73	-1.3	0	.1	none
V_s	mV	114	121	118	6	3.5	.001	.01
V_n	mV	14	15	14	7	0	.1	none
Θ	m/sec	2	2.21	2.11	10.5	5.5	.001	.001

From values V_r to V_{on} : Membrane Potential from Zero Voltage Reference
V_s V_n :   "   "   " Resting Membrane Potential

TABLE IV Parameters of the Action Current Associated with the Propagated Action Potential

Parameters Symbols and Units		Control (C)	Irradiated (I)		Difference (I−C)/C, in %		"t" test significance Level	
			Winter	Summer	Winter	Summer	Winter	Summer
V^*	mV	−53	−53	−53.1	0	0	none	none
V_i	mV	−25.5	−23	−24.1	−10	5.4	.02	.02
$\dot{V}+$	V/sec	537	594	557	10.6	4	.001	.02
V_j	mV	−6.5	−2	−5.6	−69	−14	.001	.001
$\dot{V}-$	V/sec	140	134	136	−4	−2.8	.05	.05
I_i	ma/cm²	−5.4	−6.6	−5.9	22	9	.001	.001
I_o	ma/cm²	1.09	1.18	1.13	8	3.6	.001	.02
q_i	μC/cm²	0.888	0.982	0.932	10.5	5	.05	.05
q_o	μC/cm²	0.646	0.897	0.700	38	8.3	.001	.001
Δq	μC/cm²	0.242	0.085	0.232	−64	−4	−	−

TABLE V Slope Parameters and Limiting Conductances

Parameters Symbols and Units		Control (C)	Irradiated (I)		Difference (I-C) / C, in %		"t" test Significance Level	
			Winter	Summer	Winter	Summer	Winter	Summer
k_r	$msec^{-1}$	11.8	13	12.2	10	3.4	.001	.02
k_{Na}	$msec^{-1}$	15.7	15.8	15.6	0.6	-0.6	none	none
k_K	$msec^{-1}$	2.6	3.1	2.8	19	7.7	.001	.001
$-g_{Na}$	$mmho/cm^2$	-140	-172	-150.4	23	7	.001	.01
g_{Na}	$mmho/cm^2$	265	301	289	13	9	.001	.001
g_K	$mmho/cm^2$	25	33	28	32	12	.001	.001

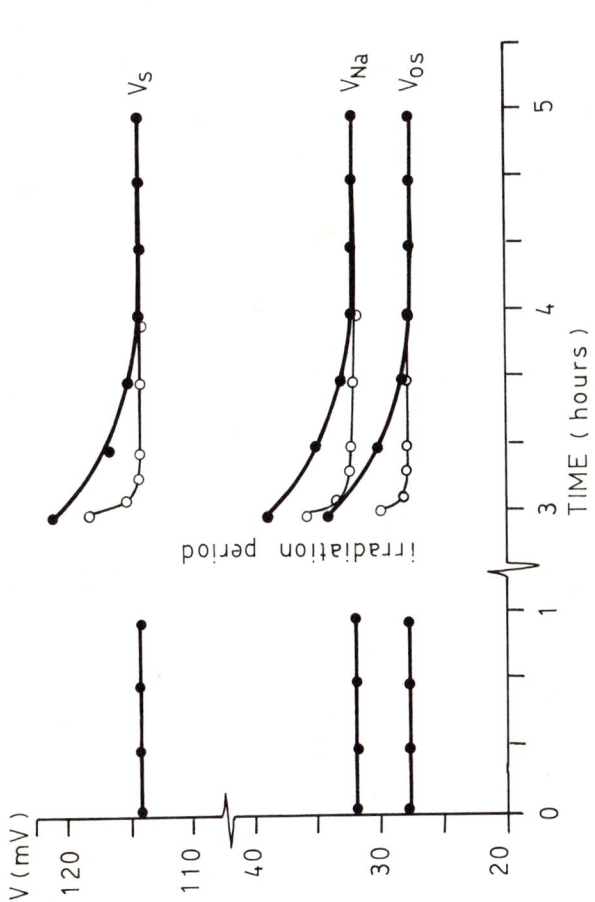

FIGURE 3. Transient effects due to microwave irradiation on the membrane action potential parameters V_s, V_{Na} and V_{os}.

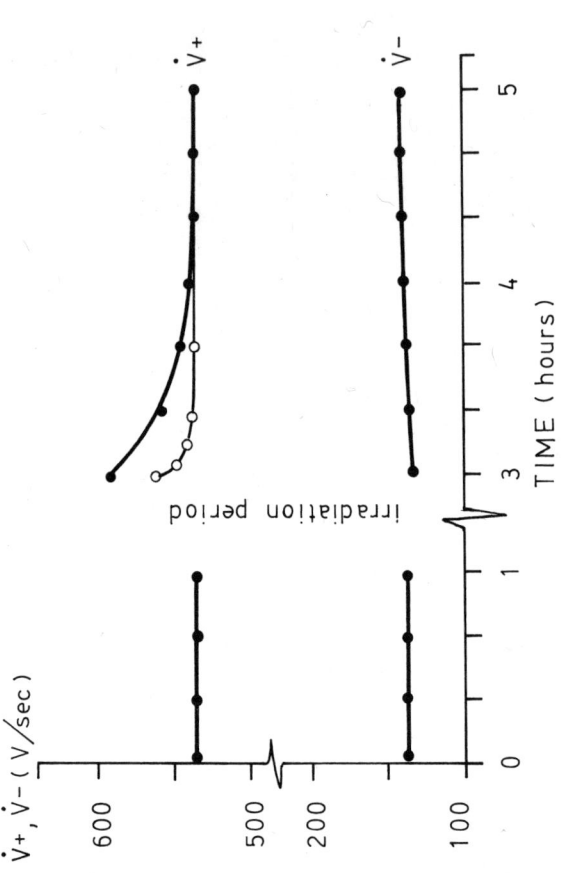

FIGURE 4. Transient effects due to microwave irradiation on the maximum rate of rise $\dot{V}+$ and maximum rate of fall $\dot{V}-$ of the propagated action potential.

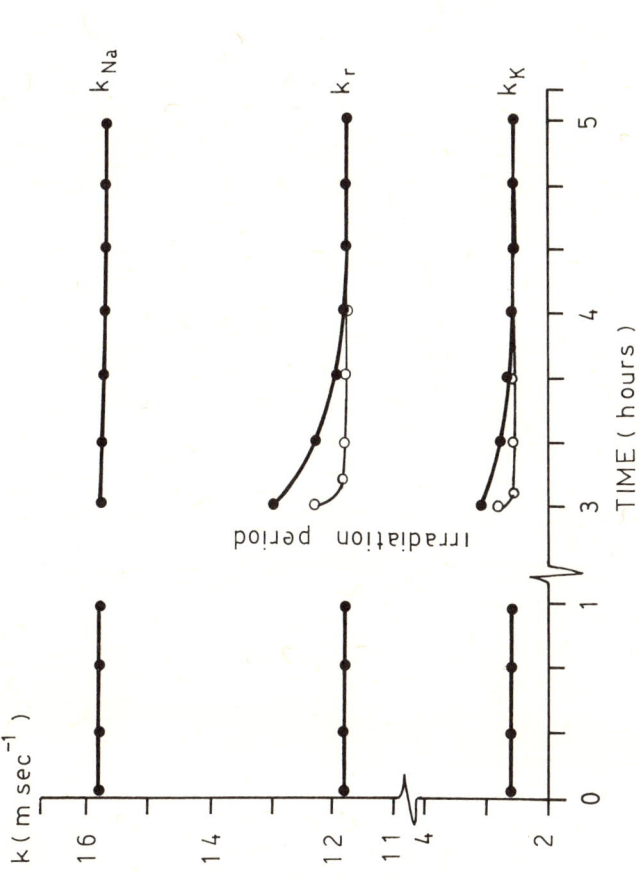

FIGURE 5. Transient effects due to microwave irradiation on the rate constants k_r, k_{Na}, and k_K for the exponential regions of the action potential.

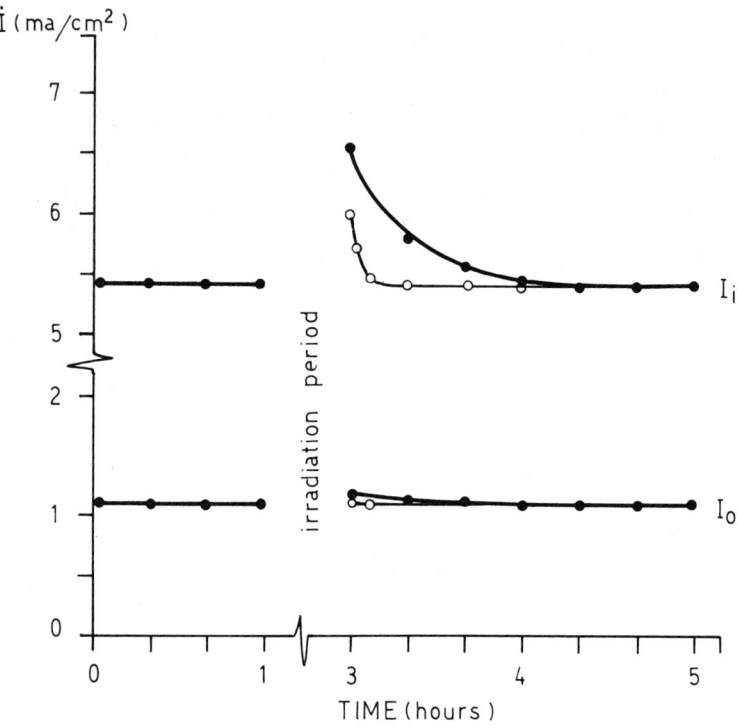

FIGURE 6. Transient effects due to microwave irradiation on the peak sodium inward and outward potassium ionic currents of the action potential I_i and I_o.

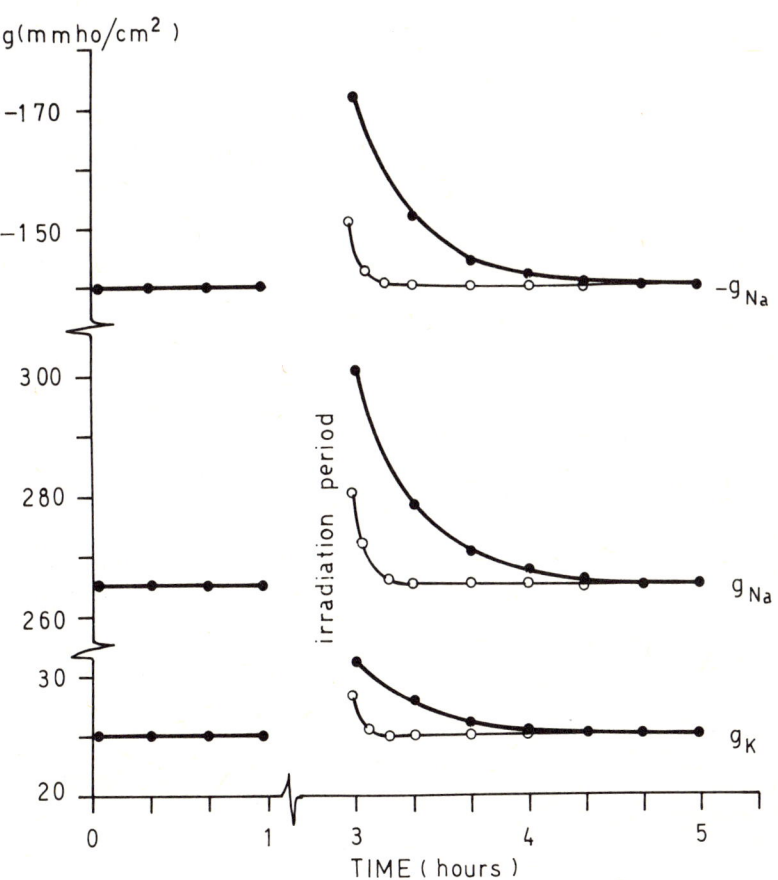

FIGURE 7. Transient effects due to microwave irradiation on the negative conductance $-g_{Na}$ and the limiting membrane conductances g_{Na}, g_K of the action potential.

virtually no effect on excitation. The excitation potential V*
refers to the value of the membrane potential at which regenerative activity begins. It must be noted that the magnitude of
the predicted sodium equilibrium potential V_{Na}, is probably a
function of the magnitude of the early inward sodium and outward
potassium ionic currents. The observed shift in V_{os} and V_{Na} may
be expected as a consequence of the net inward sodium current
associated with a transient increase in the magnitude of the
limiting sodium conductance, gNa (Table V and Figure 7). The
time to peak Na current was decreased as previously reported
(Portela et al. 1974) and related to the magnitude of $\dot{V}+$.

The outward ionic current, corresponding to movement of
potassium ions, reaches the observed transient increase of the
peak outward ionic current I_o shortly after the second inflection point V_j of the action potential as shown in Table IV.
However, I_o bears a close relationship to the maximum rate of
fall of the action potential ($\dot{V}-$). The magnitude of I after I_o
was attained is related to the increase value of the limiting
potassium conductance gK. The initial value of the negative
after potential V_{on} (or V_n) had not been altered by microwave
irradiation (Tables III and V).

The rate constants for the foot of the action potential k_r,
and for the terminal region of the action potential k_K were
transiently increased. However, the rate constant k_{Na} was not
altered (Table V and Figure 5). Therefore, the transient
increase in gK is given by the observed changes in k_r, k_K and
C_m and in gNa by changes in k_r and C_m.

From the analysis of the above studied parameters, it is
shown that the transient changes evoked by microwave radiation
are higher in those experiments done during the winter season.
The net charge accumulation Δq, associated in the generation of
an action potential is diminished by microwave irradiation
exposure, even though both q_i and q_o were increased.

Data show that the transient changes in the magnitude of
passive or active electrical membrane parameters were higher in
muscle cells from "winter" frogs than from "summer" frogs,
indicating that seasonal environmental changes may bring
about adaptive alterations in living organisms, causing changes
in cellular control functions. The transient changes produced
by microwave irradiation on the indicated parameters return to
normal values following similar time courses for muscle cells
from frogs of the same season (Figures 3, 4, 5, 6 and 7).

The analysis of the striated muscle cell action current and
related parameters associated with the propagated action

potential indicates the influence of microwave energy absorption on the cell molecular structures responsible for excitation and cell conduction.

Muscle Cell Water Parameters

The averaged results of experiments on a group of 60 single cells from muscles of 20 "winter" frogs and on a group of 80 cells from 27 muscles of "summer" frogs, are presented in Table VI. Figure 2 illustrates typical results from "winter" frogs. As it was recently reported, reducing the bathing solution osmolarity to half-normal, results in a significant volume increase (Portela et al. 1971, 1974). Noteworthy is the constancy of cell volume after equilibration and its reproducibility on repeated changes as given in Figure 2. Since the cell volume repeatedly achieves constant values in both normal and test solutions, all fluxes of ions, water and non-electrolytes into and out of the cell must be in balance under both these conditions, either in control or irradiated systems. Furthermore, since the values are reproducible over a number of normal⇌test⇌normal cycles (Figure 2), there must be no loss of solute in the test solution conditions. In addition, the fact that the different volume changes in muscle cells immediately following irradiation are equally reproducible means that, whatever the internal transient compartmental changes resulting in microwave energy absorption (indicated by the increase of both, the W_{eff} and P_W), they are not apparently affected by the cell stretching experienced in the test solution.

The magnitude of W_{eff} as well as of P_W are transiently increased by microwave irradiation, having less effect on "summer" frogs than on "winter" frogs, as shown in Table VI. These transient effects on cell water parameters disappeared within 10-20m after irradiation. However, the time courses for recovery are similar for experimental data of muscle cells from either "summer" or "winter" frogs. However, the magnitudes of W_{eff} and P_W of irradiated muscle cells from both "summer" and "winter" frogs are significantly different from those of the respective controls. Those values of W_{eff} and P_W corresponding to cells of "winter" frogs are significantly lower (P less than 0.001, by t-test) suggesting a consolidation or "tightening up" of the muscle cell structure in "winter" as compared to "summer" frogs.

Moreover these data indicate that the irradiated cell has not changed its initial water content, but has transiently increased the membrane water permeability and the water fraction which is available for free exchange with extracellular fluid.

TABLE VI Cell Water Parameters

Parameters Symbols and Units		Winter (W)			Summer (S)			Control Difference (S-W)/W, in %
		C	I	Difference (I-C)/C, in %	C	I	Difference (I-C)/C, in %	
Vc/Vco	—	1.52	1.67	10	1.70	1.76	3.5	12
Weff	—	0.52	0.67	29	0.70	0.76	8.6	34.6
Pw	cm^4/osmole-sec	0.42	0.52	24	0.55	0.61	11	31
Number of muscle cells measured			60			80		

These transient changes must be correlated to the observed increase of ion conductances and ionic currents (G_m, gNa, gK, etc.). Data suggest that the observed seasonal differences in the magnitude of microwave irradiation effects may be related to specific cellular mechanisms of physiological adaptation, depending on adaptive changes which affect specific molecular controlling properties of membrane cellular structures and cellular metabolic regulations.

Cellular living processes cannot proceed normally unless a relatively constant temperature is maintained, either in the environment or internally; in this regard, in poikilotherms, the thermal properties of the water should be of fundamental importance. The metabolic processes in "summer" frogs Amphibious Leptodactilus ocellatus result in generation of heat, but because actively metabolizing tissues (i.e., muscle) are at least 75-80% water, the resultant temperature increase is minimized. The observed summer values of cell water parameters may be consistent with mechanisms for protection from fluctuations in temperature. Therefore, from the comparative analysis of data from animals collected during both seasons, and assuming that the effects of microwave radiation mainly results from heat generation, it is inferred that the observed lower response to microwave exposure during summer may be due to the high ability of water to dissipate heat, reinforced by seasonal adjustments of water parameters.

ACKNOWLEDGEMENT

We are indebted to Dr. Atila Gosztonyi, Margarita Brennan and Marcelo Campi for their assistance in the electrical and permeability measurements and statistical work. We are also indebted to Professor Dr. Lev Kayushin, from the Institute of Biophysics Academy of Sciences of the USSR, Moscow, and members of the Institute for stimulating discussions with Professor Portela during his visit in 1974 under the auspices of the Academy of Sciences of the USSR.

NOMENCLATURE

Passive Electrical Membrane Parameters

2a — Fiber diameter, microns

ϕ — Total effective resistance, k-ohms ($1/2 \sqrt{r_m \cdot r_i}$)

λ — Length constant, mm ($\sqrt{r_m/r_i}$)

Rm — Membrane specific resistance, k-ohms-cm^2

Jm — Time constant, msec

Cm — Membrane capacitance, $\mu F/cm^2$

Gm — Membrane conductance, mmho/cm^2

d — $\sqrt{4Ri/\pi r_i}$

r_m — $2\phi\lambda$

r_i — $2\phi/\lambda$

r_m — Membrane resistance per cm of fiber length, ohms-cm

r_i — Resistance of the internal fluid per cm of fiber length, ohms/cm

R_i — Specific resistivity of the intracellular fluid, 250 ohms-cm

Dynamic Electrical Parameters of the Active Membrane

V — Membrane potential, mV

Transmembrane Potential from zero voltage reference:

Vr — Resting membrane potential, mV

Vos — Overshoot potential, mV

Von — Initial value of the negative after potential, mV

V_{Na} — Sodium equilibrium potential, determined from k_{Na} rate constant in the V axis intercept, mV

V* — Excitation potential, mV

Vi — Potential at the first inflection point, mV

Vj — Potential at the second inflection point, mV

Transmembrane Potential, from reference resting potential:

Vs — Maximum amplitude of the action potential, mV

V_n — Initial value of the negative after potential, mV

Time Derivatives:

dV/dt (or $\dot{V}$) First time derivative of the action potential, V/sec

$\dot{V}+$ Maximum rate of rise of the action potential, V/sec

$\dot{V}-$ Maximum rate of fall of the action potential, V/sec

Rate Constants:

k_r - Rate Constant for the "initial" phase of the action potential, msec^{-1}

k_{Na} - Rate Constant for "second" linear region of the action potential, msec^{-1}

k_K - Rate Constant for the "terminal" linear region of the action potential, msec^{-1}

Ionic Currents:

I_t - Total membrane current mA/cm^2

I - Membrane ionic current, mA/cm^2

I_i - Peak sodium inward ionic current, mA/cm^2

I_o - Peak potassium outward ionic current, mA/cm^2

Other parameters:

t - Time, msec

θ - Velocity of propagation of the action potential, m/sec

gNa - Limiting membrane conductance for Na inward current, mmho/cm^2

gK - Limiting membrane conductance for K outward current, mmho/cm^2

q_i - Ionic charge uptake during generation of action potential, μC/cm^2

q_o - Ionic change loss during generation of action potential, μC/cm^2

-gNa- Negative Membrane Conductance

Water cell parameters

V_c - Volume per unit length of the muscle cell in a steady state in T=0.5 Ringer

V_{co} - Volume per unit length of the muscle cell in a steady state in standard Ringer

T - Tonicity of perfusing solution. By definition, standard Ringer (osmolarity 0.222) is assigned a value of T=1.0, so that the solution of osmolarity 0.111 has a value of T=0.5

and is referred to as T=0.5 Ringer

V_c/V_{co} - Relative cell volume

Weff - The fraction of the cell volume per unit length which is osmotically available water, when the cell is in a steady state in standard Ringer. It is usually expressed as a percentage.

Pw - Cell membrane permeability coefficient, per unit area of membrane, cm^4/osmol-sec

LITERATURE CITED

Chan, K. A., R. A. Sigelmann, A. W. Guy and J. F. Lehmann, Calculation by the Method of Finite Differences of the Temperature Distribution in Layered Tissues. IEEE Transactions on Biomed. Engineering. Vol. BME-20 N°2: 86-90 (March 1973).

Fan, L. T., F. T. Hsu and C. L. Hwang. A Review on Mathematical Models of the Human Thermal System. IEEE Trans. Biomed Engineering. Vol. BME-18 N° 3: 218-234 (May 1971).

Fatt, P. and B. Katz. An Analysis of the Endplate Potential Recorded with an Intracellular Electrode. J. Physiol. 115: 320-370 (1951).

Florkin, M. and E. Schoffeniels. _Molecular Approaches to Ecology_. Academic Press, New York (1969).

Frank, K. and M. C. Becker. Microelectrodes for recording and stimulation, in _Physical Techniques in Biological Research_. 5: 22-87. Academic Press, New York (1964).

Gordon, Z. V. The problem of the biological action of UHF. In A.A. Letavet and Z. V. Gordon (Eds.), The biological action of UHF, Moscow Acad. Med. Sci. U.S.S.R., P. 2 (OTS62-19175, R. 816) (1962).

Gordon, Z. V. Biological Effect of Microwaves in Occupational Hygiene. Israel Program for Scientific Translations, Jerusalem (1970).

Graham, D. and D. McRuer. Analysis of non-linear control systems. John Wiley and Sons, Inc. N. Y. (1961)

Hodgkin, A. L. and W. A. H. Rushton. The electrical constants of a crustacean nerve fiber. Proc. Roy. Soc. B 133: 444-479 (1946).

Hodgkin, A. L. and A. F. Huxley. A quantitative description of membrane current and its applications to conduction and excitation in nerve. J. Physiol. 117: 500-544 (1952).

Hubbard, S. J. The electrical constants and the component conductances of frog skeletal muscle after denervation. J. Physiol. 165: 443-456 (1963).

Jenerick, H. Phase Plane Trajectories of the Muscle Spike Potential. Biophys. J. 3: 363-377 (1963).

Jenerick, H. Analysis of the striated muscle fiber action current. Biophys. J. 4: 77-91 (1964).

Kamenskiy, Yu I. The influence of microwaves on the functional conditions of the nerve. Biofizika 9: 695-700 (ATD Report T-65-39, Library of Congress, Transl.) (1964).

Kamenskiy, Yu I. "Effect of Microwaves on the Kinetics of Electric Parameters of a Nerve Impulse" Trans. Moscow Society of Naturalists 28: 164-172 (1968).

Katz, B. The electrical properties of the muscle fiber membrane. Proc. Roy. Soc. B. 135: 506-534 (1948).

Lobanova Ye. A. Survival and development of animals with various intensities and durations of the influence of UHF. In A. A. Letavet and Z. V. Gordon (Eds.), Biological action of UHF. Moscow: Acad. Med. Sci. USSR: 68-74 (OTS 62-19175 R.816) (1962).

Lobanova Ye. A. and Gordon, Z. V. Investigation of the olfactory sensitivity in persons subject to UHF Fields. In A. A. Letavet and Z. V. Gordon (Eds.), Biological Action of UHF. Moscow: Acad. Med. Sci. USSR: 50-56 (OTS 62-19175 R.816) (1962).

Lobanova Ye. A. and M. S. Tolgskaya. Change in the higer nervous activity and interneuron connections in the cerebral cortex of animals under the influence of UHF. In A. A. Letavet and Z. V. Gordon (Eds.), Biological Action of UHF. Moscow: Acad. Med. Sci. USSR: 68-74 (OTS 62-19175 R.816) (1962).

Lobanova Ye A. and A. V. Goncharova. Investigation of conditioned reflex activity in animals subjected to the effect of ultrashort and short radio-waves. Inst. Occupational Hygiene, Acad. Med. USSR; Medicine 1: 29-33 (1971).

McAfee, R. D. "Neurophysiological effect of 3cm microwave radiation." Amer. J. Physiol. 200: 192 (1961).

McAfee, R. D. "Physiological effects of thermide and microwave stimulation of peripheral." Amer. J. Physiol. 203: 374 (1962).

McAfee, R. D. "The neural and hormonal response to microwave stimulation of peripheral nerves," presented at the Symp. Biol. Eff. Health Implic. Microwave Radiat., Richmond, Va. (1969).

McLees, B. D. and E. D. Finch. Analysis of Reported Physiologic Effects of Microwave Radiation, in <u>Advances in Biological and Medical Physics</u>, Vol. 14, Academic Press, New York pp. 163-223 (1973).

Michaelson, S. M. The Tri-Service Program. A tribute to George M. Knauf, USAF (MC), IEEE Transactions on Microwave Theory and Techniques, Vol. MTT 19, N° 2: 131-146 (1971).

Minorsky, N. Introduction to non-linear mechanisms. Ann Arbor, J. W. Edwards, U.S.A. (1947).

Nastuk, W. L. and A. L. Hodgkin. The electrical activity of single muscle fibers. Jour. Cell Comp. Physiol. 35: 39-73 (1950).

Plonsey, R. and Fleming, D. G. <u>Bioelectric Phenomena</u>. McGraw Hill Book Co., New York, pg. 78-201 (1969).

Ponder, E. The coefficient of thermal conductivity of blood and of various tissues. J. Gen. Physiol. 45: 545-551 (1962).

Portela, A., J. C. Pérez, M. Luchelli, P. A. Stewart, T. Hajduk, M. N. Parisi and A. Garrison. Potassium and Cesium Effects on Sodium Efflux and Oxygen Consumption of Muscle Cells. Biochim. Biophys. Acta 109: 495-502 (1965).

Portela, A., J. G. Vaccari, R. J. Pérez, A. Ardizzone, J. C. Pérez. Electrical Membrane Constants of Sartorius Muscle Fibers from the South American Frog Leptodactilus ocellatus. Experientia 26: 957 (1970).

Portela, A., R. J. Pérez, J. Vaccari, J. C. Pérez and P. Stewart. Muscle Membrane Depolarization by Acetylcholine, Choline and Carbamylcholine, Near and Remote from Motor End-Plates. J. Pharmacol. Exp. Ther. 175: 476-482 (1970 a).

Portela, A., J. Vaccari, P. A. Stewart, R. J. Pérez and J. C. Pérez. Cesium Effects on Muscle Membrane Responses to Quaternary Ammonium ions. J. Pharmacol. Exp. Ther. 175: 483-488 (1970 b).

Portela, A., M. Garfunkel, J. G. Vaccari, A. M. Delbue, P. A. Stewart and J. C. Pérez. Radiation Effects on Water Permeability and Distribution in Frog Muscle Cells. Radiation Res. 47: 704-715 (1971).

Portela, A., M. Brennan, J. Vaccari, J. C. Pérez and P. Stewart. Denervation Effects on Water Permeability and Distribution in Frog Skeletal Muscle Cells. Studia Biophysica, in press, August 1974.

Portela, A., J. G. Vaccari, S. M. Michaelson, O. Llobera, M. Brennan, A. E. Gosztonyi, J. C. Pérez and J. Jenerick. Transient Effects of Low-Level Microwave Irradiation on Bioelectric Muscle Cell Properties and on Water Permeability and Its Distribution - Studia Biophysica, in press, September 1974.

Presman, A. S. and N. A. Levitina. The nonthermal effect of microwaves on the systolic rhythm of animals. II The effect of pulsed microwaves. Byull. Eksp. Biol. Med. 52: 39-43 (1962).

Presman, A. S. and N. A. Levitina. Nonthermal action of microwaves on the cardiac rhythm. Byull. Eksp. Biol. Med. 53: 36-39 (1963).

Presman, A. S. Electromagnetic Fields and Life (Translated from Russian) New York-London: Plenum (1970).

Prosser, C. L. (editor). Physiological Adaptation. American Physiological Society, Washington, D. C. (1958).

Schwan, H. P., A. Anne, M. Saito and O. M. Salati. "Relative microwave absorption cross sections of biological significance" in Biological Effects of Microwave Radiation, Vol. 1, New York, Plenum Press, pgs. 153-176 (1961).

Shitzer, A. Addendum to "A Review on Mathematical Models of the Human Thermal System." IEEE Trans. Biomed. Eng. Vol. BME-20, N° 1: 65-66 (January 1973).

Thesleff, S. The mode of neuromuscular block caused by acetylochline, nicotine, decamethonium and succinylcholine. Acta Physiol. Scand., 34: 218-231 (1955).

Vernberg, W. B. and F. J. Vernberg. Environmental Physiology of Marine Animals. Springer-Verlag, New York Inc. (1972).

-DISCUSSION-

LOTZ - When you irradiate with microwaves, do you keep the temperature constant?

PORTELA - Yes, at 25°C.

LOTZ - And, you got significant changes in such things as sodium permeability?

PORTELA - Well, that is the change I was mentioning.

LOTZ - And then when you heat your single cell. do you get changes all in the same direction?

PORTELA - We have determined reversible changes having different recovery time constants. Those "heating" effects on the indicated parameters provoked by non-electromagnetic heating - for instance, by means of an electric heater - were minimized by laboratory cooling, with time constants ranging from 1 to 2 minutes.

However, these calculated cell parameters have indicated that the transient changes induced by microwave irradiation are higher, in muscle cells from winter frogs than from summer's: for winter, you remember, the time constant was 20 minutes, for summer it was in the order of 3 minutes. Then we learnt that an explanation of this phenomenon is related to cell water metabolism: W_{eff} and P_W of muscle cells are higher in summer than in winter.

The conditions of heat exchange by a dynamic adjustment of cell water, P_W and extracellular fluid volume, are functionally dependent on seasonal influence upon the system. This is one of the interpretations of the phenomena described in this symposium.

LOTZ - What is the range of field intensity and frequency in your microwave exposures?

PORTELA - The microwave source generating at 2.88 GH_z is delivering energy in the far field region, 10mW/cm^2, corresponding to about 194 volts/meter.

POLSON - Can I go back to your theoretical model? You talked about the Hodgkin-Huxley equations. No where did I see any mention of the leakage component, the chloride ion, in your version of the Hodgkin-Huxley equations. What happened to that in your model?

PORTELA – We determined the membrane ionic current of muscle cell, conducting an action potential following the H-H mathematical model. The contribution of chloride ionic current did not alter the relationships between the membrane voltage and ionic current as reflected in the phase plane trajectory ($\dot{V}$, V) and I-V. The areas under the inward and outward ionic current plots correspond to the uptake and loss of charges, assuming the capacity (Cm) to be constant during the entire process. The influence of chloride current leakage was determined by using radioactive chloride and sulphate.

The ionic current was shown to be due mainly to an early transient component of the intensity, carried by sodium ions, and a delay component carried by potassium ions. It was shown that chloride does not contribute significantly to the ionic current. The magnitude of this ionic current leakage was also estimated from the current required to maintain the membrane at the observed potential. The total membrane current was demonstrated to correspond basically to the capacity current (C$\dot{V}$) and the well known ionic current (i_{Na}, i_K, i_1). In the H-H mathematical model the leakage current i_1 did not account for the observed displacement between $\dot{V}$+(from $\dot{V}$, V) and I-V analysis (see Studia Biophysica, Springer Verlag; in press).

Other ions contribute as shown by V_1, but this effect may be neglected in muscle cells from Leptodactillus ocellatus, and we did not take them into account. Applying H-H circuit, we deduce the total current into the cell as given by

$$I = C_m \frac{dV}{dt} + gNa\ (V - V_{Na}) + gK\ (V-V_K)$$

where V is the membrane potential, neglecting the effect of other ions. In the steady state, I = 0 for dV/dt = 0, and

$$V = (gNa\ V_{Na} + gK\ V_K)/(gNa + gK)$$

THERMAL FACTORS IN ULTRASONIC FOCAL DESTRUCTION IN ORGANIZED TISSUES

P. P. Lele
Departments of Mechanical Engineering and
  Nutrition and Food Science
Laboratory of Experimental Medicine, 26-023
Massachusetts Institute of Technology
Cambridge, Massachusetts 02139

Introduction

A topic of some considerable interest among investigators concerned with biomedical applications of ultrasound is that of the mechanisms by which ultrasonic irradiation can effect reversible and irreversible changes in organized tissues. Such changes include the formation by focused ultrasound of lesions, located deep within tissues (1,2); functional changes, such as paralysis of limbs (3) (and accompanying histological lesions, if any); electrochemical changes (4); and possibly chromosomal damage (5,6). For certain phenomena, especially those observed in tissue rather than in suspension(7), the concept of a purely thermal mechanism has been frequently mentioned (1-4,8-18) - in many instances if only to discard it. The present paper attempts to define explicitly the author's concept of thermal mechanisms of ultrasonically induced damage.

The viewpoint adopted here is that the phrase "thermal mechanism" is as yet only implicitly defined and that this lack of definition has often led to some ambiguity in the interpretation of experiment. Thus, it is necessary to state completely the specific hypothesis so that it may be subjected to rigorous experimental proof. In this regard, the context in which the phrase "thermal mechanism" is often used in ultrasound literature would appear to suggest the hypothesis that a certain specified change is initiated once the local temperature exceeds a certain threshold temperature characteristic of the biological medium. In order to distinguish this hypothesis from one subsequently discussed, it is here referred to as the threshold temperature hypothesis.

A practical implication of this threshold temperature hypothesis, once substantiated, would be that it affords a means of estimating whether a specified ultrasonic dosage is sufficient to create a specified effect. With only modest supporting assumptions, plus a knowledge of the ultrasonic absorption coefficient and the thermal properties of the medium, one may compute the temperature at any point as a function of time, and thereby determine whether it exceeds the threshold value. As is well known, there are a number of experimental observations (3,4,10,11,15) which at least partially contradict the threshold temperature hypothesis. Some workers interpret this data as requiring a total rejection of a thermal mechanism of tissue damage and the postulation of purely mechanical mechanisms without special regard to ultrasonic dosage or biological conditions.

The threshold temperature hypothesis, however, does not completely or accurately state the thermal hypothesis which so far has been only implicitly considered in discussions of "thermal mechamisms" in the literature (12,15). It is our contention that the thermal hypothesis deserves more careful consideration by workers in the field than it has received in the past. This, we believe, will lead to a lesser polarisation of views among the investigators.

Subject to a number of qualifications which are detailed subsequently, the thermal hypothesis may be briefly stated as follows:

For a wide class of reversible and irreversible effects caused by the action of ultrasound on tissues, the same effects can equivalently be produced by nonacoustic localized heating of the tissue, provided the temperature history during heating and cooling duplicates the quasi-steady (averaged over a cycle) temperature history during irradiation.

Physical Basis of the Thermal Hypothesis

Since present knowledge of intra- and intercellular processes is still relatively rudimentary, it appears infeasible to prove or disprove the thermal hypothesis on the basis of fundamental physical considerations. However, in any discussion of mechanisms, the following facts deserve consideration. For brevity, we here limit the discussion to the formation of lesions by focused ultrasound in brain tissue.

One important consideration is the relatively large magnitude of an ultrasonic wavelength (0.5mm for 3 MHz irradiation) to any representative cell size in mamalian tissue (for example, neuron body diameters are invariably less than 100μ or 0.1mm, may be as small as 0.004mm). Another consideration is the fact that the energy absorbed per unit volume during a wave period is but a small fraction of the total energy per unit volume associated with the wave. For example, the fraction is found to be only 0.04 if one takes it as $\mu\lambda$ with the value $\mu=0.8$ cm^{-1} for the intensity absorption coefficient as measured by Pond for $\lambda=0.05$ cm wavelength radiation in rat brain (15). A third consideration is that, at ultrasonic intensities generally achieved (less than 10,000 watts/cm^2) in ultrasonic irradiation experiments, irreversible effects are not observed unless the pulse constitutes a large number of cycles (1-3,10-18). Two more considerations are the facts that the pressure changes, per cycle and accumulative, are generally small during irradiation and that the accumulative temperature change may be substantially larger than the change in any one cycle.

The relatively small absorption per cycle suggests that any small element of tissue during irradiation passes through a sequence, of "quasi-equilibrium" states. Thus one would expect temperature (which, strictly speaking, is not a fundamental physical quantity, but a parameter characterizing a probability distribution (19))to remain a viable concept. In particular, we expect the characteristic time for relaxation of a localized perturbation to be small compared to a wave period. Here, the phrase "localized" implies small compared to a wavelength, which could still be appreciably larger than a representative cell dimension.

The supposition one could conceivably make is that any small element (small compared to a wavelength) of tissue during irradiation passes through a sequence of states, which in a statistical sense could be predicted from a knowledge of just the time history of the pressure and temperature of the immediate environment of the element. While the element is not, strictly speaking, in thermal equilibrium, it responds to changes in temperature T or pressure p in statistically definable ways providing comparable changes take place sufficiently slowly. Here, it is argued that the criterion of being sufficiently slow is almost always applicable. During the irradiation one may consider p and T to have a cyclic part and, in addition, one may expect T to tend to increase monotonically unless heat generation and dissipation are in equilibrium. Typically, at intensities of the order of 1,000 watts/cm^2, the temperature cycling is of the order of $0.1°C^{20}$ while the monotonic rise of temperature is of the order of 20°C. during the formation of lesions. Note that there is no such steady rise in pressure.

Nonthermal Mechanisms

The principal competing mechanisms to thermal mechanisms most frequently mentioned in the literature are cavitation (11) and acoustical streaming (21). Both of these mechanisms lie outside the context of the discussion presented in the previous section in that they may create large magnitude inhomogeneities with scale sizes considerably less than a wavelength. The appropriate scales are, respectively, the diameter of a bubble and the thickness of a streaming boundary layer. It is difficult to assess just when such mechanisms should be dominant. In the case of cavitation, we have little knowledge on the statistics of possible cavitation nuclei in tissue. Whether the occurrence of lesion formation at points other than the location of the beam focus is a sufficient proof of cavitation damage is debatable. Examples of such occurrences at intensities above 2,000 watts/cm have been recently exhibited by Fry, Kossoff, Eggleton, and Dunn (11). Thus, while cavitation cannot be ruled out at these power levels, the demonstration of a positive correlation between the location of such lesions and the distribution of cavitation nuclei, or a positive proof of the occurrence of cavitation within the tissue during its irradiation would be more convincing. In the absence of cavitation, the present author would tend to believe that the lesion formation should be consistent with the thermal hypothesis; and furthermore, that more than one mechanism may be operative under certain dosage and state conditions.

Streaming, while perhaps important in vitro, appears to be an unlikely mechanism in tissue. In order to achieve the necessary large velocity gradients, it would appear necessary to have relatively large streaming velocities, which in turn would require large scale circulation. Such circulation, however, would be largely impeded in any structured tissue. It is also relevant that studies of ultrastructure of early (20 seconds after irradiation) threshold lesions by electron microscopy reveal that the lesions have the characteristics of coagulative necrosis with no evidence of disruption (as by cavitation) or displacements (as by streaming) of the cellular structure (22).

Some Comments on Apparently Negating Experiments

There are a substantial number of well-known experimental results which are commonly cited (3,4,10,15) as evidence that various specific ultrasonically induced effects are not caused by a thermal mechanism. While it is infeasible to discuss here each of these results in detail, it should be pointed out that most of these results may actually not be inconsistent with either the thermal hypothesis or the more specific hypothesis that temperature time histories with consistently higher temperatures lead to a greater probability of an irreversible effect.

For example, in experiments with poikilothermic animals under hypothermia, paralysis may have been induced by purely thermal mechanisms even when the peak temperature in the spinal cord at irradiation did not exceed 37°C. Whether or not such an effect occurs would depend on the pre-irradiation adaptation temperature of the animal. It is well known that the metabolic state of an animal or tissue, as reflected in its oxygen consumption (or requirement) is dependent on its (steady state) temperature and changes exponentially with temperature (23). The changes in the heart rate and the cardiac output follow closely those in the oxygen consumption (or requirement) of the animal. Thus, in an animal adapted to a given temperature, the oxygen supply is just barely adequate to meet the oxygen requirements of the tissues at that temperature. If, in a hypothermic animal, the local oxygen requirements of an organ or a small volume of tissue within an organ were suddenly raised (for example, by localized heating with ultrasound) the circulation will be unable to meet the increased local requirements and the tissues involved will be subjected to hypoxia and may be irreversibly damaged. The central nervous system is well known for its extraordinarily high metabolic rate (greater than that of a powerfully contracting muscle (24));and extreme vulnerability to interruption or inadequacy of oxygen supply (cerebral ischaemia of longer than 2 minutes causes irreparable damage (25). Since the oxygen requirements of the brain increases by 300 percent with a rise of its temperature from 10°C. to 20°C.(23) irreversible hypoxic damage may occur even at this "low"temperature. Hypoxia has recently been shown to decrease the exposure time required for ultrasonically produced injury by 40 percent (26). And, it may be added that the occurrence of brain damage secondary to transient cardiac arrest or fibrillation in patients at normal temperature is a well-known fact in clinical medicine (23). Under such conditions, one may expect that the site of the initial damage will be the synaptic terminals with their high metabolic activity (24) and the earliest manifestation of such damage will be disruption of the synaptic transmission processes (resulting, for example, in paralysis). Such damage though functionally obvious immediately, will not be detectable by the usual histological techniques for several minutes until the whole neuron is involved. The only implication of the hypothesis that is pertinent here is that if one started with the same animal in the same ambient state, the occurrence of paralysis could be induced by any nonacoustic process which duplicated the temperature history of a paralysis-inducing ultrasonic dose.

In the experiments referred to above, the adoption of a functional end-point, viz. the induction of paralysis, (as opposed to a morphologic end-point, viz. measurement of dimensions of necrotized tissues determined histologically) as an index of the occurrence of ultrasonic damage also deserves some consideration. In the central nervous system, which is a phenomenally complex neuronal network, a miniscule damage at the proper site may result in a complete paralysis of the related muscles; but extensive damage to adjacent areas may produce no observable functional effects whatever. Thus the occurrence of paralysis may be misleading as an indicator of threshold damage to the nerve tissue itself.

Moreover, paralysis of limbs may result even without any direct ultrasonic damage to the neuronal tissues. Non-neural tissues adjacent to the spinal cord may have lower thresholds to ultrasonic damage and may secondarily produce effects on the spinal cord. The blood supply of the spinal cord is notoriously tenuous and variable from one segment to another and easily jeopardized in certain regions. Vascular insults to meninges on the posterior surface of the cord may result in necrosis of an entire segment and produce neurological symptoms comparable to complete cord transection (27). Ultrasonic irradiation of the spinal cord shows a predilection to hemorrhage (26,28) which can result in paralysis of lower limbs. The adoption of a morphological end-point of ultrasonic damage in a larger organ, such as the brain, may obviate much of the uncertainty in these types of experiments.

Similarly, in experiments where repeated apparently identical ultrasonic doses achieve a given effect, while a single dose does not, the threshold temperature hypothesis may be negated but the thermal hypothesis may not be. The distinction is that the thermal hypothesis ascribes a possible significance to the history of the temperature cycle as well as to the maximum value it achieves. Just which features of the temperature history are most important remains a topic for experimental investigation. Undoubtedly, the maximum temperature reached is an important parameter, but it is apparently not sufficient by itself to characterize the experimental outcome. This would certainly seem to be the case when one considers results of experiments in which the temperature was cyclically raised and lowered, as in multiple pulse irradiation.

Discussion of Experiments by Robinson and Lele and by Pond

There is some recent evidence which suggests that under restricted circumstances the threshold temperature hypothesis may be a useful means of estimating lesion sizes. Robinson's (1968) doctoral thesis (13) and Robinson and Lele(14) report on experiments with 2.7 MHz focused ultrasound in cat brains where local temperature was measured with a calibrated 0.002-inch-diameter chromelconstantan thermocouple threaded through the brain. With a dosage (2.0-second pulse duration) sufficient to create lesions, a succession of experiments were performed in which the lesion center was placed at successively closer radial distances to the thermocouple. In this manner the peak temperature was determined as a function of radial distance. In particular, the peak temperature at the lesion boundary was measured and found to be approximately 56°C. for both grey and white matter, as contrasted to an initial brain temperature of approximately 36°C. Given that all such lesions have the same maximum boundary temperature (consistent with the threshold temperature hypothesis), and given empirically determined specific heats and coefficients of ultrasonic absorption, the authors developed a general mathematical model which allowed them to estimate lesion length and diameter for any specified (that is, given pulse duration and peak intensity) ultrasonic dosage. Such calculations were then compared with experimental results, also with 2.7 MHz focused ultrasound in cat brain, previously reported by Basauri and Lele (29). The results appear to support the threshold temperature hypothesis (and accordingly, also support the thermal hypothesis).

Pond (15,16) reports on similar experiments in which lesions were created in rat brains. In one such series with focused 3 MHz ultrasound, he determined the minimum continuous pulse duration necessary to create a lesion for various specified peak intensities. For each such intensity-pulse duration threshold, he computed the maximum temperature from heat transfer fundamentals using a procedure similar to that mentioned above in the discussion of Robinson and Lele's work. The results are summarized in table 1. For two such cases, Pond reports that the maximum temperature at threshold was measured, and these appear to agree well with his calculations. The spread in calculated temperatures at threshold is not markedly inconsistent with the threshold temperature hypothesis, when one considers the variances in the experimental data on which the estimates of the threshold dosages were based. Presuming the rat brains were initially at a temperature of 37°C., one might also conclude that the results are consistent with Robinson and Lele's threshold temperature of approximately 58°C. In another series of experiments reported by Pond, an electric current heated a conducting wire passed through rat brains. (Similar experiments are also described by Robinson (13). The radius of the resulting lesion was subsequently measured and then the maximum temperature experienced at this radius was computed from heat transfer fundamentals. The resulting derived threshold temperature increments were between 21°C. and 25°C., which, with some allowance for experimental variations, would appear consistent with the results shown in table 1, and with the threshold hypothesis. Pond also points out that these results are consistent

Table 1

Intensity time conjugates for ultrasonic focal lesion thresholds as given by Pond (1968)

Peak intensity (watts/cm^2)	Time duration (sec)	Calc. ΔT (°C)	Meas. ΔT (°C)	Coeff. var. ΔT
75	300	17.0		
100	100	18.3		
200	9	21.0	21.2	0.20
315	3	24.1	24.7	0.18
500	1.6	27.2		
1500	0.14	18.5		
2500	0.040	19.4		

with experiments of Moritz and Henriques (30) in which the surface of the skin of pigs and humans were suddenly raised and held at specific temperatures and the minimum time required for tissue damage was measured. In particular, they found a temperature increment of 23°C. causes damage after 3 seconds while 33°C. causes damage within less than a second. These latter experiments, incidentally, are quite clearly inconsistent with the threshold temperature hypothesis, but not necessarily with the thermal hypothesis. As the present author has previously stated (12), "The inactivation (and destruction) of nerve fibers by heat is a complex process governed by many interdependent factors, e.g., the extent, the duration and the rate of elevation of the temperature of the tissue of the species and even the temperature to which the individual of the species was acclimatized at the time of the experiment."

In the discussion near the end of his thesis, Pond cites the smaller values (relative to say 23°C.) of calculated threshold temperature increments for the higher peak intensities of 1,500 watts/cm^2 and 2,500 watts/cm^2 which appear here in Table 1 as evidence of the existence of a "nonthermal effect". He notes that the experiments of Moritz and Henriques would not indicate that such relatively low temperatures could cause damage in such relatively short time.

The present author tends to believe the latter conclusion may be somewhat premature. Comparison with Moritz and Henriques' experiments would seem inconclusive since the environments (surface of skin as contrasted with interior of brain) and natures of the tissue were considerably different. One might perhaps conclude that the restricted hypothesis that consistently higher temperatures give a greater probability of damage is negated by the fact that calculated peak temperatures for threshold doses in table 1 for I_o=1,500 and 2,500 watts/cm^2 are less than those for I_o=500 watts/cm^2. However, before one could accept this, it would be necessary to ascertain whether the calculated peak temperatures at the higher intensities are those which are actually present. Unfortunately, Pond found that thermocouple measurements are somewhat suspect for transient phenomena shorter than 1 second. We might tend to disbelieve the computed values of peak temperature at higher intensities for a number of reasons. One possibility is that for a given acoustic power output, the intensity at the focus may tend to decrease with time because when the medium becomes heated by the wave, the medium tends to refract wave energy away from the higher temperature, and hence, higher sound speed regions (31). Pond evidently was primarily monitoring power output and the intensities in table 1 were calculated from an independently derived (constant) proportionality between peak intensity and power output. According to the theory, for very short pulse durations, the temperature increment ΔT at the focus center should vary roughly as $I_o t$. If I_o should vary with time this product would be replaced by an integral or by as $(I_o)_{ave} t$. Since $(I_o)_{ave}$ decreases with time, estimates of peak temperatures based on Pond's model could actually be too low for short duration pulses, or, alternately too high for long duration pulses. A more detailed quantitative examination of this possibility would seem desirable.

A second reason for suspecting the theoretical values is the temperature dependence of the absorption coefficient. Robinson and Lele(14) have found that the absorption coefficient of the brain increases precipitously at a temperature of approximately 50°C. and point out that calculations of peak temperatures based on an assumption of a constant absorption coefficient may be misleading. Further work in these areas is clearly warranted.

One further point that should be made concerns the caution necessary in comparing the results obtained in different experimental animals. Lele (12) quotes Hodgkin and Katz in relation to the effects of heat on nerves:"The factors which determine the physiological range of temperature in different types of nerve are not yet completely understood. The squid axon ceases to conduct at a temperature which is optimal for mamalian fibers." In addition to the species differences, the age of the animal and the size of the target organ appear to be of consequence in quantitative ultrasonic studies. Thus, in the brain of growing rabbits Young and Lele (32) found that for the same ultrasonic dosage there was greater variability in the results in the smaller brains than in the larger, leading them to suspect the existence of interference from the brain-bone-air interfaces at the base of the skull. In the larger brains the reflected beam would be more highly attenuated due to longer path lengths and would thus cause less interference. Such interference would be expected to become pronounced at higher intensities - particularly in spinal cords and in small brains as those of rats. In such situations, therefore, deduction of the existence of "non-thermal" mechanisms from departures from monotonic trends in dose-effect relationships may be untenable.

REFERENCES

1. FRY, W.J. "Intense Ultrasound in Investigations of the Central Nervous System", in "Advances in Biological and Medical Physics" Vol VI (Academic Press, New York, 1958),pp.281-348.

2. LELE,P.P. Production of deep focal lesions by focused ultrasound: Current status. Ultrasonic 5:105-112(1967).

3. DUNN, F. Physical mechanisms of the action of intense ultrasound on tissue. Amer.J. Phys. Med. 37:148-151(1958).

4. FRY, W.J., V.J.WULFF, D. TUCKER, and F.J. FRY, Physical factors involved in ultrasonically induced changes in living systems:I Identification of non-temperature effects, J. Acoust. Soc. Amer. 22:867-876(1950).

5. MACINTOSH, I.J.C. and D.A. DAVEY, Chromosome aberrations induced by ultrasonic fetal pulse detector, Brit.Med. J. 4:92-93(1970).

6. COAKLEY, W.T., D.E.HUGHES, J.S.SLADE, and K.M.LAWRENCE, Chromosome aberrations after exposure to ultrasound, Brit. Med. J.1: 109-110(1971).

7. CLARK, P.R. and C.R. HILL, Physical and chemical aspects of ultrasonic disruption of cells, J. Acoust. Soc. Amer., 47:659-653 (1970).

8. HERRICK, J.F., Temperatures produced in tissues by ultrasound: Experimental study using various techniques, J. Acoust. Soc. Amer.,25:12-16(1953).

9. LEHMANN, J.F., The biophysical mode of action of biologic and therapeutic ultrasonic reactions, J. Acoust. Soc. Amer., 25:17-25(1953).

10. BARNARD, J.W., W.J. FRY, F.J. FRY, and R.F.KRUMINS, Effects of high intensity ultrasound on the central nervous system of the cat, J. Comp. Neurol., 103:459-484(1955).

11. FRY,F.J., G.KOSSOFF, R.C.EGGLETON, and F.DUNN, Threshold ultrasonic dosages for structural changes in the mammalian brain, J. Acoust. Soc. Amer.,48:1413-1417(1970).

12. LELE, P.P., Effects of focused ultrasonic radiation on peripheral nerve, with observations on local heating, Exper. Neurol. 8:47-83(1963).

13. ROBINSON,T.C.,An analysis of lesion development in plexiglas and nervous tissue using focused ultrasound, Doctoral Thesis, Massachusetts Institute of Technology(1968).

14. ROBINSON, T.C. and P.P.LELE, An analysis of lesion development in the brain and in plastics by high intensity focused ultrasound at low megahertz frequencies, J.Acoust.Soc. Amer.51:1333-1351(1972).

15. POND,J.B., A Study of the Biological Action of Focused Mechanical Waves(Focused Ultrasound),Doctoral Thesis, University of London(1968).

16. POND,J.B., The role of heat in the production of ultrasonic focal lesions, J.Acoust. Soc.Amer.,47:1607-1611(1970).

17. WELLS,P.N.T., Some Biological Effects of Ultrasound, Doctoral Dissertation, University of Bristol(1966).

18. DUNN,F. and F.J.FRY, Ultrasonic threshold dosages for the mammalian central nervous system, IEEE Trans.Biomed.Engin. BME-18: 253-256(1971).

19. LANDAU,L.D. and E.M.LIFSHITZ, "Statistical Physics",(Pergamon Press,London,1959)p.33.

20. MATISON,G. and P.P.LELE,Scattering of Ultrasonic Plane Waves by Ultrasonic Focal Lesions in Tissues,I.E.E.E. Ultrasonics Symposium Proceedings,72 CHO 708-8 SU:446-452(1972).

21. NYBORG, W.I.,Mechanisms for nonthermal effects of sound, J.Acoust. Soc.Amer. 44:1302-1309(1968).

22. LELE,P.P.,"Ultrasound in Biology and Medicine"in "BioMedical Physics and BioMaterials Science",H.E.Stanley Ed. M.I.T. Press (1971).

23. BLAIR, E.,"Clinical Hypothermia", MCGraw-Hill,1964.

24. KEELE,C.A. and E.NEIL,(revised),"Samson Wright's Applied Physiology",Oxford University Press, London(1966).

25. BELL,G.H.,J.N. DAVIDSON,and H. SCARBOROUGH,"Textbooks of Physiology and Biochemistry",E.&S.Livingstone Ltd.,London(1968).

26. TAYLOR,K.J.W.,Ultrasonic damage to spinal cord and the synergistic effect of hypoxia,J.Path.,102:41-47(1970).

27. TRUEX,R.C. and M.B.CARPENTER,"Human Neuroanatomy",Williams & Wilkins Co.,(1969).

28. LELE,P.P.,Progress Report to U.S.Public Health Service,Project No.NB03222(1965).

29. BASAURI,L. and P.P.LELE,A simple method for production of trackless focal lesions with focused ultrasound:Statistical evaluation of the effects of irradiation on the central nervous system of the cat,J. Physiol.,160:513-534(1962).

30. MORITZ,A.R. and J.HENRIQUES,Studies of thermal injury;conduction of heat to and through skin and the temperatures attained therein; theoretical and experimental investigation,Amer.J.Path., 23:695-720(1947).

31. MATISON,G.G.,Sonar Detection of Ultrasonic Lesions During and After Production, Doctoral Thesis,Massachusetts Institute of Technology(1971).

32. YOUNG,G.F. and P.P.LELE,Focal lesions in the brain of growing rabbits produced by focused ultrasound,Exper.Neurol.,9:502-511 (1964).

-DISCUSSION-

TAYLOR - Pond did some experiments like this some years ago (Pond, 1968). The same temperature cycle which had been produced by a focused beam of ultrasound was simulated by heating wires passed through the brain substance. Up to an intensity of 650 W/cm^{-2}, he found that heat alone was sufficient to account for ultrasonically induced lesions. I think Professor Dunn would probably go down lower than that, but these two independent workers agree that there is a region of dose parameters below 650 W/cm^{-2} in which a thermal change is sufficient to account for the production of a lesion. Both workers also agree that cavitation accounts for the disruptive effects observed at intensities above 1200 W/cm^{-2}. This leaves a range of 650 - 1200 W/cm^{-2} in which Pond found that he was unable to simulate the same lesion by thermal change that was produced by ultrasonic means. This was strong evidence for a non-thermal mechanism in this dose range.

LEHMANN - What you found on the occasion of this lesion and demonstrated by histologic means is an ultrasound effect which may be produced by the same temperature elevation, as was produced by the ultrasound. I think this applies to various types of reactions to ultrasound. At low intensities within the range of less than 10 W/cm^2, peak; and for longer exposures, this holds true for muscle metabolism, for permeability of membranes, for changes in the mechanical properties of collagen tissues, and so on. I think it is very likely that we get biological effects from ultrasound which are temperature mediated. On the other hand, there are effects which are also very well described in the literature, however, the mechanism of those still escapes us and we can not define the therapeutic use or the dangers. Some of them are clearly destructive, but at higher intensities than would be used in a therapy. I remember experiments we did some time ago with high intensities, 100 W/cm^2 or more where the tissue looked like mashed potatoes, virtually destroyed, and yet we could not reproduce this type of lesion by temperatures with similar rate of rise and peak levels. It was not cavitation because simultaneous application of adequate pressures did not prevent it either. To this day I don't know what the mechanism of damage was. Similar effects were seen on diffusion processes through biologic membranes. The question is, in all these things, what significance they do have, and under what conditions do they occur?

LELE - The experimental conditions are really germane to these results. We have some other tissues we are looking into. We don't turn our back on experiments to extrapolate these results to a wide range of tissues or to other irradiation schedules.

LELE - At the moment, I agree completely that the effects of either reversible blockage of nerve conduction or a diffuse impairment of nerve conduction can be exactly mimicked by a localized heat source.

PHYSIOLOGICAL RESPONSES TO HEAT

John Bligh

Agricultural Research Council Institute of

Animal Physiology, Babraham, Cambridge, England

ABSTRACT

The characteristics of the mammalian processes of thermoregulation have been likened to that of a physical regulator in which a signal representative of the controlled variable is compared with a reference or set-point signal. A sustained thermal disturbance elicits sustained thermoregularity responses, but the steady level of core temperature may be displaced. This shift is interpreted by some as a shift in the biological set-point, but others argue against the set-point concept and suggest that the controlled variable may not be a body temperature. An alternative proposal discussed in this paper is that the regulatory process is essentially a servomechanistic maintenance of the balance between heat production and heat loss. Since the mechanisms of thermoregulation are fundamental to the understanding of the nature of disturbances to the system, these alternative concepts are discussed.

Introduction

When non-ionizing radiation impinges on an organism most of it is converted into heat within the tissues where the progress of the waves is arrested. The extent to which temperature rises during the exposure, and the effect this rise in tissue temperature has on the total organism, depends on the intensity and duration of the exposure to radiation, and the ease with which this heat can flow away to other parts of the body or to the environment. The consequences of this rise in local tissue temperature may be physical, physiological or pathological. The purely physical consequences of changes in thermal gradients, and the pathological consequences of local tissue damage are outside the scope of this paper. Discussion is restricted to the physiological responses which basically are twofold: i) the direct effects on local tissue functions, and ii) the reflex effects operating via sensors, the central nervous system and effectors.

The local effects of heat on tissues are complex since virtually every biophysical and biochemical process is temperature dependent. Mention is made here of only two of the more salient effects: changes in the rate of tissue metabolism, and in the local circulation of the blood. An increase in metabolic rate resulting from a rise in local temperature further increases the rate of generation of free energy and is additive in its thermal consequences. Peripheral vasomotor tone is regulated reflexly via the central nervous system, when ambient conditions are close to thermoneutrality, but at low ambient and skin temperatures, the peripheral vessels tend to be kept constricted, and at high ambient and skin temperatures the vessels may be kept dilated by local thermal effects on them. Thus except in a narrow range of ambient temperatures, within which the ear vessels were responsive to sympathetic control, centrally administered substances which activated or inhibited other thermoregulatory effector functions, were without action on peripheral vasomotor tone (5). Very likely, therefore, local thermal disturbances due to non-ionized radiation exert a direct effect on the peripheral blood vessels which render them unresponsive to any influence from higher neuronal integrative centres.

Mainly, however, a thermal disturbance will act on local thermosensors and reflexly induce appropriate corrective changes in the pattern and intensity of thermoregulatory effector functions. Apart from peripheral vasomotor tone, which may be influenced mostly by direct thermal effects, these comprise of shivering and non-shivering thermogenesis which increase the heat content, and evaporative heat loss by panting or sweating, which decrease the heat content of the total organism. These functions are controlled by central regulatory processes in neuronal pools in the hypothalamic region, and possibly other regions, of the brain.

The basic concern in this paper will be the search for the closest analogy to the mechanism by which body temperature, and its regulatory processes are controlled under sustained thermal load such as during prolonged exposure to non-ionizing radiation.

The regulation of body temperature in mammals has been compared with the regulation of the water temperature in a laboratory waterbath, and it is accepted by some as the most reasonable hypothesis, and by others as a fact, that some biological equivalent to a thermostat must exist in the mammalian brain. Evidence, or even the notion of how neurones in the brain perform a thermostat-like function has still to be realized, and from time to time an alternative hypothesis to the thermostat analogy is put forward. Basically, the alternative hypothesis is based on the proposition that the regulated parameter is not a body temperature, but the balance between the rates of heat production and heat loss. According to the protagonists of this thesis, the relative stability of body temperature is a consequence of this balance, and small variations in the level of body temperature can be attributed to a time delay in the re-establishment of the balance. Since shifts in the level of body temperature which may stem from the action of a pyrogen, physical exercise, the administration of drugs, changes in ambient temperature or exposure to non-ionizing radiation, are now often described as 'shifts in the set-point', rather than in more guarded terms, it may be useful to review briefly the evidence and discuss the interpretations placed on them.

The set-point concept

A comparison of the temperature disturbance-response characteristic of man and other mammals with those of different types of physical regulators (12) revealed that the characteristics of the mammalian pattern of thermoregulatory responses to thermal disturbances are very similar to those of a proportional controller, in which a signal derived from the controlled variable (a core temperature (T_c)) is compared with a set-point of reference signal, and in which the intensity of the thermoregulatory effectors is proportional to the error between the signals representative of the set-point and the controlled variable (Fig. 1A). If it is hypothalamic temperature (T_{hy}) or any other specific T_c which provides the controlled variable, then the convergence of other signals from variously located extra-central sensors would seem to be theoretically inadmissable. Thus the thesis that temperature regulation can be explained as the relation between a set-point and a controlled variable signal derived from a T_c takes no account of the roles of the extra-hypothalamic temperature sensitive structures such as those at peripheral surfaces and in the spinal cord. There is also the problem of the apparent variability of the level at which T_c is regulated in different circumstances.

A 'solution' to both these problems (11) was that extra-hypothalamic and non-thermal influences act on and modify the set-point signal (Fig. 1B). Thus emerged the concept of changes in the thermoregulatory set-point not only in the particular case of fever, but also as a result of many other disturbances. The basic experimental evidence for a shift in the set-point during thermal stress is expressed in Fig. 2 If the set-point were fixed, it might be supposed that after an initial disturbance, T_c would

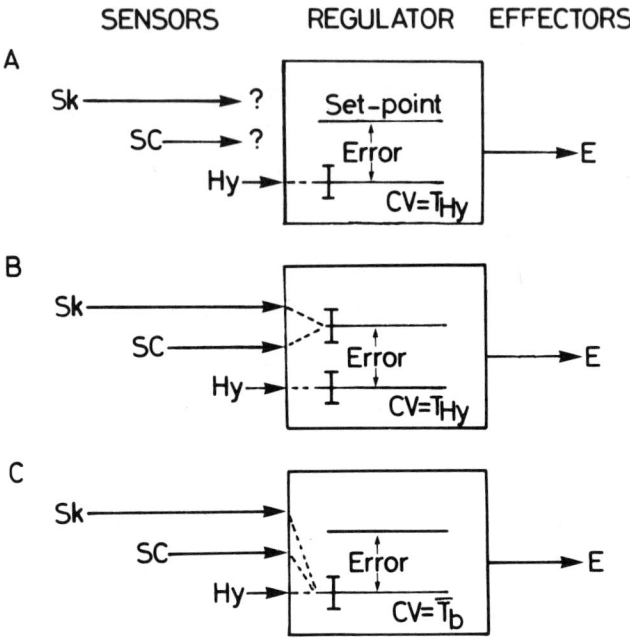

Figure 1. Diagram expressing the role of the central regulator of body temperature. A) The controlled variable (CV) is hypothalamic (Hy) temperature. A signal derived from Thy sensors is compared with a fixed reference or set-point (SP) signal. The difference between SP and CV signals, the error, determines the drive to the thermoregulatory effectors (E). The influence of extra-hypothalamic thermosensors, represented by skin (Sk) and spinal cord (SC) sensors are undefined. B) The signals from Sk and SC sensors are presumed to influence the SP which is thereby varied. C) The signals from Sk, SC and Hy thermosensors are considered to converge and all contribute to the CV which is thus representative of something close to mean body temperature ($\bar{T}b$). The SP may be fixed.

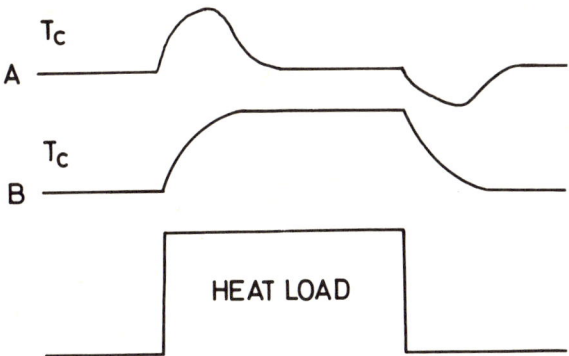

Figure 2. A diagram expressing the possible effects on core temperature (Tc) when an animal or man is subjected to a square-wave heat load. A) The expected pattern of Tc if the set-point is fixed; Tc returns to its pre-disturbance level while the disturbance remains. B) The general observed pattern of an elevation of Tc to a new steady level.

return to its controlled level while the disturbance persisted (Fig. 2A), but the typical pattern is the curve illustrated in Fig. 2B: Tc rises to a new steady state which is sustained for so long as the disturbance persists. This pattern of events has been interpreted as evidence of a shift in the set-point, although the sustained rise in Tc could be interpreted as a sustained increase in the <u>error</u> necessary to maintain the drive to heat loss effectors.

The servomechanism concept

An earlier proposal that thermoregulation operates as if the heat content of the body, rather than a particular body temperature, is the controlled variable (8) was revived by Snellen (22) who argued that if a signal representative of average body temperature (Tb) were fed to the central regulator, the assumption of a changed set-point of Tc would be unnecessary. Houdas et al. (14,15,16) also argue against the set-point concept and for a controlled balance between heat production and heat loss. On the basis of their analyses of the relationship between thermal disturbances and thermoregulatory responses they propose that the relative stability of Tc can be based on another physical model - a servomechanism

- and that the shift in Tc during a sustained thermal load can be attributed to the time lag in the rebalancing of heat production and heat loss. An analogy with a water cistern was used (18) to illustrate the set-point concept (Fig. 3A). A similar analogue (Fig. 4B) has been used to explain the operation of a servomechanism to maintain the balance between water inflow (= heat production) and water outflow (= heat loss) (14).

The basic thesis here is that an increase in signal intensity from 'cold' sensors is an indication of an increase in heat loss or reduction in heat production, while an increase in the signal intensity from 'warm' sensors is an indication of an increase in heat gain or production, or a reduction in heat loss. Thus the signal from thermosensors could provide information about heat balance rather than temperature which is a derivative of heat content. The relation between thermosensors and thermoregulatory effectors

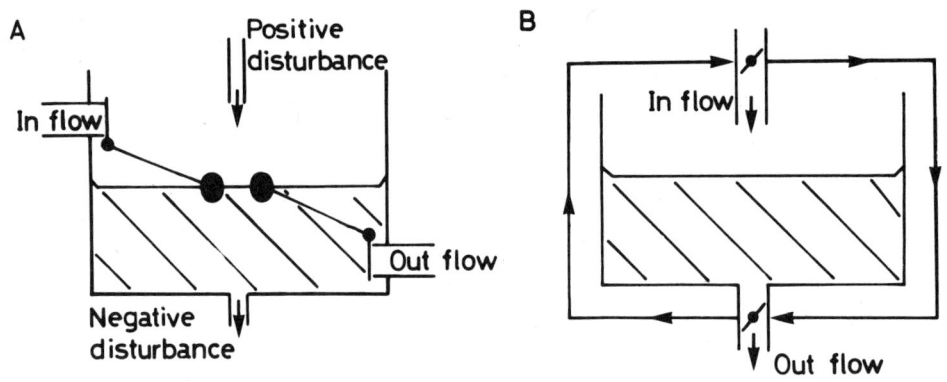

Figure 3. Physical analogies (water cistern level) of the process by which homeothermy may be achieved in a biological system. A) the water level is regulated by ballcock (disturbance sensor) control of water inflow (= heat production) and outflow (= heat loss). A positive disturbance (= a non-thermoregulatory increase in heat production) or a negative disturbance (= a non-thermoregulatory increase in heat loss) changes the water level (= controlled variable) and activates the regulatory valves. (from Mitchell et al. 1972). B) the water content of the tank (= heat content of the body) is held at an approximately constant level by servomechanisms which continuously equate inflow (= heat production) with outflow (= heat loss). A sustained change in outflow elicits a sustained and balancing change in inflow, but because of the lag in response via the servomechanism, the water level (= temperature) will shift in the direction of the disturbance (modified from Houdas & Guieu, 1974).

could be essentially that of a servomechanism which effects the re-establishment of the balance between heat production and heat loss during a disturbance. Such a system would have the effect of maintaining Tc at a more-or-less fixed level although Tc is not actually the controlled quantity.

Convergence of pathways from thermosensors

Evidence has now accrued, from both hypothalamic unit activity studies (6,9,13,20,23) and from studies involving chemical interference with synaptic events in the hypothalamus (4,17) of the convergence of pathways from peripheral, spinal cord and hypothalamic temperature sensitive structures (Fig. 1C). This evidence reduces the acceptability of Thy as the controlled variable, and could support the proposition (22) that the controlled variable might be $\bar{T}b$ (or something very near to $\bar{T}b$, depending on the distribution of thermosensors in the different thermal compartments of the body, roughly distinguished as core and shell). Consistent with this interpretation of neuronal studies is the evidence from disturbance response analyses that the intensity of shivering in the guinea-pig (7) and sweating in man (21) correlate better with $\bar{T}b$ than with Tc or T skin.

Hitherto, the reasons for preferring the concept of the regulation of a Tc close to a set-point, rather than the maintainance of a balance between heat production and heat loss on a servomechanistic basis have been i) the relative constancy of Tc, and the relative lability of shell temperatures, and ii) the absence of any clear idea of how heat production and heat loss could be sensed and equated. However, as has been pointed out by Houdas (personal communication), an integrated signal derived from thermosensitive structures in different compartments of the body is representative of the heat content of the body as well as of $\bar{T}b$. Such information, Houdas suggests, could be the basis for the maintenance of a balance between heat production and heat loss by a servomechanism: a reduction in heat content → a reduction in $\bar{T}b$ → a signal to increase heat production and reduce heat loss; conversely, an increase in heat content → a rise in $\bar{T}b$ → a signal to decrease heat production and increase heat loss.

Neuronal models of the central controller

Neuronal models have been constructed, quite independently, of each other, on the basis of a) disturbance-response analyses; b) studies of electrical events in temperature sensitive neurones in the hypothalamus, and c) the effects of thermoregulation of drugs which influence synaptic events in the hypothalamus. Bligh (1,2,3) has noted that when represented in a common format, these

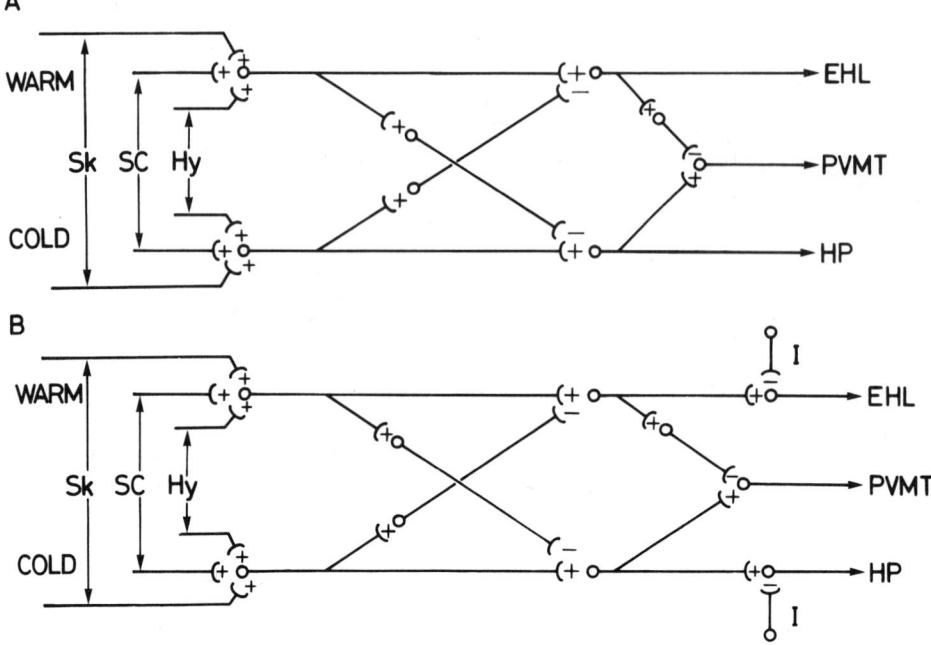

Figure 4. Models of the neuronal arrangements in the hypothalamus relating the input from temperature sensors in the skin (Sk); spinal cord (SC) and hypothalamus (Hy) to the output to the evaporative heat loss (EHL), peripheral vasomotor tone (PVMT) and heat production (HP) effectors. A) is based on synaptic interference studies. B) includes the addition of inhibitory influences (I) on the EHL and HP pathways to provide a theoretical explanation for the thresholds for EHL and HP (modified from Bligh, 1974).

models are remarkably similar to each other, having the following common features: i) convergence of the signals from peripherally and deeply located temperature-sensitive structures; ii) two main neural pathways through the hypothalamus - one from warm sensors to heat loss effectors, and the other from cold sensors to heat production effectors, and iii) crossing inhibitory influences of each pathway on the other (Fig. 4A).

An interesting distinction between the physical models of the central thermoregulator based on engineering analogues, and the neuronal models, is that the latter apparently contain nothing to represent a set-point mechanism. On the other hand, Fig. 4A could

PHYSIOLOGICAL RESPONSES TO HEAT

well be regarded as a neuronal equivalent of a servomechanism, which does not require a reference signal generator.

Comment

Neither of these concepts, in the simple forms presented here, can adequately explain the observed relations between thermal disturbance and thermoregulatory responses. Fig. 5 is a diagrammatic representation of the relations between a $\bar{T}b$ (which may be Tc or $\bar{T}b$) and the intensity of the thermoregulatory effector functions. Thermoregulatory heat production first appears at a threshold Tb and increases as Tb falls further. Similarly, evaporative heat loss appears at another threshold Tb and increases as Tb rises further. Peripheral vasomotor tone operates between these thresholds. This pattern indicates a need to hypothesize distinct and separate set-points for each of the three principal thermoregulatory effector functions.

The set-point concept does not explain these thresholds, but neither does the servomechanistic concept or the neuronal model.

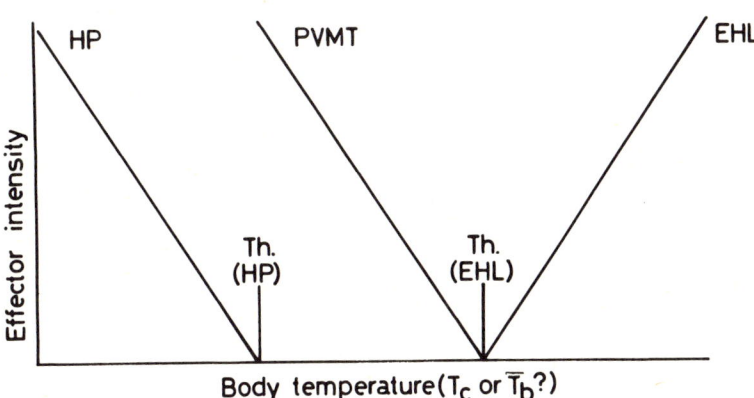

Figure 5. A diagrammatic expression of the general relations between the intensity of thermoregulatory effector functions and a body temperature (Tb) which may be a specific core temperature (Tc) or mean body temperature ($\bar{T}b$). Head production (HP) increases when Tb falls to or below a threshold (Th) level. Evaporative heat loss (EHL) increases when Tb rises to and above a threshold level. Peripheral vasomotor tone (PVMT) operates between these two thresholds.

A hypothetical neuronal device to operate these threshold activities was incorporated into the neuronal model by Bligh (Fig. 4B). The only findings to support this embellishment, apart from the observed disturbance-response relations, are the reports of temperature insensitive neurones in that area of the hypothalamus concerned with temperature control (10, 19). These could be the source of a steady level of synaptic inhibition on the pathways to evaporative heat loss and heat production effectors, and could block the passage of excitatory signals from the thermosensors to the effectors until the excitatory influence at the synapse becomes sufficient to exceed the fixed inhibitory influence.

On the face of it, this modified neuronal model has the functional characteristics which express the central nervous processes necessary to create the observed relations between thermal disturbance and thermoregulatory response. For example, when thermoregulatory effector intensities are plotted against Tc, a peripheral thermal disturbance modifies the threshold Tc for heat production or evaporative heat loss. If, however, there is a summation of the signals from the differently located warm and cold sensors, and the signal strength from the sensors at one of these locations is changed, there will be a change in the activity in the post-convergence pathways relative to any particular Tc. Therefore the threshold, expressed in terms of that particular Tc rather than Tb, will have changed. Such apparent variations in the set-point or threshold temperatures could relate to this arbitrary choice of a particular Tb as the controlled variable.

This continuing analysis and debate on the nature of the central regulation of body temperature, and the effect on it of a thermal stress, is directly relevent to what happens to the regulation of body temperature during exposure of the body to a level of non-ionizing radiation which has completely reversible thermoregulatory effects. The local accumulation of heat, mostly in peripheral tissues, changes the input from peripheral thermosensors, and this may change the level of core temperature at which heat production and heat loss come back into balance. But has a thermoregulatory set-point been changed? We cannot say with certainty; we can only say that it is as if a set-point has been changed.

REFERENCES

1. BLIGH, J.: Neuronal models of mammalian temperature regulation, Ch. 9, Essays on Temperature Regulation (Bligh, J., Moore, R.E. Eds.). North Holland, Amsterdam (1972).

2. BLIGH, J.: Temperature Regulation in Mammals and Other Vertebrates. North Holland, Amsterdam (1973).

3. BLIGH, J.: Neuronal models of hypothalamic temperature regulation, in Recent Studies of Hypothalamic Function (Cooper, K.E., Lederis, K. Eds.) Karger, Basel (1974)

4. BLIGH, J. and BACON, M.: Interaction of spinal heating and cooling and intracerebral ventricular injection of monoamines and carbachol in sheep. (Int. Symp. Temperature Regulation, Jerusalem, 1974 - submitted).

5. BLIGH, J., COTTLE, W.H. AND MASKREY, M.: J. Physiol., Lond. 212 (1971) 377.

6. BOULANT, J.A. and BIGNALL, K.E.: Am. J. Physiol. 225 (1973) 1371.

7. BRÜCK, K. and WÜNNENBERG, W.: Arch. Ges. Physiol. 293: (1967) 215.

8. GLASER, E.M. and NEWLING, P.S.B.: J. Physiol., Lond. 137 (1957) 1.

9. GUIEU, J.-D. and HARDY, J.D.: J. Appl. Physiol. 29 (1970) 675.

10. HAMMEL, H.T.: Neurones and temperature regulation, Ch. 5, Physiological Controls and Regulation (Yamamoto, W.S., Brobeck, J.R. Eds.). Saunders. Philadelphia (1965).

11. HAMMEL, H.T. et al.: J. Appl. Physiol. 18 (1963) 1146.

12. HARDY, J.D.: Physiol. Rev. 41 (1961) 162.

13. HELLON, R.F.: Arch. Ges. Physiol. 321 (1970) 381.

14. HOUDAS, Y. and GUIEU, J.-D.: Physical models of human thermoregulation. Proc. 2nd. Int. Symp. Pharm. Thermoreg., Paris 1974. Karger, Basel (in press).

15. HOUDAS, Y. et al.: J. Physiol., Paris 66 (1973) 137.

16. HOUDAS, Y. et al.: Trans. Am. Soc. Mech. Engrs. Series G. 95 (1973) 331.

17. MASKREY, M. and BLIGH, J.: Int. J. Biometeor. 15 (1971) 129.

18. MITCHELL, D., ATKINS, A.R. and WYNDHAM, C.H.: Mathematical and physical models of thermoregulation, Ch. 5, Essays on Temperature Regulation (Bligh, J., Moore, R.E. Eds.). North Holland. Amsterdam (1972).

19. NAKAYAMA, T., HAMMEL, H.T., HARDY, J.D. and EISENMAN, J.S.: Am. J. Physiol. 204 (1963) 1122.

20. NUTIK, S.L.: J. Neurophysiol. 36 (1973) 250.

21. SNELLEN, J.W.: Acta Physiol. Pharmacol. Neerl. 14 (1966) 99.

22. SNELLEN, J.W.: Set-point and exercise, Ch. 11, Essays on Temperature Regulation (Bligh, J., Moore, R.E. Eds.). North Holland. Amsterdam (1972).

23. WIT, A. and WANG, S.C.: Am. J. Physiol. 215 (1968) 1151.

PHYSIOLOGICAL RESPONSES TO HEAT

-Discussion-

LELE - Which anatomical location do you think is most sensitive to warming in the hypothalamus, if there is a discrete target site?

BLIGH - This could depend on which species you are talking about.

LELE - The cat or the rhesus monkey.

BLIGH - There seems to be a far greater number of warm sensors deep within the body and in the hypothalamus than in the periphery, and there is some evidence that there are more cold sensors in the periphery than in the hypothalamus. But, one would have to qualify this. In the sheep and in some other species, but not in the dog, warming the scrotum will activate evaporative heat loss and will drive body temperature right down. The resultant depression in hypothalamic temperature shows no tendency to counteract the effect of the peripheral warm stimulus.

One must conclude that there is a very powerful drive from the peripheral warm sensors which predominates over any concurrent drive from central cold sensors.

CARSTENSEN - What is the control in that case? Is there any kind of local control over vasomotor tone, or is it all from the hypothalamus?

BLIGH - Well, the main drive that brings body temperature down when you heat the scrotum in the sheep is the evaporative heat loss by panting. Undoubtedly the signals from the scrotal warm sensors are passing through the hypothalamus, or at least through the CNS. Again, one must be a little cautious, because there is some evidence that there may be other non-hypothalamic pathways from thermosensors to thermo-regulatory effectors.

CARSTENSEN - If you have a single organ or a region in a body that's irradiated -- let's say you are getting diathermy treatment over the muscle in the back, would there be a local control over the blood flow in that region, or is it all from the central nervous system?

BLIGH - Well, again, I can only speak with experimental certainty about the sheep. A locally applied temperature affects peripheral vasomotor tone, and causes the vessels to dilate in that particular area. But the responses in other areas would be via the hypothalamus as a result of the drive from temperature sensors.

BEISCHER - I would like to draw attention to some experiments which Dr. Houk did with Dr. Michaelson where rats were irradiated in a field of about 10 mW/cm^2 for 4 or 5 hours (Michaelson, S. M., W. M. Houk, N. J. A. Lebda, S. T. Lu and R. L. Magin. Biochemical and neuroendocrine aspects of exposure to microwaves. New York, Ann. N.Y. Acad. Sci., 247: 21-45, 1975). The core temperature of the animals increases and stays at an elevated level during several hours exposure. There is obviously no attempt of the animal to counteract this small increase in core temperature.

BLIGH - I think this is the general observation. If you cause a thermal disturbance such as a heat load, then body temperature is not regulated as might be expected from the set point hypothesis; an initial disturbance to body temperature being nullified. A sustained displacement is the more likely effect of a sustained thermal load. Is this what you are saying?

BEISCHER - Yes, that's what I am saying.

BLIGH - Now, I should like to hear Dr. Hardy's views on this because Houdas considers this sustained displacement to be indicative of a servo mechanistic form of control. In such a system there would be a delay before heat production is increased in response to a cold stimulus, and heat loss is increased in response to a warm stimulus. When, as a result of these responses to disturbances, heat reduction and heat loss were again in balance, the heat content of the body would have changed a little and would remain displaced from its previous level for so long as the disturbance persisted. And, Houdas says that this pattern of core temperature during a thermal disturbance is more representative of a "servo mechanism" than the "set point" type of control. On the other hand, it seems to me that if there were a "set point" type of control and if a thermal disturbance persisted, there would have to be an increase in the "error" in order to maintain a corrective response and re-establish thermal stability. I would like to know whether Dr. Hardy considers this sustained deviation in core temperature to be a maintained error in terms of the set point hypothesis.

HARDY - I suppose so. I have always thought the physiological thermoregulator probably was a proportional controller; there must be an error signal to set the regulator in operation. As far as the Houdas, Colin and Guieu hypothesis is concerned, they do not accept a set point as just a fixed element, which itself is temperature insensitive. They substitute for the set point a non linear function, but in the non linearity they have, in a sense, built in what amounts to both a set point and thermo-sensitivity. "Set point" is something like the dosimetry that

Hardy - you gentlemen were talking about this morning. It is a term that comes easily off the tongue, but pretty soon it refers to so many things that we wish one hadn't thought of it. Basically, I think, it doesn't make any difference. Going over Houdas' et al. and also Wyndham and Mitchell's hypotheses, which are not really very different since they both arrive at the same conclusions, I think I have the feeling that the significant question is, as the result of convergence, do the incoming signals add up or do they multiply. The more recent work from my laboratory would indicate that they may appear to be a sum, but, in fact, the convergence probably is a multiplication. And, it also appears that the set point or reference point is a biologically important point in that it is the threshold of noxious thermal stimulation (45°C). The body operates, let's say, at 37° or 38° or 39°C while the regulator is really referencing itself at all the times to 45°C.

Dr. Bligh, would you give us your ideas on how fever is produced? I mean, is fever a change in the set point or is it a change in the cerebral mechanism or what?

BLIGH - I don't know. But we were able to study fever and to express its action in terms of our neuronal model. I must admit that fever is crucial to this whole question of "set point" versus "servo mechanism." Clearly we have first to decide what we really do mean by "set point." There has been a tendency on the part of some to consider it as an injected signal in the strict engineering sense of a "reference signal." Whereas some pharmacologists working on drug effects on body temperature deny that by using the term "set point" they imply the existence of anything analogous to the engineer's set point. When they speak of the set point, they refer only to the fact that ordinarily body temperature remains fairly stable at some point.

The model that we are examining has two main pathways one from the warm sensors to heat production effectors, and one from cold sensors to heat loss effectors. What we have found in the sheep is that intraventricularly injected 5 hydroxytryptamine acts as if it were the transmitter on the pathway from warm sensors to heat loss effectors; it activates heat loss effectors and inhibits heat production effectors. Acetylcholine acts as if it were a transmitter on the pathway from cold sensors to heat production effectors; it activates heat production effectors and inhibits heat loss effectors. Noradrenaline exerts an inhibitory influence on both pathways. A pyrogen-induced fever has been used to test the actions of these putative transmitter substances. When a pyrogen is administered, heat production is activated, and heat loss is inhibited. This suggests that the pyrogen is acting at or near the point at which acetylcholine seems to act. We

BLIGH - could test this. If during a pyrogen-induced rise in body temperature, adrenaline is injected into the cerebral ventricles it should inhibit the shivering and halt the rise in temperature. That is exactly what happened when we did the experiment. On the other hand, an intraventricular injection of 5 hydroxytryptamine during the rising phase of fever should inhibit shivering and activate panting, so the change in body temperature should be reversed. Again, that is exactly what happened when we did the experiment. I can't prove this now, but we plotted what we thought would happen to body temperature beforehand, did the experiments, and after the experiments all we had to do was put in the standard errors. So, in our neuronal format we know where a pyrogen acts. But if a pyrogen acts on the set point mechanism where in this neuronal format is the set point machinery. That I can't answer.

HARDY - You had indicated a neuron located off to the side in one of your diagrams; I assume you were using it as kind of a "set point" neuron.

BLIGH - Yes, that is so. It is a purely theoretical embellishment added to allow for "threshold" in the activation of heat production and heat loss effector function. We have no neuronal or synaptic evidence for this at all, but I agree that a "threshold generator" might be indistinguishable from a "set point" generator.

Much of the model is descriptive of experimental findings, but that doesn't make it true. However, the threshold generator is purely theoretical. There is no evidence whatever. It is not a description of any experiment. It simply expresses the neuronal concept of threshold which is that intensity of synaptic excitation at which excitation exceeds inhibition and signals are generated in the post pathway. So, when I use the word "threshold" I am thinking neuronally, but whether I am really talking about the same thing as you are when you use the word "set point" I don't know.

HARDY - I have used "set point" as a concept rather than to refer to any specific physiological entity, because it is the simplest way to make my theoretical system work. Thus, if one wishes to set up an analog model or digital model of thermoregulation, it is most convenient to enter a fixed temperature into the equations and call this the "set point." So, that's the way I really use it. However, in terms of responses of single units in the hypothalamus, I was impressed with the fact that a lot of these neurones that I observed were completely temperature insensitive. Of course, this is the kind of element one would need, if one wanted a "set point" -- a kind of signal, something

HARDY - that wasn't influenced by temperature.

BLIGH - I quite agree that if these threshold generators, which I have put into my model without any supporting evidence, should turn out to be the temperature insensitive neurones described by you, then I think we would suddenly discover that we were talking about the same thing, and that my concept or preference for the term "threshold," and your preference for the term "set point," would be purely semantic. It would mean that a beginning had been made in the understanding of biological temperature set point in neuronal terms.

FLOOR - Can Dr. Bligh give me an example of internal -- that is to the body -- transducer, that could measure heat content?

BLIGH - I will put it this way. That this is an old problem. When body temperature is being kept at a more-or-less steady level, heat production and heat loss are being held in balance perhaps by set point or by servo mechanistic form of control. Suppose it is heat content rather than core temperature that is being regulated. This would require the sensing of heat content, and the question is "how could this be achieved?" Because we know that there are temperature sensors in the hypothalamus, and we know that the deep body temperature is stable and the peripheral temperature is fairly labile, we tend to regard the hypothalamic temperature sensors as having a particular significance. It now seems possible that afferent pathways coming in from the cutaneous and from the spinal cord thermosensors and from the hypothalamus converge synaptically. Now, Houdas has suggested to me that if this thermal information is summated in this way the resultant signal conveys information about mean body temperature (or something like it). It also carries information about the heat content of the body, because if you had a structure of several different compartments and you wanted to keep the total heat content of the structure constant, then, presumably you would place temperature sensitive devices into the different compartments in proportion to the mass of each compartment. The summed signal would then be proportional to the heat content of the total structure and could be used in a control system to maintain the relative constancy of the heat content of the system. Now, I haven't yet thought out whether this Houdas thesis is a reasonable one, but I am sure Dr. Hardy, who is much better versed in physics than I am, will be able to make some useful comment. The point that is being made is: whether the summated signal from variously located temperature sensors gives information only on body temperature, or also gives information on body heat content.

FLOOR - The point is then that the body has some sense of its

FLOOR – own mass?

BLIGH – Has some concept of it.

FLOOR – Has some sense of its own mass. It needs to know what its mass is.

BLIGH – As I said, it will only have sense of its own mass if the signals from variously located thermosensors are appropriately weighted.

HARDY – That is the point of the question as to whether the body really knows what weight it wants to be.

FLOOR – It must know that if it is to know its heat content.

HARDY – It does, but I don't think that heat content makes very much difference to the thermal regulator.

FLOOR – I agree.

HARDY – For example, consider a baby, and how much it increases its mass over its lifetime. The body temperature remains pretty much the same, so that its heat content increases by a factor of 20 or 30 times.

BLIGH – That is a very valid objection. I agree with you.

HARDY – I do not see any rationale to the idea of regulation of heat content of the body.

VOGELMAN – What happens when you use drugs to vasodilate the skin? You change the surface temperature, but you are changing the blood circulation at the same time. What kind of reaction do you get in the core?

BLIGH – The change in body temperature, you mean?

VOGELMAN – Do you get a body temperature change when you use a vasodilator?

BLIGH – I don't know. I just don't remember.

HARDY – I would say no. A man who is chronically vasodilated in a neutral or cool environment may shiver all the time. In certain exfoliating dermatoses you have this condition, if the skin is red and flushed all the time. Such an individual has to stay under the covers to keep warm, or he is shivering all the time.

VOGELMAN - What is his core temperature?

HARDY - Normal.

BLIGH - On the other hand, I think it only fair to point out that there are cases cited in the literature of human patients with wide-spread skin lesions and with very large loss of thermo-sensivitity in the skin. Their body temperatures are reported to vary considerably and to be much more labile as a result of this peripheral information. So, it is very difficult to be sure about the relation between peripheral thermosensitivity and the stability of core temperature.

CZERSKI - May I ask a question? A specific characteristic of microwave heating is non-uniform heating. If you expose a rat to an incident field in the far field zone, along the direction of propagation, you will get certain temperature increases in the brain which may differ in, say, an incident field of about 100 mW/cm^2, from the temperature in the liver by 5°C after five minutes of exposure. Such differences in temperature at the end of very short exposure, between brain, hypothalamic region, heart, liver and kidney, may be between 1 and 7°C. And of course, if you turn the rat around, the distribution of such temperature gradients, will be quite different. You will expect some heating around the kidneys, around the liver or in the liver, and much less in the hypothalamic region. When you expose the rat laterally, then you get still another temperature gradient distribution at the end of the exposure of the same duration. I am just wondering if you would care to comment on how far would you expect all these situations to be physiologically different in activating thermal regulatory mechanisms? Do you feel that such situations may be compared from the point-of-view of thermal regulation?

BLIGH - I do not know how far; we are still within the realm of normal physiology. Do you know, Dr. Hardy? But you can raise hypothalamic temperature as much as 5°C.

CZERSKI - Oh yes, without any difficulty.

BLIGH - But this is way outside normal physiological variations in hypothalamic temperatures, isn't it? Everything depends, it seems to me, upon whether this convergence is true or not, and whether certain receptors in the deep body have a special function or whether their signals are integrated with those from elsewhere in the body. Until we are certain about the answer to this, I think we are just speculating.

CZERSKI - Yes. May I ask, is any information available about

CZERSKI - the distribution of temperature sensors in various organs, anatomically speaking?

BLIGH - No. Some people still assume that thermosensors occur in only two areas -- that is: the skin and the hypothalamus. There is now very, very strong evidence, indeed, of thermoreceptors in the spinal cord. There is evidence from Dr. Hardy's laboratory, in the sheep at least, of receptors in the stomach. Even if you locally heat a particular area or tissue of the body, and get no response, one cannot be certain that there are no thermoreceptors there. It is possible that thermoreceptors are widely, but sparsely, distributed in the body. Their presence in low concentration cannot be demonstrated by the ordinary techniques of locally heating or cooling. But nonetheless, when the temperature of the whole body is changed, they could be contributing quite a sizable input to the central controller. So, at the moment, it would be extremely dangerous to assume that there are any areas of the body that contain no thermosensors and which are totally temperature insensitive. I come back to my statement that everything now depends upon whether the signals from the variously located thermosensors are integrated or are not.

CZERSKI - Taking into account the present situation, would you say that it would be extremely difficult at present to state the thermal balance characteristics of an animal exposed to non-uniform internal deep body heating?

BLIGH - I think it would be extremely difficult. Yes.

HARDY - Thank you very much, Dr. Bligh.

BEISCHER - I think microwaves offer an excellent means to check the validity of some of the theories presented here. The caloric input to the animal can be well localized and problems of dosimetry should be solved in time.

LELE - Dr. Bligh, is anything known about the local vascular response to local heating? If one heats, let's say, some part of the body, raises the temperature of a cubic centimeter of tissue by 4 or 5°C, what will the time course and the magnitude of the local vascular response be?

PORTELA - I should like to give the answer if Dr. Bligh will allow me. Let us consider an organ of a system localized in some part of the body which raises the temperature under a specific physiological action. Suppose that this component of such a system is a muscle where there is a vascular response to local heating, with heat transfer within the medium. The question

PORTELA - is whether this "working" muscle can experience significant temperature rise due to an increase of its metabolic activity, associated with glucose utilization related to insulin interaction with specific receptor sites at the level of muscle cell membranes. One will discover that this heat transfer - the transfer of thermal energy resulting from temperature differences - provokes a transient rise of the extracellular space of muscle, the cellular fraction of water osmotically active (W_{eff}), and the water permeability (P_w), correlated to an increased exchange of water. The magnitude of the local vascular response to lose organ heat will be facilitated by these basic phenomena.

This is an interesting complex mechanism for thermal regulation of organs. In our example the muscle contributes to raise the temperature as a result of an hormonal effect on the metabolism. You have experience with insulin, acting at determined levels of muscle cells. We have studied this problem in intact muscles and isolated cells from Tibialis anticus, semitendinosus and gastrocnemius. Insulin evokes a quick increase of metabolism and consequently of temperature (2°C to 3°C), W_{eff}, P_w, extracellular space, and again, the rate of water exchange. Thus, the resultant increase in temperature is minimized, due to the high capacity of water to remove heat from these complex systems that is cell and muscle.

These parameters apparently play a critical role in the regulation of organ and system temperature. In regard to the vascular response to local heating, there must be an inherent coupling effect between the internal and external heat exchange mechanisms. There is convincing evidence for the existence of localized control through vasomotor action in a specific organ.

LELE - There is a lot of vascular perfusion in muscle.

PORTELA - That is right. There is a significant mixed extravascular space, forming capillary networks.

LELE - What is the time course?

PORTELA - The time course depends on the species. Perhaps in dogs with time constants between 60 seconds to 160 seconds, this is an important parameter. We are talking in terms of tissues, and we must not forget some specific parameters, such as the extracellular space, the surface of the muscle membrane and the active surface of other tissues that are involved in the exchange of water or of other metabolites besides water. That is important from the point-of-view of heat exchange.

LEHMANN - We have tried to get at this question in another way.

LEHMANN - The final results are not in, but we know how much energy you put into the muscle and at which rate. In man as well as in animals, the temperature goes up first then apparently a vascular change occurs and the temperature goes down and seeks a new equilibrium, even though the energy input and its rate remain the same. Once you know how much you put in, you can also calculate the losses using these measurements. Now, we are calculating how the losses are achieved, assuming that blood is the substance which removes the heat energy.

LELE - We are trying to do something very similar such as trying to measure the local perfusion rates by sonic heating.

Microwave—Biological Effects

ELECTROPHYSIOLOGICAL EFFECTS OF ELECTROMAGNETIC FIELDS ON ANIMALS

Arthur W. Guy, James C. Lin and C.K. Chou
Bioelectromagnetics Research Laboratory
Department of Rehabilitation Medicine RJ-30
University of Washington School of Medicine
Seattle, Washington 98195

ABSTRACT

In vivo and in vitro studies of nervous tissues exposed to microwave radiation are reported. It is shown that the conduction and transmission latencies and amplitudes of evoked potentials in both the CNS of anesthetized cats, isolated nerves of cats, and ganglia of rabbits are affected by CW microwaves in a manner very similar to that of localized conduction heat. Temperature rises are always associated with any observable changes of the measured characteristics in the nervous tissues exposed to CW irradiation. The threshold of occurrence of latency changes falls between 2.5 - 5.0 W/kg of absorbed power density in the affected tissues which is about one-quarter to one-half of the normal metabolic rate of brain tissue. This absorbed power corresponds to 5 - 10 mW/cm^2 and 10 - 25 mW/cm^2 incident upon a cat head and a human head, respectively. Pulsed microwaves of high peak intensity and low average power induce a hearing phenomenon in man. Electrophysiological studies on cats indicate that pulsed microwaves interact with mammalian auditory systems in a manner similar to that of conventional acoustic perception. A possible mechanism of microwave interaction is the acoustic energy release from rapid thermal expansion due to power absorption in the gross structure of the head.

I. INTRODUCTION

The literature is abundant with reports on the effects of electromagnetic fields (EMF) on the central nervous system (CNS) and on

the peripheral nervous system of both man and animals. A large
number of papers indicate that these effects can take place at
relatively low incident field levels (2, 11, 14, 33, 41) while
others show effects only with high incident fields (21, 28, 40).
As a result, there has been considerable controversy regarding the
actual EMF strength at which these effects can occur. Additional
controversy has revolved around whether the observed phenomena are
thermal or athermal in nature (11, 29, 36). The issue is confused
by the fact that absorption of microwaves by human and animal bodies
is intimately related to body configuration and size. Furthermore,
exposure facilities such as cavities and instrumentation involving
metal leads used frequently in the past can produce fields in the
tissue far greater than one would expect from simple extrapolations
of field measurements made in the incident wave or in the subject's
immediate surroundings (18, 21). The results of behavioral studies
do not give a clear indication of what tissues or portions of the
CNS are being affected when exposed to EMF. Electrophysiological
measurements, on the other hand, can provide a method for directly
quantifying the reactions of the nervous system exposed to EMF and
appears to be more useful for establishing thresholds for EMF interactions. Early electrophysiological work consisted of measurement
of EEG signals from exposed animals (1, 3, 27, 31) and humans (24,
38). Such work has continued and is still ongoing in many laboratories (2, 4, 11, 13, 30, 38). In addition, recordings have been
made to determine the effects of EMF on the extracellular activity
of neurons within the CNS (5, 20, 21, 40), and in peripheral nerves
(6, 22, 34). Most of the above studies, however, involved standard
off-the-shelf physiology equipment and techniques which can introduce problems such as EMF fringing effects on wire terminals and
electrode tips and can produce pronounced perturbation of the incident fields by interconnecting wires. These problems, coupled with
the uncertainty of the magnitude of the EMF induced in the tissues,
make it very difficult to assess the significance of the results in
terms of human exposure to a myriad of possible EMF sources. To
resolve these problems, a coordinated interdisciplinary investigation involving physiology and electromagnetic engineering was undertaken to quantify field interaction levels on the CNS and peripheral
nerves of laboratory animals. Exposure, dosimetry and instrumentation techniques and a number of animal preparations were developed
to provide a means for determining the thresholds of various electrophysiological phenomena observed in terms of actual absorbed EM
power levels and possible temperature increases at the sites of
interaction. Methods were devised for providing equivalent energy
depositions by thermal sources to ascertain whether observed effects
were thermal or athermal in nature. Redundant methods of dosimetry
including both theoretical and experimental approaches were used to
minimize chances for errors or artifacts.

EFFECTS OF ELECTROMAGNETIC FIELDS ON ANIMALS

II. EXPOSURE, METHODS, DOSIMETRY AND INSTRUMENTATION

A. Cat CNS Exposure System

In order for the results of EMF exposure of animals to be related from animal to animal, from experiment to experiment and from animal to human exposure, it is necessary to relate observed effects to the actual fields and absorbed power densities in the tissues exposed rather than an incident power density to the subject since the power coupling to the animal is related to many factors including source, frequency, animal size, animal shape and presence of other objects. This can be clearly illustrated theoretically by considering a spherical model of the exposed body or part of the body. If we consider a sphere representing the head of an animal or man consisting of a core of brain material surrounded by five concentric layers of tissue material, the power absorption density patterns in the model due to exposure to an incident plane wave can be theoretically computed (21, 25, 37).

Figure 1 illustrates such absorption patterns for a 7.0 cm diameter model representing a cat or monkey head and a 20 cm model representing a human head. The figures illustrate the power absorption density patterns along the major axes of a cartesian coordinate system whose origin coincides with the center of the sphere. The incident plane wave of 1 mW/cm^2 power density is assumed to have an E field parallel to the x-axis with direction of propagation along the z axis. The peak and the average power absorption densities are noted in each figure. The thickness and the complex dielectric constants for the concentric layers are given in Table I. The figures clearly illustrate the higher absorbed power density directly at the center of the model for the human head and 0.6 cm off

TABLE I The Six Layered Model of Cranial Structure

		SKIN	FAT	BONE	DURA	CSF	BRAIN
DIELECTRIC CONSTANT	918 MHz	51.4 -j 25.08	5.56 -j 0.856	5.56 -j 0.856	51.40 -j 25.08	80.85 -j 14.05	34.42 -j 15.49
	2450 MHz	47.52 -j 11.42	6.0 -j 0.856	6.0 -j 0.856	47.52 -j 11.42	77.0 -j 13.94	32.78 -j 15.37
THICKNESS (cm)	ANIMAL	0.10	0.07	0.2	0.05	0.2	2.88
	MAN	0.15	0.27	0.7	0.8	1.1	6.98

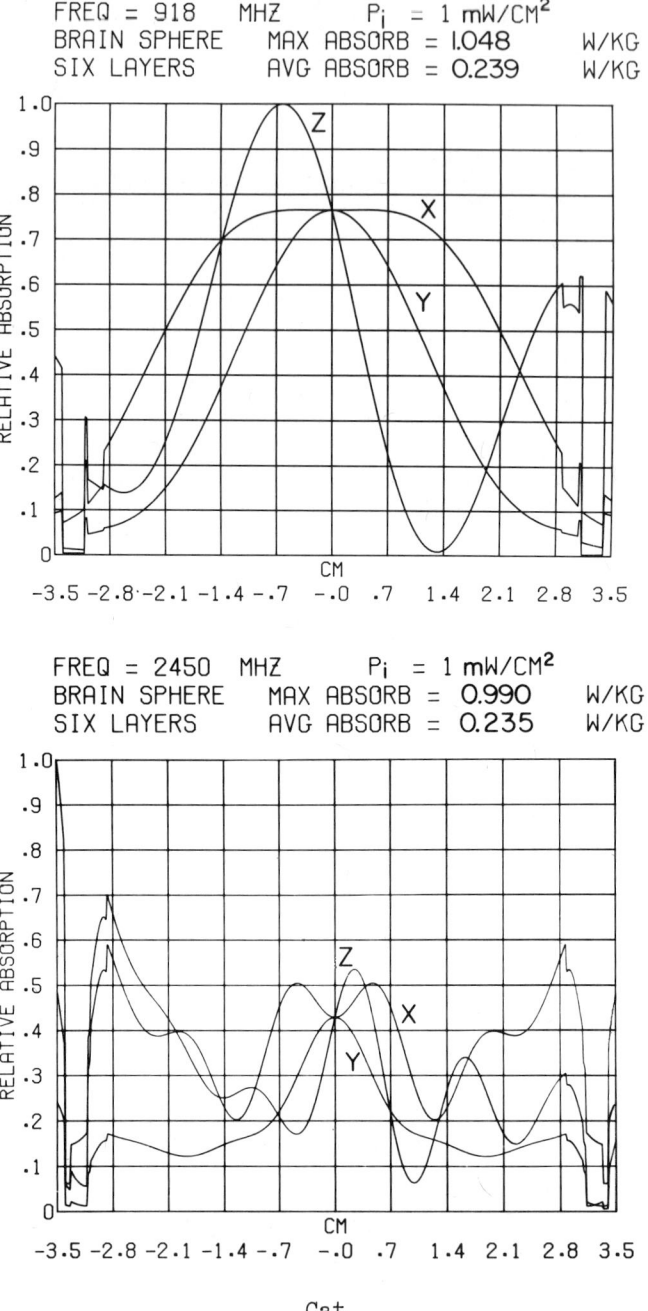

Figure 1 Theoretical power absorption density patterns in spherical phantoms of human and cat heads exposed to a plane wave source (propagation along z axis with E field polar-

EFFECTS OF ELECTROMAGNETIC FIELDS ON ANIMALS

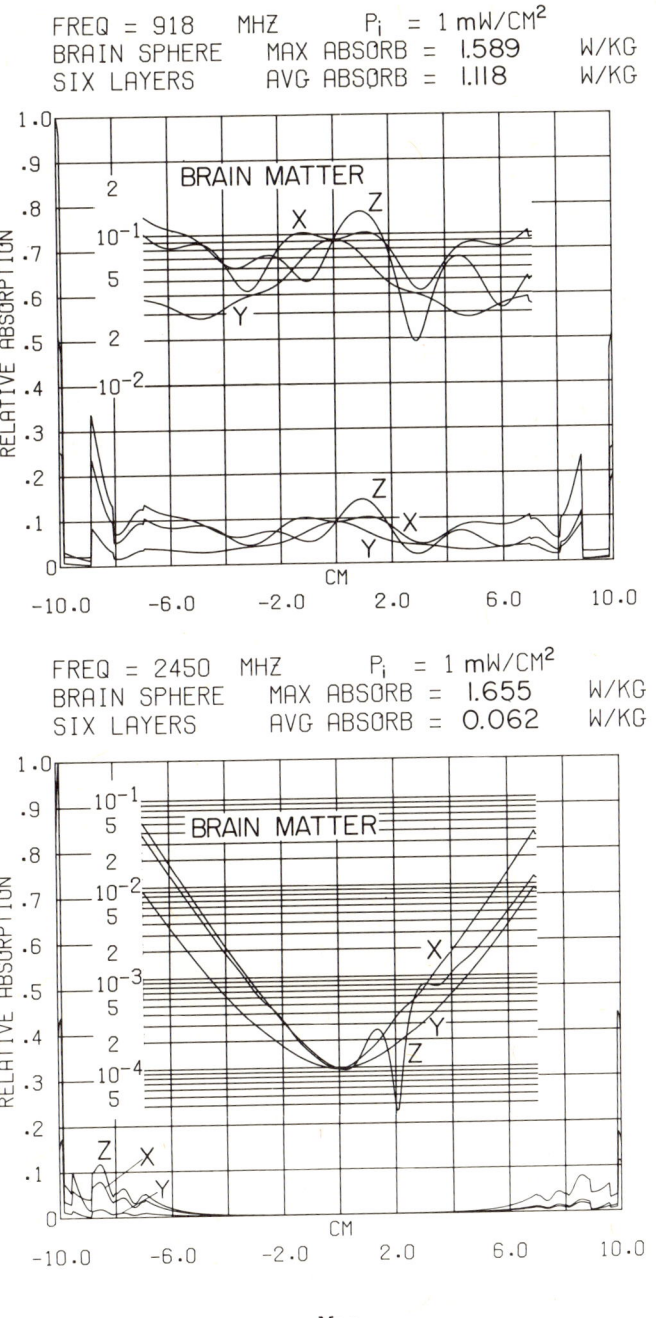

Man

ized along x axis with origin at center. Tissue boundaries are located at absorption discontinuities).

center for the model of the animal head for 918 MHz exposure. Also, one may note that the maximum absorption in the model for the animal head is larger by a factor of five than in the model for the human head. With 2450 MHz exposure, on the other hand, there is maximum absorption in the anterior portion of the simulated human head while there is strong absorption in the center portion of the simulated animal head. The regions in the center of the simulated heads with the more intense absorbed power density are due to a combination of high refractive index and the radius of curvature of the model producing a strong focusing of power toward the interior of the sphere which more than compensates for transmission losses through the tissue. These results are significant in view of the many reported CNS effects at frequencies near 1 GHz. Since greater similarities existed in the power absorption patterns in the simulated cat and human heads with incident EMF at 918 MHz than at 2450 MHz, most of the exposures on the actual animals were conducted at the former frequency. An experimental protocol was developed to irradiate the animal's head with EMF at frequencies of 918 MHz or 2450 MHz, while measurements were being made of responses in the brain due to various stimuli. Whole body irradiation of the animal with plane waves could not be meaningfully used in this case since a metal stereotaxic apparatus had to be used in the vicinity of the animal's head to accurately place recording electrodes in the brain. This apparatus, along with other instrumentation needed in the vicinity of the animal would have seriously perturbed any applied plane wave field to the point where it would be difficult to determine what type of field the cat's head was actually being exposed to. Instead, either a 918 MHz, 13 cm square aperture source or a 2450 MHz, 12 cm square horn source was applied at 8 cm from the dorsal surface of the head, as shown in Figure 2. In some experiments, a 2450 MHz diathermy "C" director corner reflector source was used. The 918 MHz square aperture with a TE_{10} mode aperture distribution, developed in previous studies, was designed originally to provide optimum depth of EMF penetration in exposed tissues for diathermy use. Both the 918 MHz and the "C" director sources have been described previously in the literature (19). Cats averaging approximately 2.3 kg were anesthetized intravenously with alpha-chloralose (55 mg/kg) dissolved in Ringer's solution (20 cc) and 0.2 mg of atropine sulfate was administered intramuscularly after induction of anesthesia. The animals were paralyzed with Flaxedil (20 mg) and placed on artificial ventilation. The body temperature of the animal was held constant at 38°C by a heating pad connected to a rectal temperature control unit. A pair of wood ear bars instead of metal was used to hold the cat in a stereotaxic instrument in order to minimize the distortion of the fields around the cat's head. For the same purpose, all metal pieces for fixing the inferior orbit and the upper jaw were replaced by wood fixtures. An electrode was placed in the area of the brain to be analyzed by the standard Horsley Clark's method. The electrode consisted of a

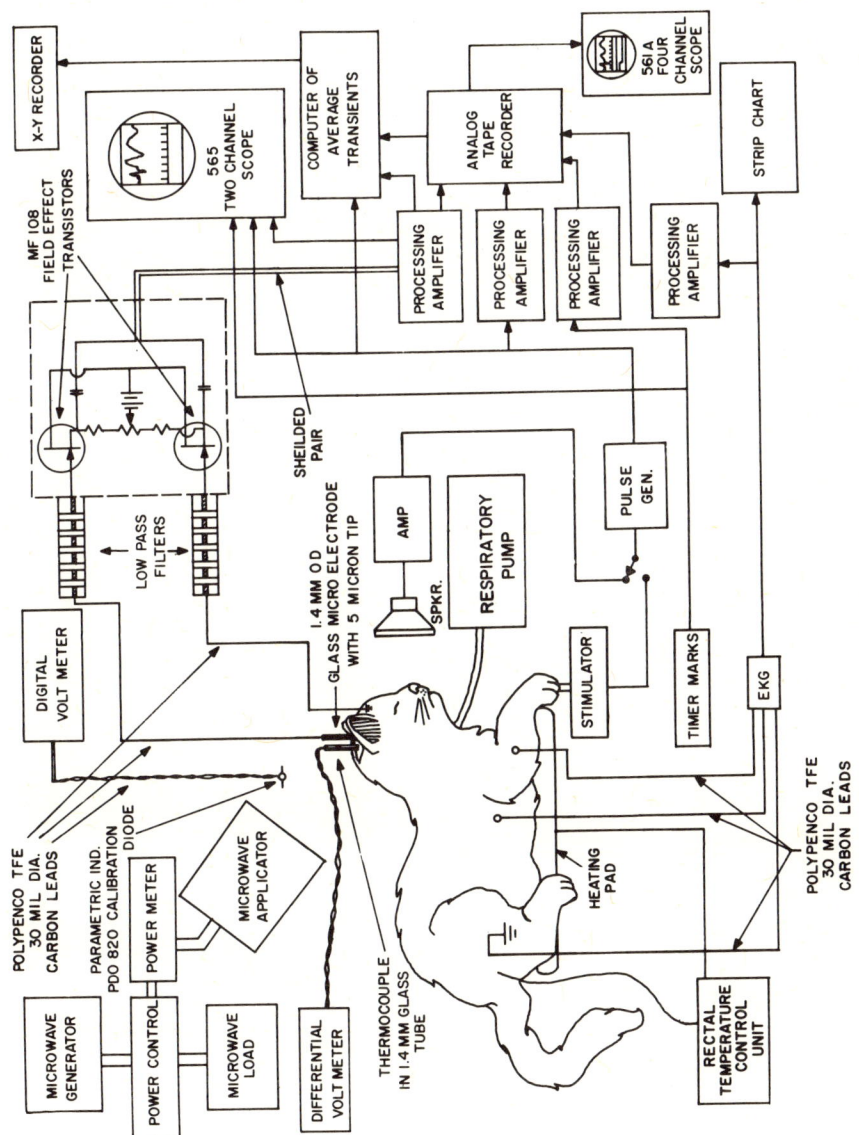

Figure 2   Block diagram of instrumentation used to quantify CNS effects of microwave radiation (21).

Ringer's solution-filled glass pipette with an 80-100 μ-tip diameter. Since the electrical conductivity of the fluid-filled electrode was close to that of tissue, no EMF perturbation was expected when the animal was exposed. Also, in order to prevent perturbation of the fields, the electrode and accompanying ground connections were coupled via high resistance 1000 ohms/cm teflon-carbon polymer conductors (transparent to EMF) through a low pass EMF filter to a high impedance input physiological signal processing amplifier, oscilloscope, computer of average transients, and an x-y plotter. The conductors are of the type used by the National Bureau of Standards for preventing field perturbations when coupling voltages from EMF sensing probes to amplifiers in field survey meters. The microwave filter was designed to provide more than 150 db attenuation of the coupled EMF currents with less than 20 pF of shunt capacitance presented to the amplifier input. The EMF effects and threshold EMF level for effects on the CNS of the cat were made by the following types of recordings: (1) The response of the thalamic somatosensory area of the cat's brain to electrical stimulation of the skin of the contralateral forepaw with an electric shock and the response of the thalamic auditory (medial geniculate) area due to stimulation of the contralateral ear with an acoustic click was recorded both with and without the presence of EMF applied for intervals of 15 min on and 15 min off, varying in incident power density between 1.3 mW/cm^2 and 52 mW/cm^2 corresponding to a peak absorbed power density of 1 W/kg to 40 W/kg in the thalamic area. Proper placement of the electrode tip was assumed when a proper latency period was observed for the evoked responses. The electrode placement was verified in some of the animals by histological examination of the brain tissue. (2) The above was repeated using a constant incident energy density of 1 mW-hr/cm^2 corresponding to the American National Standards Institute C95.1 Safety Guide for Radio Frequency Radiation Hazards. The initial exposure was 10 mW/cm^2 incident power density or a peak absorbed power density of 7.8 W/kg applied for 6 min. This corresponds to 0.78 W-hr/kg of absorbed energy for each successive exposure. The power density level was set to twice the preceding power density level but applied for one-half the time duration. (3) The response of the cochlea, VIII nerve, medial geniculate nucleus, auditory cortex, and thalamic somatosensory area of the cat to direct acoustic and microwave pulse stimulation at frequencies of 918 MHz and 2450 MHz and 9 GHz. Acoustic stimulation was applied by a loudspeaker or a piezoelectric transducer cemented to the cat's skull as described elsewhere in the literature (39). A Narda Model 8100 power monitor was used to measure the average incident power density to the head of the cat (without the cat present) and the incident and reflected power to the aperture, horn or corner reflector source was monitored by means of a Hewlett-Packard 477 bolometer and 430C power meter combination connected to a Microlab FXR 30 db bi-directional coupler inserted between the coaxial cable and the radiator. A 1.5 mm glass pipette with a 0.3 mm diameter sealed tip of the same stock used for the microelectrodes was inserted in the homologous point in the opposite side of the brain at the same depth as the electro-physiological recording electrode. A thermocouple was kept in the pipette at all times during the experiment when the EMF fields were off to monitor brain temperature. The thermocouple was removed during the exposure times to prevent any fringing field effects. The dosimetry was based on the following methods of calibration which have been discussed previously in the literature (21): (a) the incident power density measured at the position of the cat's head by the Narda Model 8100 electromagnetic radiation monitor, (b) the electric field at the surface of the cat's brain was measured by means of a calibrated microwave diode coupled by high resistance wire to a digital voltmeter shunted by a 10 kilo ohm resistor, (c) the temperature was measured with the thermocouple before and after a short term exposure to high power EMF as a function of position in the cat's brain and the changes were converted to absorbed power density information, and (d) after each experiment, the sacrificed animal was frozen in the stereotaxic support, cast in a polyfoam block, bisected, returned to room temperature and rejoined back into the stereotaxic support for a short-term, high intensity

TABLE II   Calibration of 918 MHz Microwave Applicator (8cm away, 1 watt in) for Typical Cat

REGION	ABSORBED OR INCIDENT POWER	METHOD
FREE SPACE	2.6 mW/cm^2 (8 cm AWAY) 1.7 mW/cm^2 (11 cm AWAY)	NARDA 8100 MONITOR
IN STEREOTAXIC SUPPORT WITHOUT CAT	2.48 mW/cm^2	NARDA 8100 MONITOR
MAXIMUM HEATING AREA OF THALAMUS OF CAT	2.0 mW/cm^3	THERMOGRAPHIC
MAXIMUM HEATING AREA IN PHANTOM SPHERE	1.5 mW/cm^3	THERMOGRAPHIC (8 cm FROM APERTURE SOURCE)
MAXIMUM HEATING AREA IN PHANTOM SPHERE	2.1 mW/cm^3 1.4 mW/cm^3	2.6 mW/cm^2 THEORETICAL 1.7 mW/cm^2 (PLANE WAVE)
SURFACE OF CAT BRAIN	0.4 to 0.8 (mW/cm^3)	MICROWAVE DIODE
SURFACE OF CAT BRAIN	0.54 mW/cm^3	THERMOGRAPHIC
MAXIMUM HEATING AREA OF THALAMUS OF CAT	1.88 mW/cm^3	THERMOCOUPLE (LIVE BRAIN)
HOMOLOGOUS POINT IN OPPOSITE THALAMUS OF ELECTRODE POSITION	1.23 mW/cm^3	THERMOCOUPLE (LIVE BRAIN)

exposure. Immediately after the exposure, the thermograph recordings of the induced internal temperature patterns in a half-section of the cat's head were made and converted to power absorption patterns. Typical calibration results are tabulated in Table II and thermograms of a typical 918 MHz exposure of the cat's head and an equivalent spherical phantom model are shown in Figure 3.

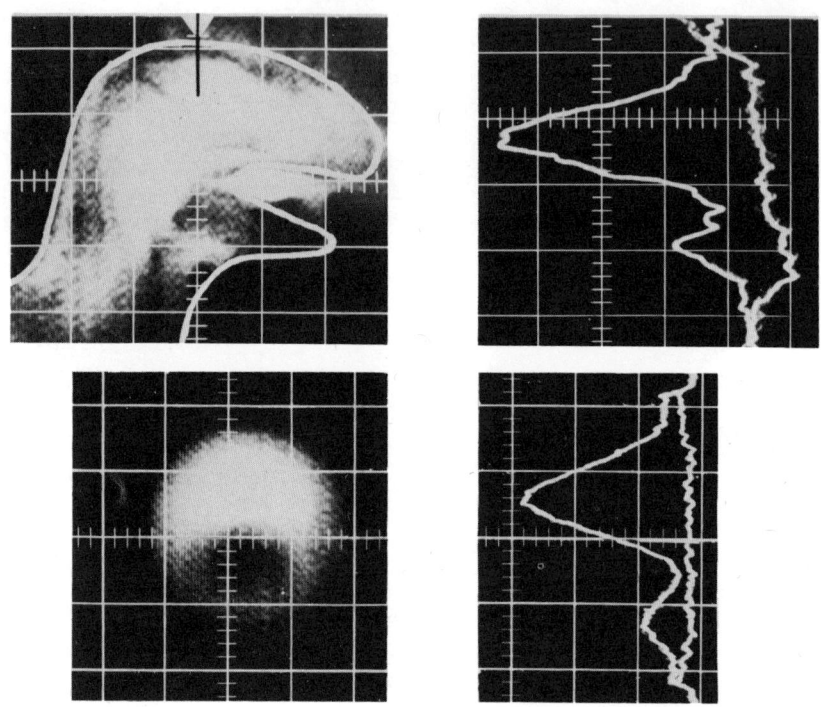

Figure 3   Comparison between cat head and 6 cm diameter phantom sphere. Subjects exposed to 918 MHz aperture source. Scale: C scan, 1 div = 2 cm; B scans, 1 horizontal div = 2 cm, 1 vertical div = 2.5°C. Subjects exposed from top with horizontal E vector in plane of paper.

The equivalent model was prepared as described in the literature (21). The EKG of the cat was recorded and monitored on a strip chart recorder and the responses of the cat's brain to various stimuli with and without the presence of EMF were processed for improved signal to noise ratio, both on-line and off-line from recorded magnetic tape by averaging up to 50 successive responses with a computer of average transients. The results were plotted by means of an x-y recorder.

B. Thermal Considerations

In order to evaluate and understand the thermal significance of a given amount of absorbed power density in the tissues to ascertain whether an effect is specific to the field or to a temperature change, one should be aware of the relationship between the absorbed energy, the tissue cooling mechanisms, and the temperature. The energy equation for the time rate of change of temperature (°C/s) per unit volume of neural tissue exposed to EMF is

$$\frac{dT}{dt} = \frac{0.239 \times 10^{-3}}{c} [W_a + W_m - W_c - W_b] \qquad (1)$$

where W_a is the absorbed power density, W_m is the metabolic heating rate, W_c is the power dissipated by thermal conduction, and W_b is the power dissipated by blood flow, all expressed in W/kg; c is the specific heat of the tissue in kcal/kg°C, and T is the tissue temperature in °C (19).

The metabolic heating rate, W_m, for the human brain is approximately 11 W/kg and the average for the entire body is 1.3 W/kg. If the EMF induced power, W_a, is small compared to W_m, we would not expect much of a thermal response. If, on the other hand, W_a is large compared to W_m, we could expect some form of thermal response. It will be shown later in the discussion on auditory system effects that this latter statement can be true, even though there is negligible temperature change associated with it. The absence of a measured temperature change does not necessarily mean the absence of a thermal response. The time rate of change of temperature directly related to W_a is an important factor, even though the total temperature change may be unmeasurable by standard means. Thus, large peak values of W_a associated with EMF pulses could be significant even though the average value of W_a may be negligibly small.

In order to ascertain whether observed changes in the CNS of the animal were EMF effects per se, tests were run on a series of animals where the microwave irradiation was replaced by either heating the animal with a heating pad or by the addition of a heat exchanger to the animal's head as shown in Figure 4. The thermal implant unit can be used for either heating or cooling by circulating fluids of different temperatures inside a brass tube in the absence of EMF or inside a pyrex tube for experiments involving simultaneous radiation and cooling. The units were applied to the base of the skull using a transpharyngeal approach with the division

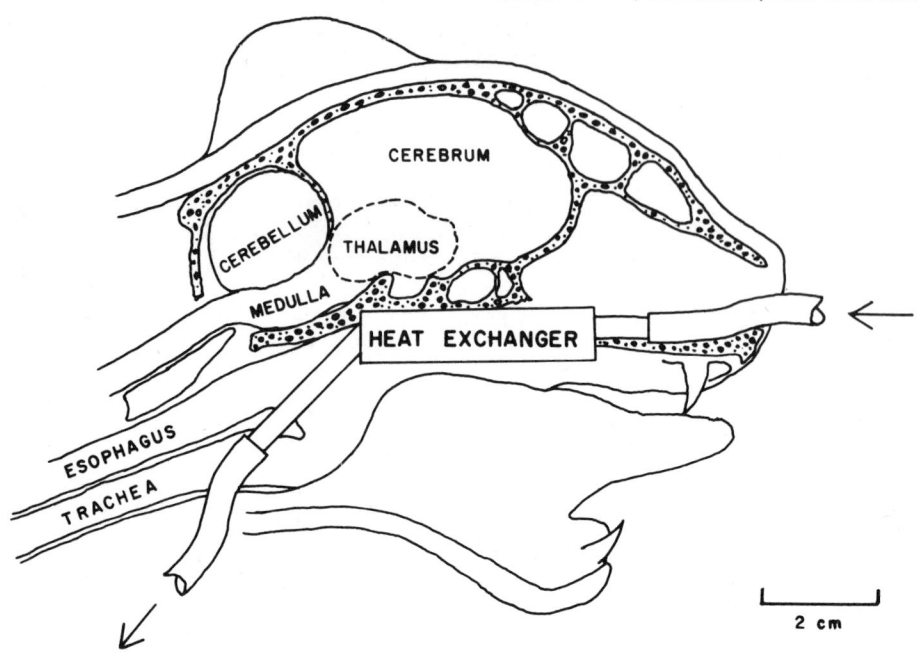

Figure 4 Detail of placement of heat exchange devices against the base of the skull for non-radiation heating or cooling, coincident with radiation, of the brain during study of thalamic evoked potentials (40).

of the posterior palate and retraction of the soft tissues. The exchangers were attached by plastic tubing to the outlet and return ports of a Brinkmann constant temperature fluid circulator.

C. Cat Spinal Cord Exposure System

A classical spinal cord preparation used in general neurophysiology for more than three decades was modified to meet the special requirements for EMF exposure. The details of the preparation have been previously reported in the literature (40). Subjects consisted of cats weighing 3.0 ± 0.5 kg. An extensive laminectomy was performed including removal of most of the bone dorsal to the vertebrae bodies of the second or third lumbar to the first sacral vertebrae in order to place a stripline EMF applicator around a portion of the spinal cord and a nerve root pair. The roots left the

EFFECTS OF ELECTROMAGNETIC FIELDS ON ANIMALS

vertebral canal at a level likely to contribute to the sciatic nerve. The stripline applicator used to irradiate the spinal cord is shown in Figure 5.

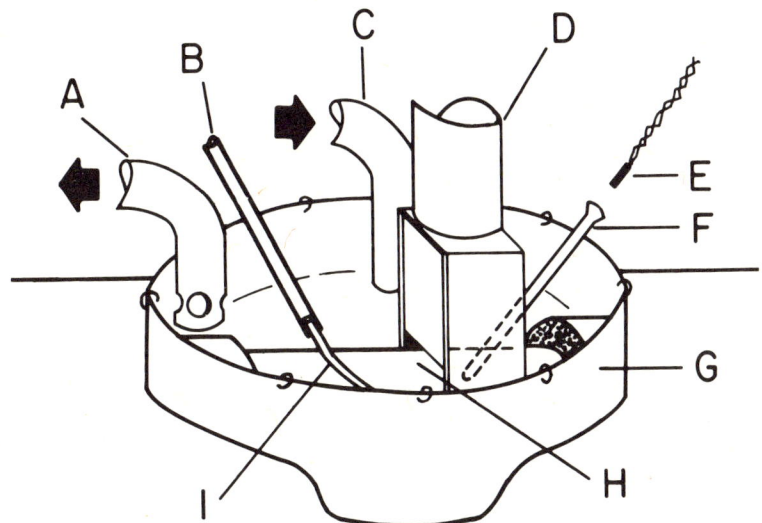

Figure 5   Semi-schematic detail of spinal cord preparation. Components are as follows: (a) circulatory outflow tube, (b) polyethylene suction electrode, (c) circulatory inflow tube, (d) microwave applicator, (e) iron-constantin thermocouple, (f) thermocouple guide tube, (g) flexible plastic dam sutured to skin, (h) spinal cord, and (i) ventral root (40).

A dam consisting of a flexible teflon ring was sutured to the dorsal skin to confine a volume of Ringer's solution over the cord and around the EMF applicator. The Ringer's solution held at a temperature of 37-38°C was circulated through the reservoir by a Brinkmann constant-temperature circulator. A Ringer's solution-filled polyethylene suction electrode was used to record from the central end of the cut ventral root which was near the EMF source. The ipsilateral sciatic nerve was dissected out and the gastrocnemius branches were elevated onto silver wire electrodes. The peripheral nerves and silver electrodes covered with warmed mineral oil were sufficiently remote from the EMF source to prevent undesirable field enhancement. The gastrocnemius nerves were stimulated by single, 0.1 msec duration pulses repeated once per sec. Spinal cord activity detected at the ventral root was recorded and processed in the same manner as the CNS recordings discussed previously. The responses consisted of typical monosynaptic spikes and late complex potentials of which the former was chosen for analysis since it required the minimum assumptions in interpretation. The power absorption density patterns were determined for the EMF stripline applicator by thermography and were found to produce in the spinal cord a peak power absorption density of 509 W/gm per watt input to the applicator. A gell material with the same properties as the Ringer's solution bath was used as a phantom model to obtain the patterns. A glass micropipette similar to that discussed for the CNS experiments was inserted into the cord between the plates of the applicator so that a thermocouple could be used to measure temperature when the 2450 MHz EMF was not applied. The spinal cord was exposed to EMF to produce up to 3.8 kW/kg absorbed power density in the tissue with the temperature of the bathing Ringer's solution held constant at and below body temperature, and with the circulator off.

D. *In Vitro* Exposure Apparatus

It is important to realize that exposure of *in vitro* preparations requires careful consideration of the exposure techniques. Since the size of such a preparation is substantially smaller than an actual body, the absorbed power can be considerably different for the former than the latter, even for the same incident field intensity. Small *in vitro* preparations do, however, allow much more flexibility in the design of the exposure apparatus and can easily simulate a whole range of low and high intensity exposure conditions even with relatively low power readily available laboratory sources, waveguides, and conventional power meters. The most practical exposure apparatus for *in vitro* nervous system studies is a waveguide designed such that the nerve can be passed through the waveguide in regions of known field configuration and magnitude while at the same time stimulating and recording electrodes may be attached to the nerve outside of the waveguide, thereby eliminating any artifacts due to field enhancements or EMF interaction with preamplifiers.

A silver-plated S band WR284 waveguide was equipped with inlet and outlet ports for circulating fluids as shown in Figure 6a. The waveguide was put in a lucite tank containing Ringer's solution 6 cm in depth (three times greater than the depth of field penetration). In each experiment, a frog sciatic or a cat saphenous nerve was pulled through the 3 mm diameter holes on the waveguide parallel to the electric field of the TE_{10} mode. Both ends of the nerve were

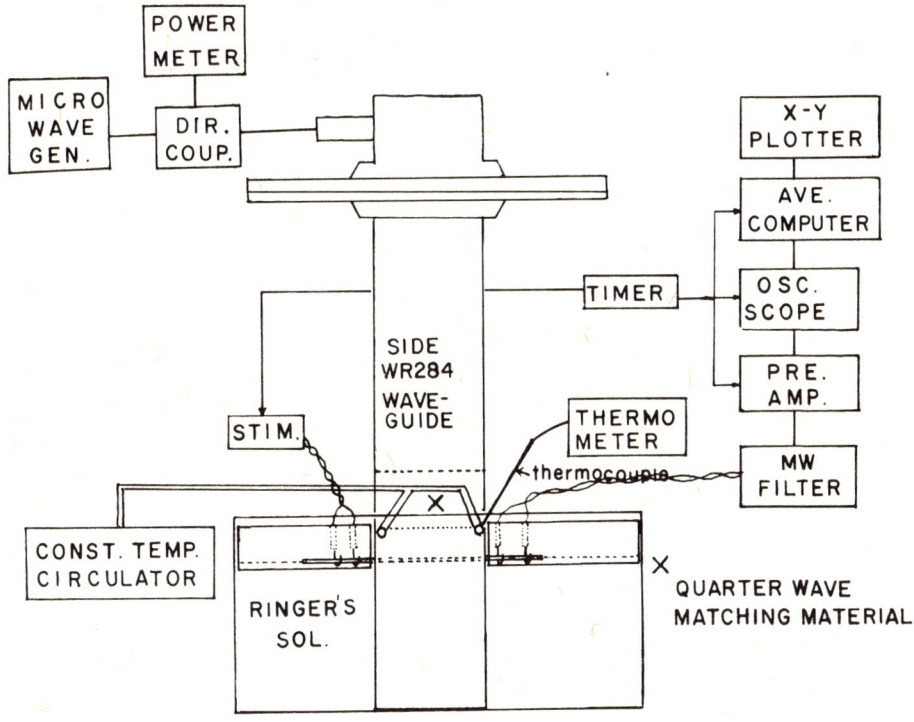

Figure 6a Waveguide exposure facility for isolated preparation with recording instrumentation, configuration for nerve.

immersed in mineral oil-filled chambers for stimulation and recording. A quarter guide wavelength of matching material with a dielectric constant of 6 was used to reduce the reflected power to less than 3% of the incident power. The temperature of the Ringer's solution was controlled within ± 0.02°C by a constant temperature circulator at a rate of 1.3 L/min. The temperature of the solution was monitored at the outlet of the waveguide with a thermocouple. CW (continuous wave) and pulsed power sources, operating at 2450 MHz with incident and reflected powers measured by means of a directional coupler and a power meter were used to feed the waveguide. In one series of experiments, the nerve was exposed to pulsed power with calculated average absorbed power densities of 3 W/kg, 30 W/kg, and 300 W/kg in the nerve. The pulse width was 10 μsec and the pulse recurrence frequency was 100 pulses per sec. The total irradiation time for each run was 20 min with the circulator being operated for the first 10 min and shut off for the final 10 min. Compound action potential data from the nerves was recorded and reduced by a computer of average transients. The absorbed power density in the nerve can be calculated by the following formula:

$$P = 4\alpha \frac{P_I - P_R}{A} e^{-2\alpha x} \qquad (2)$$

where

- P : absorbed power density in the nerve (W/gm)
- P_I : incident power (W)
- P_R : reflected power (W)
- x : depth of nerve below the surface of Ringer's solution (cm)
- A : cross-sectional area of the waveguide (cm^2)
- $1/\alpha$: depth of field penetration in Ringer's solution (1.78 cm at 2450 MHz)

The exposure apparatus was also used to irradiate the superior cervical ganglion of the rabbit. In this case, since micromanipulation had to be made on the preparation in the waveguide, the

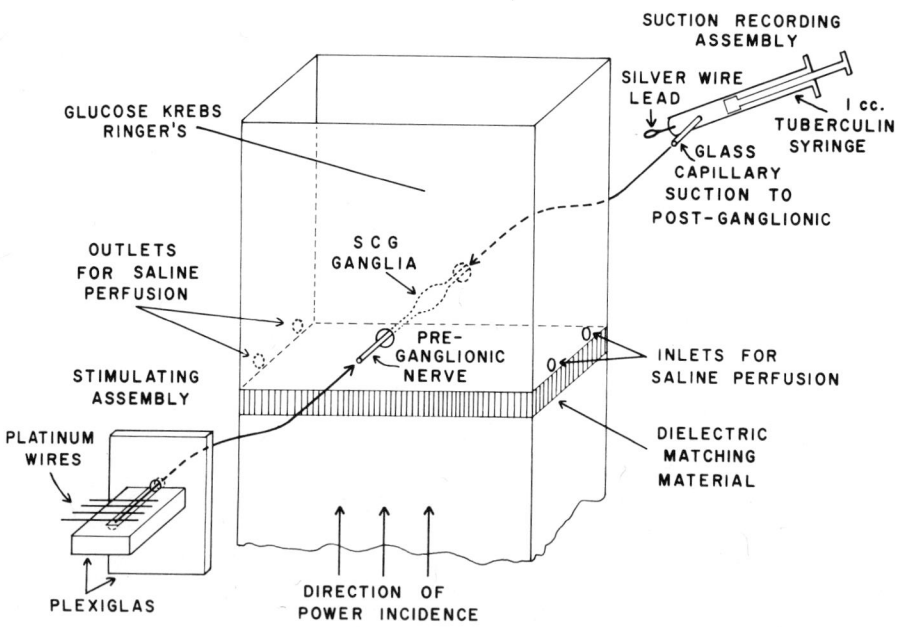

Figure 6b   Waveguide exposure facility for isolated preparation with recording instrumentation, configuration for ganglion.

waveguide was turned 180° so that the preparation could be viewed from the open end. In this case it was circulated with Ringer's solution composed of 3.5 mM KCl, 3.9 mM $CaCl_2$, 15.7 mM $NaHCO_3$, 138.3 mM NaCl, and 8.7 mM glucose, and was equilibrated with 95% O_2-5%CO_2. Figure 6b depicts the ganglion stretched across the waveguide between a set of stimulating electrodes on the preganglionic side (outside the waveguide) and a suction electrode on the other (with a glass capillary projecting into the waveguide to make contact with the postganglionic nerve inside the waveguide). Right and left superior cervical ganglia were removed from 3 to 6 kg New Zealand white rabbits under urethane anesthesia (1500 mg/kg). The postganglionic nerve (internal carotid) was difficult to dissect out with a length of more than about 5 mm. Thus, it was necessary to project the suction capillary tube into the guide so as to leave the ganglion proper in the center of the radiation fields.

III. RESULTS OF ANIMAL EXPERIMENTS

A. Effect of CW on the Somatosensory and Auditory Systems

Electrophysiological responses due to electrical stimulation of the tactile receptors of the cat's contralateral forepaw, recorded from the somatosensory thalamus area of the cat brain with and without the presence of 918 MHz EMF, have been published in the literature (21). The measurable effects of the microwaves appears to be an induced temperature rise in the thalamus with an associated decrease in latency time of neural responses within the exposed area. In attempting to study the generality within the CNS of effects observed in tactile thalamic nuclei, the auditory structures are immediately suggested. This is because of recurrent reports (8,9,12) of acoustic sensation being elicited by pulsed microwaves. Figure 7 shows the evoked thalamic nucleus activity in an area (medial geniculate nucleus) involved in auditory mechanisms, as opposed to the tactile modality which has been studied extensively (17,21), suggesting no remarkable differences in effect or mechanism. This represents an initial attempt to study the generality of CNS effects. This is not to say that some regions of the brain will not react differentially to given parameters of microwave illumination, particularly recognizing the distribution of power within the nervous mass, shown in Figure 3, but that there is some equivalence of effect and that our early investigations did not limit to some extraordinary case.

A 60 cycle notch filter was used with the processing amplifiers in order to enhance the high frequency components of each wave so that the latencies could be more accurately observed. The peak microwave power absorption density and temperature in the thalamus area are noted on each curve. The latency times between the

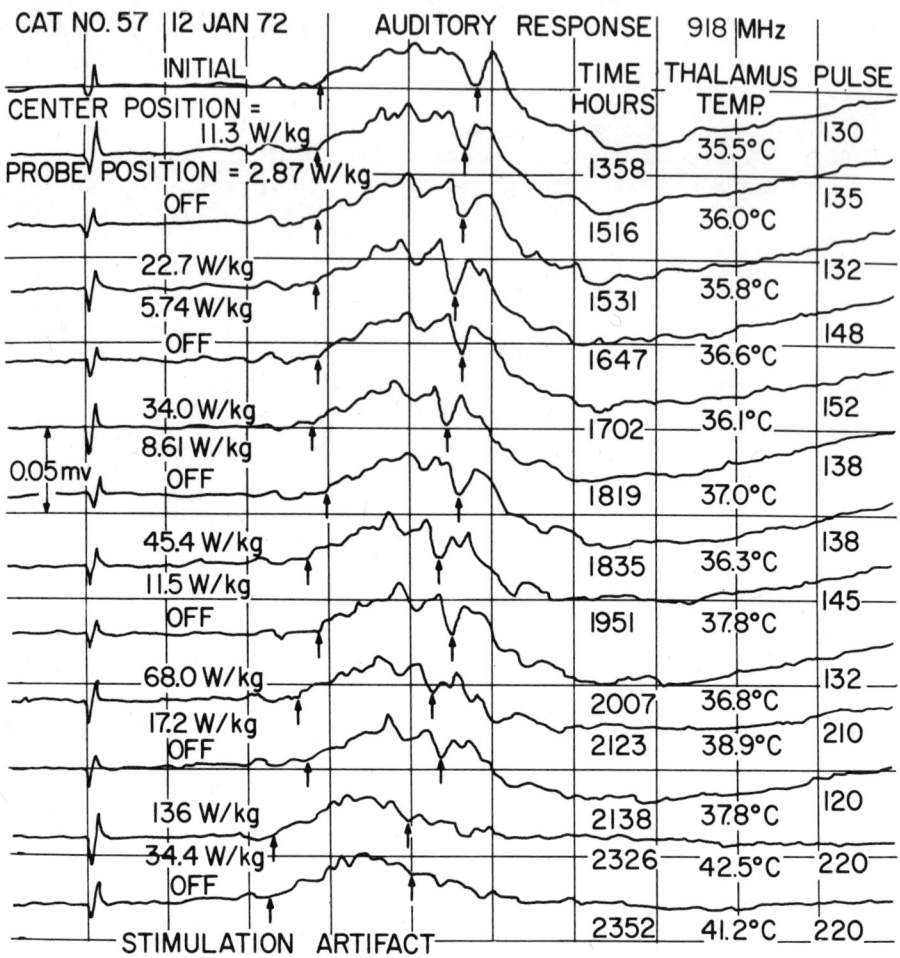

Figure 7 Effect of 918 MHz microwave power radiation on the evoked auditory response of the medial geniculate body to audio click stimulus delivered to contralateral ear of cat No. 57. 15 min exposures to continuous wave radiation at different power levels alternated with 15 min or longer rest periods.

stimulus and the initial thalamus response (denoted by the first arrow) and between the initial thalamus response and a distinguishable later event (denoted by the second arrow) are also noted. The latency between the stimulus artifact and the first arrow represents the conduction time of sound in air and the propagation time of neural impulses to the recording area, whereas, the latency between the arrows is probably more associated with pathways within the brain. Resting periods between the 15 min exposure were increased for the higher power exposure in order to allow the brain tissue to cool sufficiently. Temperature changes and heart rate are shown in Figure 8.

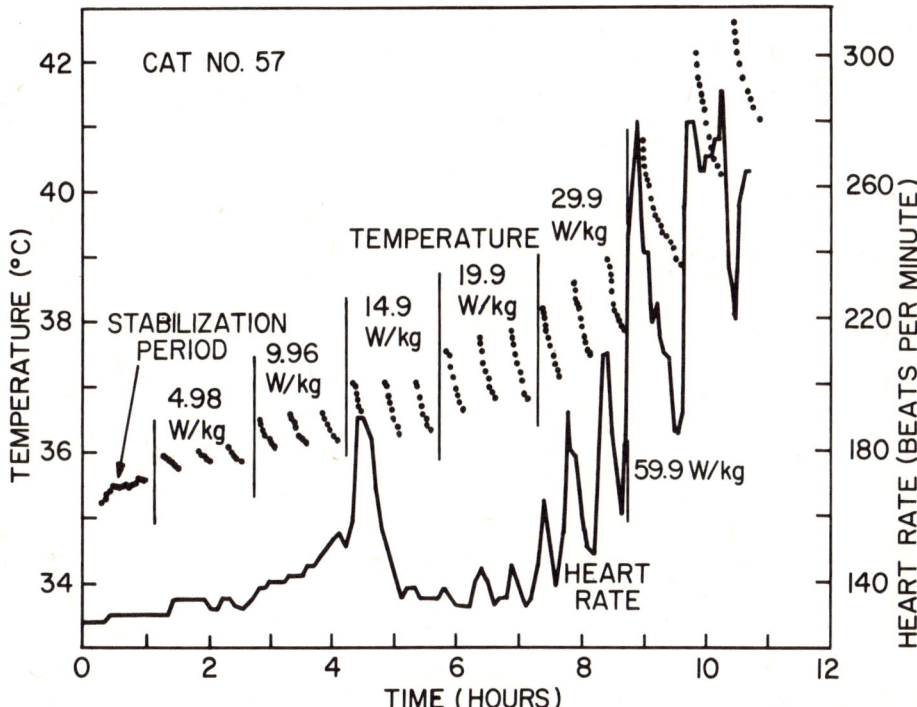

Figure 8 Thalamic temperature and heart rate of cat exposed to 918 MHz microwave radiation. 15 min exposures to continuous wave radiation of different power levels alternated with 15 min or longer rest periods (power off during temperature recordings).

Note the increase in heart rate for the higher exposure power levels. Experimental results given in Figure 9 indicate that both latencies decrease with increasing body temperature of the cat when produced by a hot pad, whereas, microwaves applied to the head have more of an effect on the later latency. The changes are reversible and have time constants that seem to be directly associated with the thermal effect of microwaves. The threshold for both temperature changes and latency changes was found to correspond to a maximum power absorption level between 2.5 and 5 W/kg at the center of the brain. Table II indicates that this corresponds to an indicated power density of 3.0 to 6.0 mW/cm^2 as measured in the unoccupied

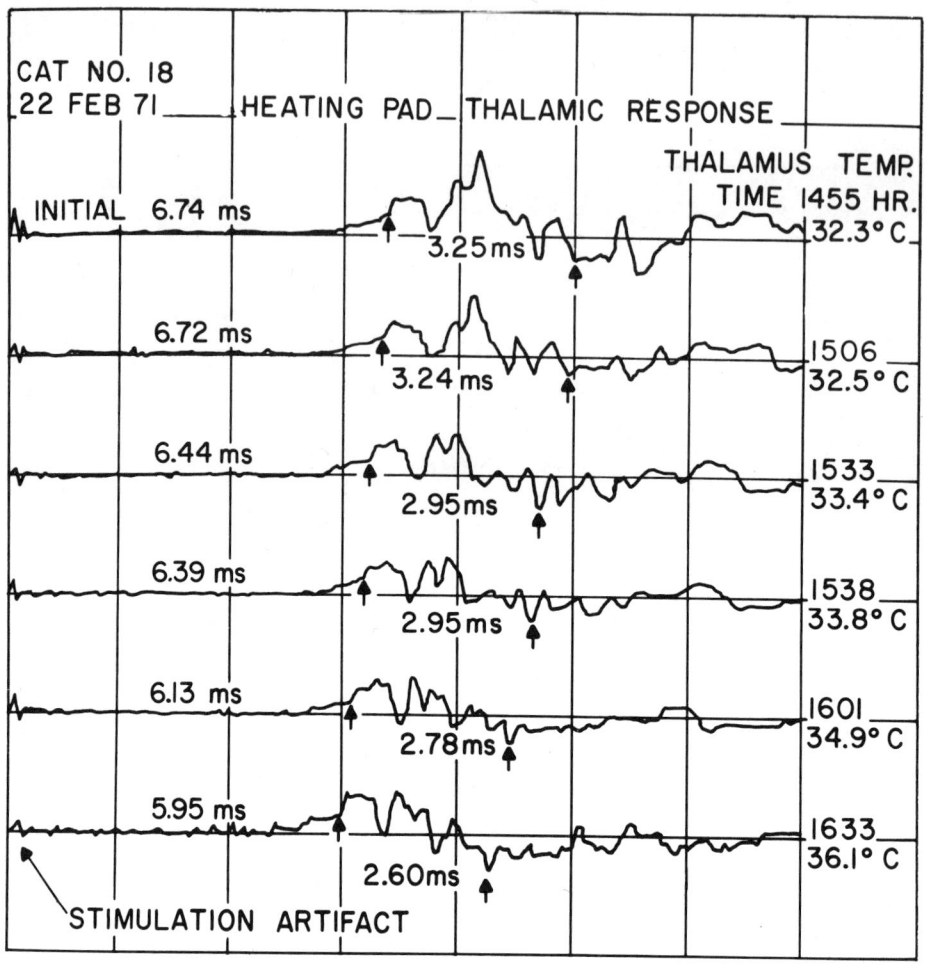

Figure 9 Effects of heating pad induced cat body temperature changes on thalamic response of cat to stimulation of contralateral forepaw.

stereotaxic support by the Narda monitor. It would take a plane wave power density of 10 to 25 mW/cm^2, however, to produce the same power absorption in the human brain assuming it can be represented by a simple model illustrated in Figure 1. Figure 10 presents the results from exposing the animal to the same total energy at different power levels and exposure times corresponding to a total incident energy of 1.0 mW-hr/cm^2. This induces a peak absorbed energy density of 0.78 W-hr/kg as calculated from thermocouple measurements. This incident energy level corresponds to the present ANSI C95.1 standard for minimum energy for human exposure for 1.0 mW-hr/cm^2 or less.

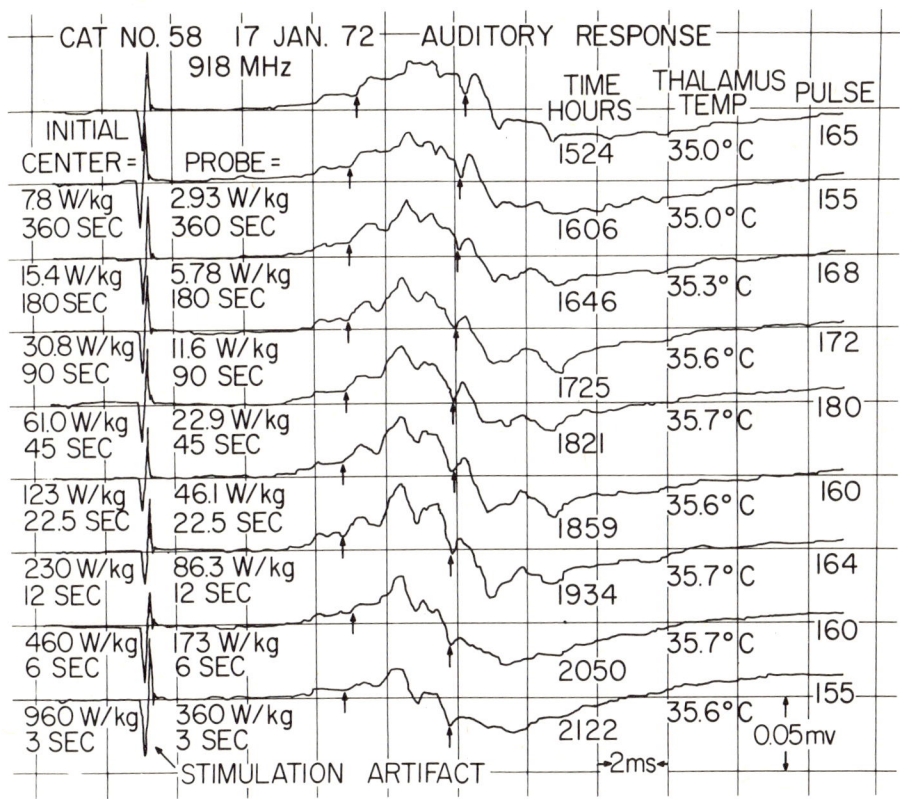

Figure 10 Effect of 918 MHz microwave power radiation on the evoked auditory response of medial geniculate body to an acoustic click stimulus delivered to contralateral ear of cat No. 58 exposed at various power levels for different time periods corresponding to total incident energy of 1 mW-hr/cm^2 (0.78 W-hr/kg peak absorption density).

The same energy was applied with increasing incident power for decreasing exposure duration. The only effect is the slight characteristic change in latency as observed with the CW exposures mentioned previously. The maximum incident power used was 1 W/cm^2 for 3.5 sec, producing 950 W/kg peak absorbed power. Figure 11 illustrates the thalamic temperature changes associated with the exposures. Note that exposures of constant energy for 3 min or less result in approximately the same temperature increase of 0.2 to 0.3°C.

Elevation of thalamic temperature by the circulation of heated fluid through the heat exchanger applied to the base of the skull as shown in Figure 4 resulted in evoked potential changes comparable

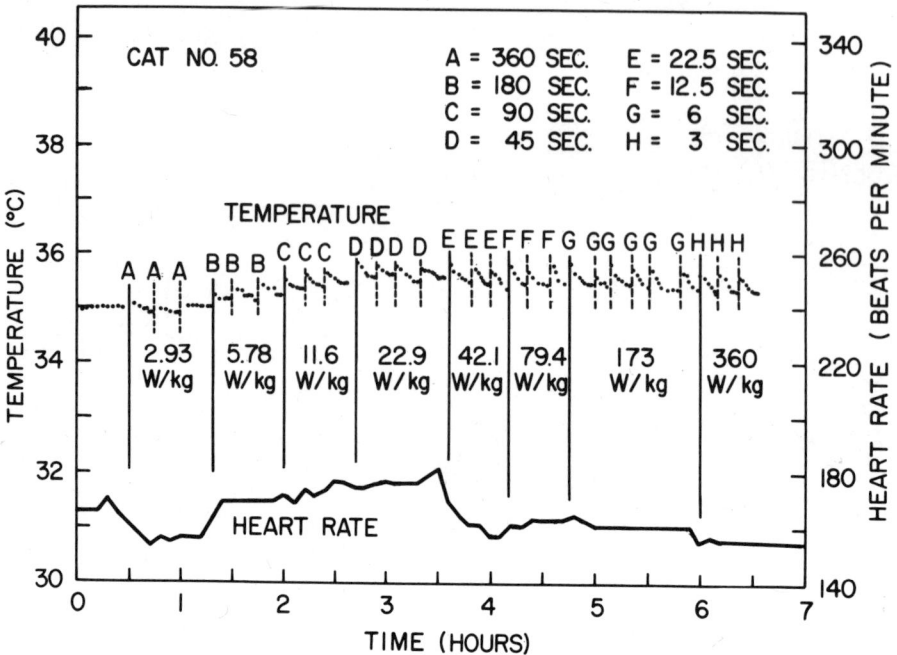

Figure 11 Thalamic temperature and heart rate of cat exposed to 918 MHz microwave radiation at various power levels and exposure periods corresponding to total incident energy of 1 mW-hr/cm^2 (0.78 W-hr/kg peak absorption density).

to those produced by microwave heating of the same magnitude. In contrast to an earlier observation that heating of the whole cat by means of a heating pad yielded decreased latency in both early and late components of the evoked potential, the exchanger-heated cats showed changes only in late components. This is directly analogous to the microwave case. This demonstration supports the contention that the above observed microwave effect is a thermal phenomenon. Figures 12 and 13 represent even more convincing support for this view. These figures show the sequential pattern of brain temperature and evoked potential latency with successive application of radiation alone, cooling alone, radiation combined with cooling, and during periods in which the animal's own temperature compensation mechanisms are in operation (i.e., during "recovery" from a radiation elevation, or a coolant depression of temperature). It is very evident that the evoked potential changes are associated with particular direction and magnitude of temperature change. With appropriate titration of coolant temperature against radiation energy, it is even possible to reverse the change that would be anticipated with radiation alone.

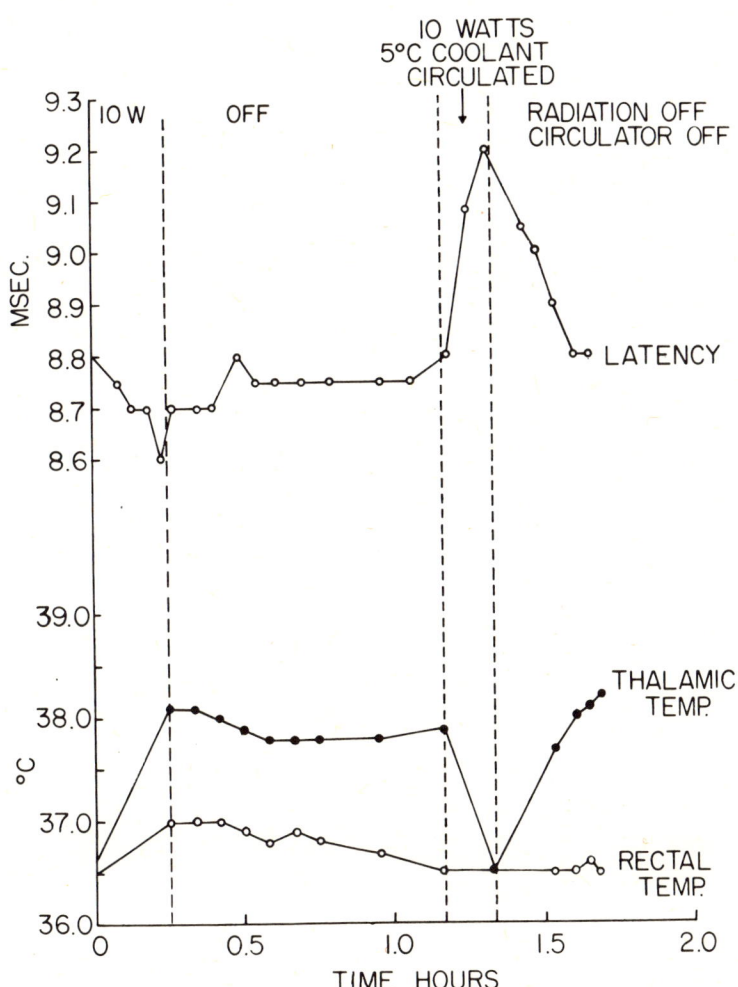

Figure 12 Sequential microwave radiation (peak power absorption density)(20 W/kg, 918 MHz CW). No treatment periods and combinations of radiation heating and active cooling with related evoked potential latencies.

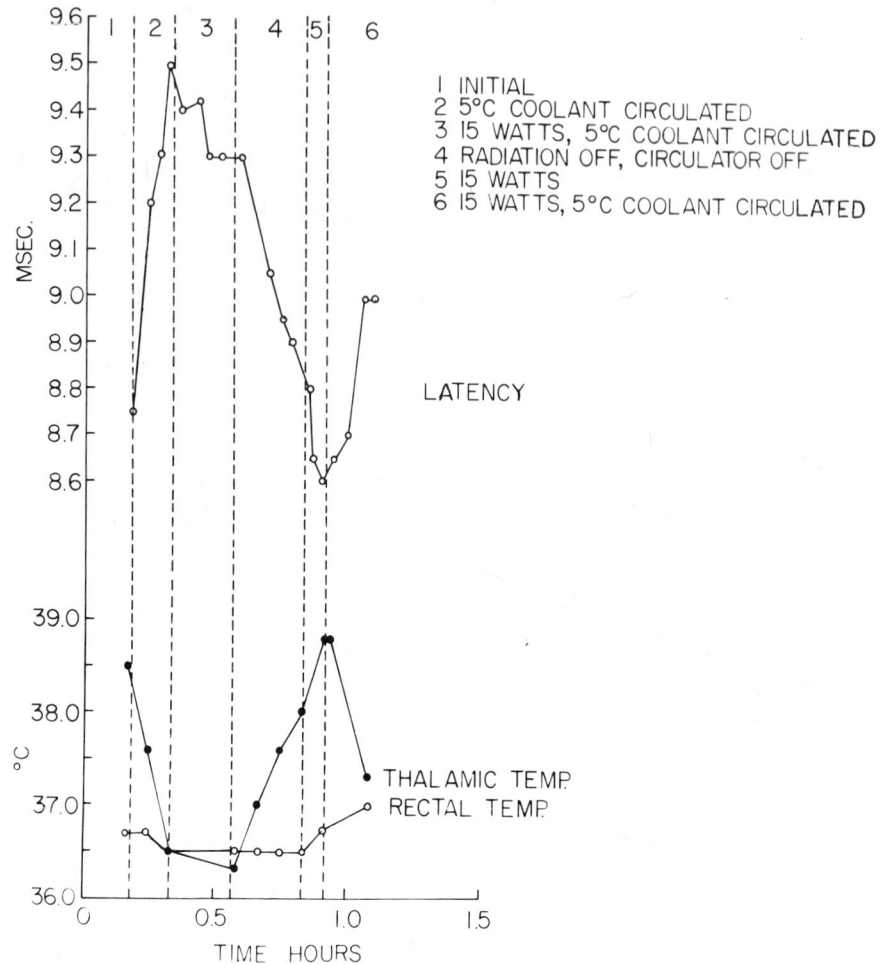

Figure 13 Combinations of 918 MHz CW radiation (30 W/kg peak absorption) and active cooling on the evoked thalamic potential latencies.

The present work supports the contention that evoked potential changes seen with CW microwave irradiation of the CNS are limited to those that can be attributed to thermal loading.

B. Pulsed Power Effects on the Auditory System

As we described in our statement on generality of thalamic effects, there appears to be no remarkable differences, at least within the resolving power of the evoked potential method, between tactile and auditory stimuli with equivalent radiation parameters. Brief pulse radiation, on the other hand, has produced, in classic auditory elements, highly specific effects. Figure 14 shows activity in the medial geniculate area of the thalamus evoked by conventional acoustic stimuli (a "click" derived from a pulsed loudspeaker, a "click" provided by piezoelectric crystal cemented on the skull) and similar activity evoked by application of microwave pulses in both the UHF and microwave bands. The effect was also found at various frequencies between the two bands.

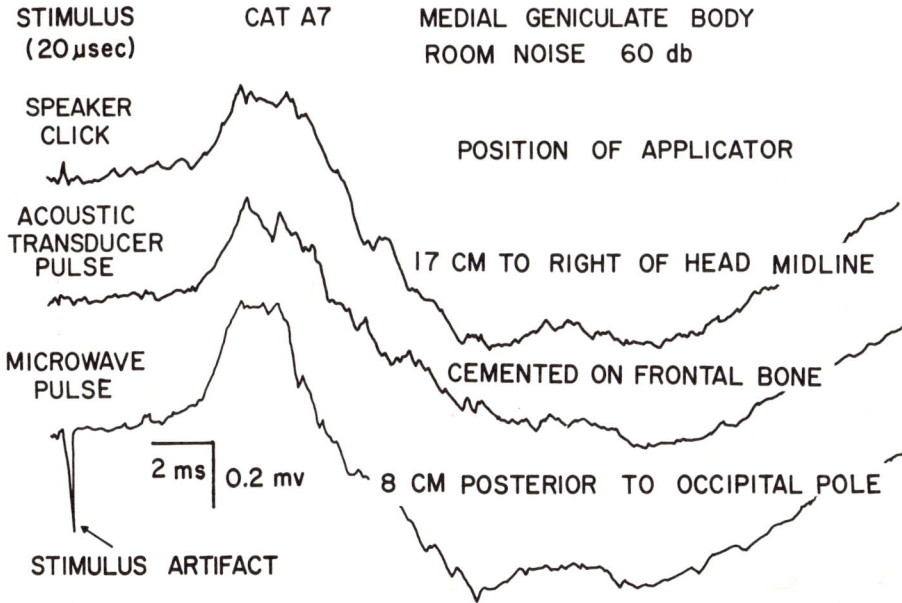

Figure 14   Evoked responses recorded from medial geniculate body of the cat (20).

Though the details of these experiments have been published elsewhere (20) the results will be summarized here. The threshold of the 2450 MHz microwave pulse evoked response as a function of pulse width is shown in Table III. The thresholds for the evoked responses with microwave pulses 0.5 to 10 μsec in duration appear to be related to the incident energy density per pulse at a level about one-half of that which produced audible sensations for the human exposure. The required threshold energy per pulse seems to increase with pulse width for 10 to 32 μsec duration pulses with the exception of the 25 μsec case. The peak absorbed energy density per pulse in the head of the cat was obtained from thermographic measurement as illustrated in Figure 3. The incident energy density per

TABLE III Threshold Evoked Auditory Responses in Cat 2450 MHz (one pulse/sec) Background Noise 64 db

PEAK INCIDENT POWER DENSITY (W/cm^2)	AVG INCIDENT POWER DENSITY (µW/cm^2)	PULSE WIDTH (µs)	INCIDENT ENERGY DENSITY PER PULSE (µJ/cm^2)	PEAK ABSORBED ENERGY DENSITY PER PULSE (mJ/kg)
35.6	17.8	0.5	17.8	10.1
17.8	17.8	1	17.8	10.1
10.0	20.3	2	20.3	11.6
5.0	20.3	4	20.3	11.6
4.0	20.3	5	20.3	11.6
2.2	21.6	10	21.6	12.3
1.9	28.0	15	28.0	15.9
1.7	33.0	20	33.0	18.8
0.6	15.2	25	15.2	8.7
1.5	47.0	32	47.0	26.7

pulse corresponding to the threshold for evoked responses recorded from the medial geniculate body due to 918 MHz radiation, differed very little from that for 2450 MHz. An evoked response from the medial geniculate body of the cat was also obtained for 2 animals using X band pulses at frequencies between 8.67 GHz and 9.16 GHz. For this case the required energy per pulse to elicit the responses was 20 to 60 times higher than required for the other frequencies. For the latter case, the X band horn had to be placed within a few cm from the exposed brain surface of the animal (through the 1.0 cm diameter electrode access hole in the skull). No response could be elicited for an animal in which the electrode access port through the skull was limited to the hole whose diameter was slightly larger than the electrode. When the skull was bared, there still was no response elicited in the above animal. When the hole in the skull was enlarged, however, a response was obtained. Recordings made of the responses to acoustic and microwave pulse stimuli were also made from the round window of the cochlea with the results shown in Figure 15. The first trace of the figure illustrates the composite cochlear microphonic and N_1 and N_2 auditory nerve response elicited by a loudspeaker pulse. The cochlear microphonic was quite strong in amplitude following the decaying oscillatory response shape of the loudspeaker (measured by optical interferometry)(20). When the

EFFECTS OF ELECTROMAGNETIC FIELDS ON ANIMALS

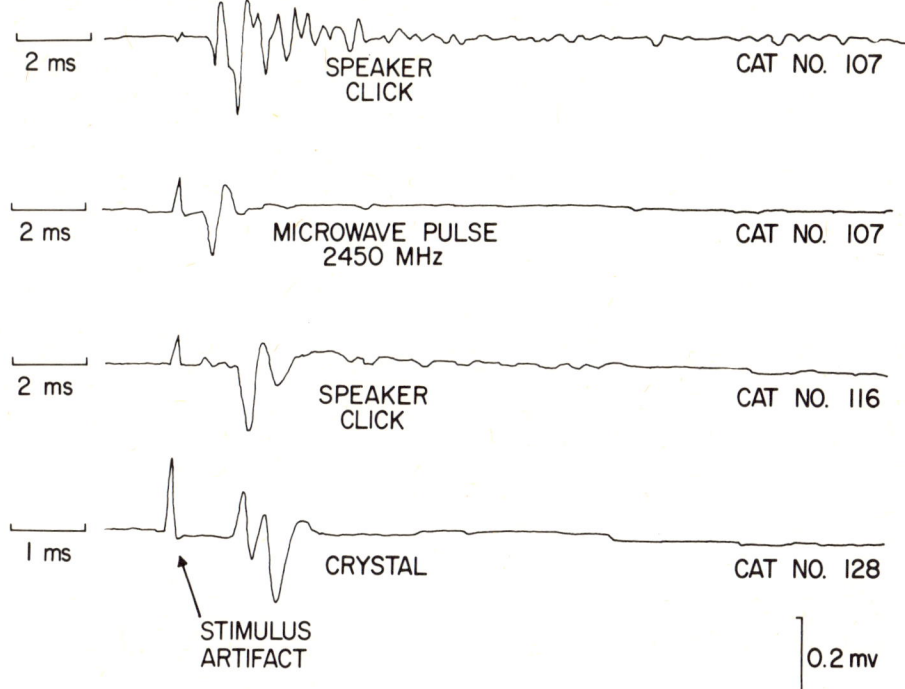

Figure 15 Responses in the round window of the cat cochlea due to acoustic and microwave stimuli (39).

auditory system of the same animal was stimulated by microwave pulses, a microwave artifact pulse and a clear N_1 and N_2 auditory nerve response was elicited, but there was no evidence of a cochlear microphonic as seen from the second trace in Figure 15. The role of the cochlea in microwave acoustic effects has been discounted partly on the basis of not observing a microphonic in either cats or guinea pigs (10). We have found, however, in some animals, that the cochlear microphonic is considerably reduced (third trace in Figure 15) or not present at all (fourth trace in Figure 15) when the auditory system of the animal is stimulated by an acoustic pulse. Furthermore, it has been pointed out that a number of factors could prevent the observance of a cochlear potential, especially when the stimulus intensity is low (43). He cites work, for example, in which auditory thresholds in cats, as determined by behavioral levels, were established as being 40 db below the stimulus level, first effective level in producing cochlear microphonic potentials of sufficient magnitude to be identified with the conventional oscilloscope display. Thus, considering the fact that the microwave pulse generator is capable of only providing a 10 to 17 db increase in pulse energy over that corresponding to the threshold of evoked responses, the absence of a microwave evoked cochlear microphonic does not rule out theories based on EM to acoustic energy transduction. The capability of the evoked auditory effect in producing potentials at CNS sites other than auditory was noted. For instance, potentials were also recorded in the somatosensory area of the brain due both to acoustic and microwave stimuli. Thus it is clear that these cross-modal evoked potentials due to microwave stimuli could be recorded at CNS sites other than those corresponding to the auditory nervous system. This leaves open the possibility that evoked potentials recorded from any location in the CNS could be misinterpreted as indicating a direct microwave interaction with the particular system where the recording is made. Cochlear disablement was found to result in total loss of all evoked potentials due to both auditory and microwave stimuli recorded from the 8th nerve, the medial geniculate nucleus, and the primary auditory cortex (20,39). The data strongly supports the contention that the microwave auditory effect is exerted on the animal in the same manner as that of conventional acoustic stimuli. The results lead us and others to examine more closely the hypothesis that the auditory effect cannot be a result of transduction of EM to acoustic energy. It has been known and observed for some time that laser and microwave pulses impinging on materials can induce audible sounds in the exposed material by rapid thermal expansion (15,44).

A theoretical and experimental analysis of the conversion of visible electromagnetic radiation from a Q-switched ruby laser to acoustic energy by thermal expansion due to absorbed energy in various liquids was reported in the literature (15). It was shown that the pressures vastly exceeded radiation pressure. This analysis was extended (7) to the case of physiological Ringer's solution exposed to microwave pulses and showed theoretically and experimentally that pressure changes far in excess of radiation pressures could produce audible, acoustic energy in the exposed medium. It is very significant to note that the audible sounds could be produced by rapid thermal expansion associated with only a 5×10^{-6}°C temperature rise in the tissue due to the absorbed EM energy. It has been estimated (20) that the acoustic pressure in the brain of approximately 2.2 - 3.0 dyne/cm^2 due to an incident 20 μsec, 2450 MHz 40 μJ/cm^2 EM pulse would be well above the computed internal threshold pressures for hearing. Gourney's equations (15) are consistent with our experimental observations that the hearing threshold is proportional to pulse energy for pulses less than 32 μsec and with Frey's observations that loudness is proportional to peak power for pulse widths greater than 50 μsec. The fact that the sounds are mediated by pulse energy levels sufficient to raise the tissue temperature of only 5×10^{-6}°C points out the extreme care that one must exercise in classifying an effect as thermal or athermal based simply on the level of temperature increase.

C. Effect of EM Fields on the Spinal Cord of the Cat

Evoked spinal cord activity at the ventral root was recorded from the preparation described in Section II-C and illustrated in Figure 5 with and without applied EMF and various amounts of cooling. The response consisted of a focal afferent signal and a variable set of spinal cord afferent potentials. These included typical monosynaptic spikes and late complex potentials. Figure 16 illustrates typical recordings. The first trace shows a typical stable response resulting from a number of averages with a circulating solution temperature of 37.5°C. When EMF was applied so that the maximum power absorption density in the cord was 3.8 kW/kg, the temperature of the cord elevated and the primary ventral root potential underwent characteristic changes; decreases in latency and attenuation in amplitude. With the radiation, the ventral root potential was essentially eliminated as shown in the second and third traces of Figure 16. With the cessation of microwave illumination, the ventral root potentials returned to their original configuration. This restoration followed the gradual reduction in temperature attributable to the circulating Ringer's solution. Stability of the early focal afferent potential in these and subsequent experiments indicates that radiation effects are exerted on a central mechanism and not on nerve conduction generally. With sufficient cooling of the spinal cord, the effect could be reduced and even eliminated indicating a temperature effect. It was also found that the effect could be produced by increasing the temperature of the cord by circulating bath. Nothing in the results indicated that a given level of CW microwave power acts upon the spinal cord in any manner except to elevate the temperature of the tissue.

IV. IN VITRO EXPERIMENTS

A. Frog and Cat Nerves

Some athermal effects on frog sciatic nerves at low power levels have appeared (22,34). It is difficult to determine, however, what the actual temperature changes were in these nerves, since the nerves were isolated in the air during irradiation. Under such conditions, power absorption density can be much higher than for the normal situation in which the nerve is buried in other tissue and the temperature inside the nerve can be higher than that measured at the surface. The aperture source used can cause hot spots in the irradiated nerve due to the discontinuities at the edge of the aperture (32). In this work, these problems are avoided using the waveguide exposure system described in Section II-D.

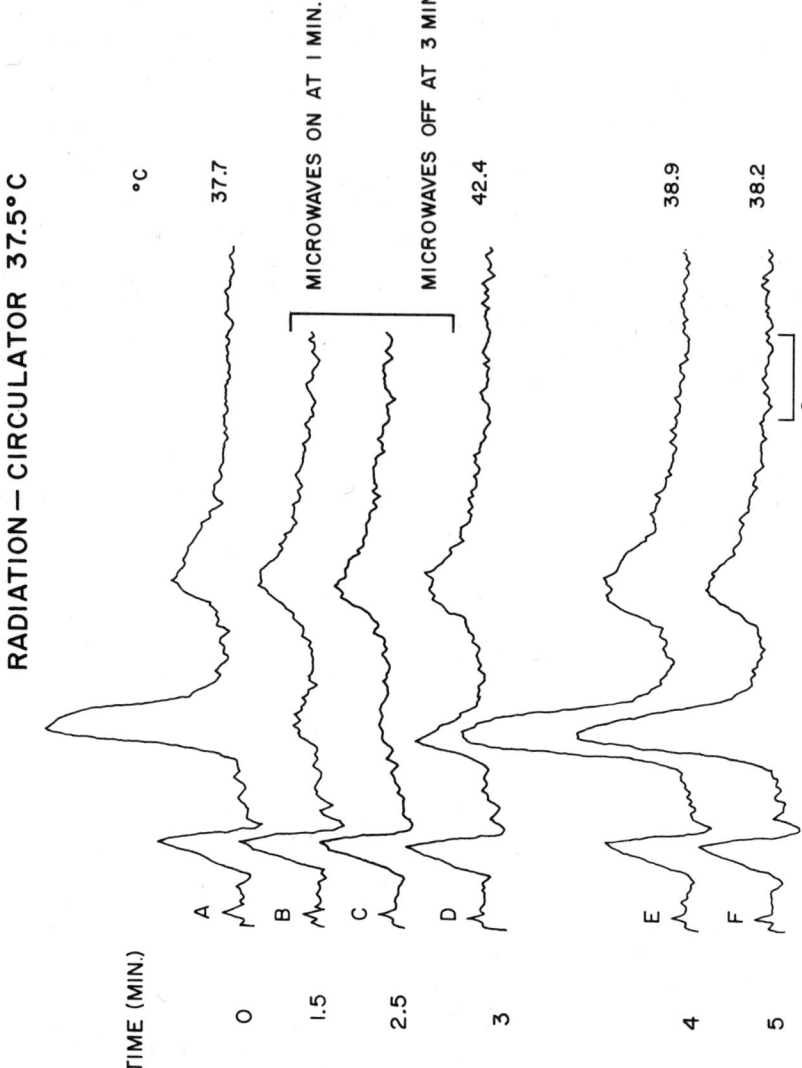

Figure 16 Spinal cord ventral root responses before, during and following exposure to electromagnetic radiation. The maximum power absorption density level in the exposed cord is 3.8 kW/kg. Circulating Ringer's solution (40).

EFFECTS OF ELECTROMAGNETIC FIELDS ON ANIMALS

A study was first performed to determine the effect of the controlled solution temperature on the compound action potential and conduction velocity of the isolated nerve. The results are shown in Figure 17 for the cat nerve.

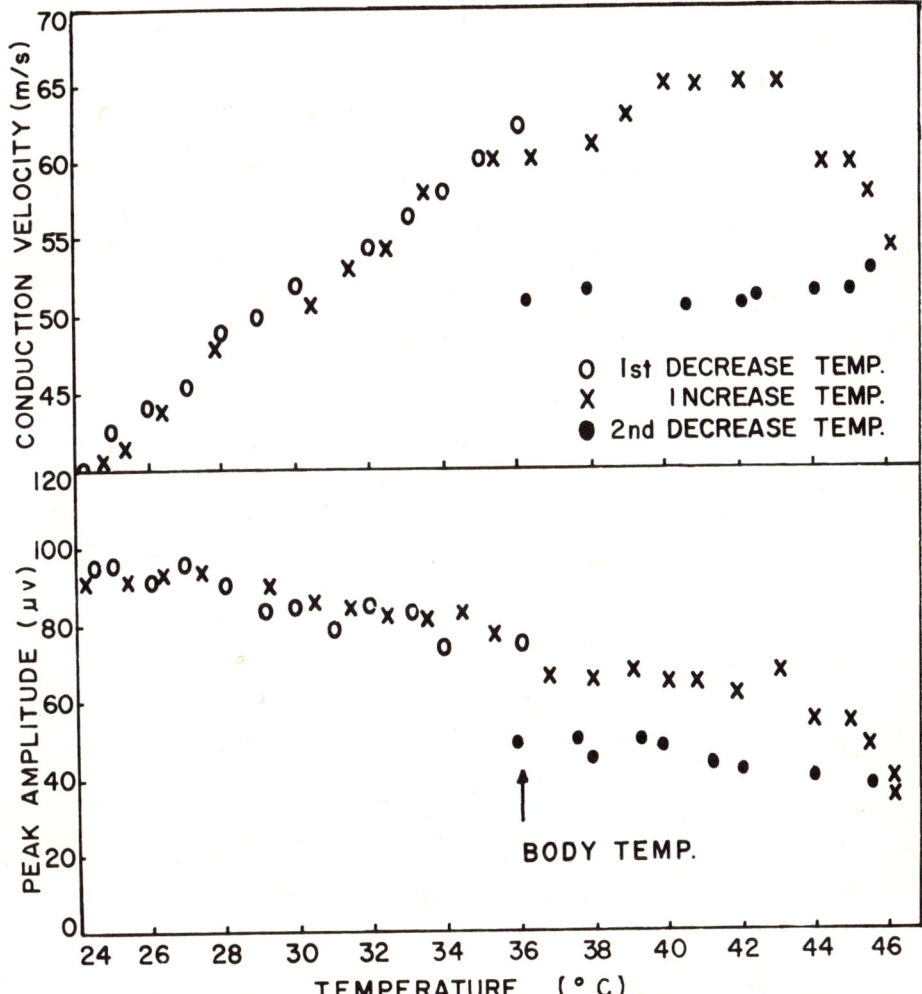

Figure 17 Amplitude of compound action potential and conduction velocity of cat saphenous nerve as a function of temperature.

In each case, the temperature was first decreased from normal to approximately 10°C below body temperature. The temperature was then increased to the point where the amplitude of the compound action potential decreased to one-half normal value. This corresponded to 46°C for the cat nerve which caused irreversible damage. The damage was evidenced by different nerve responsiveness when the temperature was again decreased. The temperature-sensitive characteristics of the frog nerve were reversible, since the maximum temperature was well below the threshold for damage when the peak amplitude decreased by 50%. These results suggest that mammalian animals rather than cold-blooded animals should be used to evaluate effects of microwaves on the nervous system for better quantitative extrapolation to humans.

Exposure of the frog and cat nerves did not result in any amplitude, conduction velocity, or excitability changes during the time that the constant temperature circulator was on for any level of either CW or pulsed irradiation. Slight increases in excitability and conduction velocity were detected during the time that the circulator was off for power absorption levels of 30 and 300 W/kg. These effects can be attributed to temperature changes according to Figure 17. Figure 18 shows that the absorption (1.7 kW/kg) due to very high power irradiation caused a slight increase of conduction velocity during the circulator on-phase. This was due to the limited pumping rate of the circulator and consequent temperature rise. A much greater increase in conduction velocity was observed when the circulator was shut off and the temperature was allowed to increase due to radiation. Frog action potential amplitude and latency were degraded at a temperature around 35°C, which was consistent with the frog temperature versus nerve velocity measurements. After irradiation, the response either recovered or degraded, according to the degree of temperature elevation. In a third series of experiments, Figure 19, double stimulation pulses 1.2 msec apart were used. The response due to the second pulse was in the relative refractory period of the first response. The characteristics of the second response were more sensitive to temperature changes. This indicates that a slight microwave-induced temperature rise in the peripheral nerve could possibly alter firing patterns of the neurons in the CNS.

B. Rabbit Superior Cervical Ganglion

This work was designed to extend the above study from simple nerves to synaptic junctions represented by rabbit superior cervical ganglion. The superior cervical ganglion has been well studied by physiologists (26), and pharmacologists (42) and is known to have synaptic systems utilizing both acetylcholine and catecholamines as transmitters. This ganglion is of interest in its own right as a major output relay for sympathetic nervous system functions related to cardiac acceleration, vasoconstriction, and secretion in the salivary glands, and pupillary dilation and vasoconstriction in the eye. The waveguide exposure system used above was inverted as shown in Figure 10.

EFFECTS OF ELECTROMAGNETIC FIELDS ON ANIMALS 199

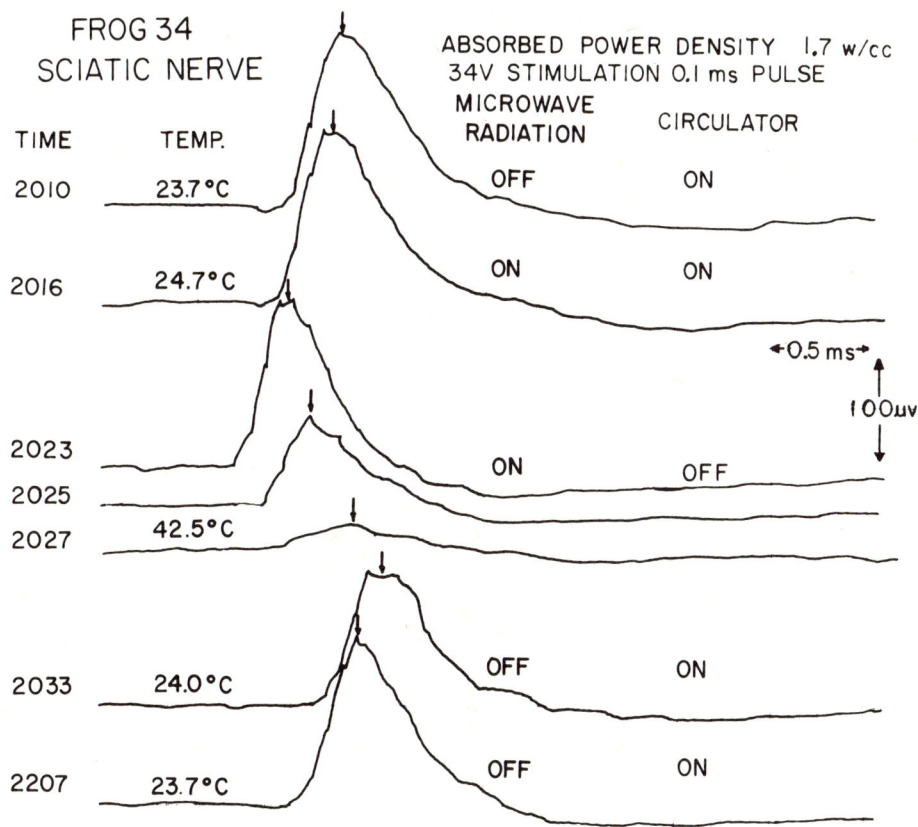

Figure 18 Heating effect on nerve action potentials due to 1.7 W/gm CW radiation when the circulator was off.

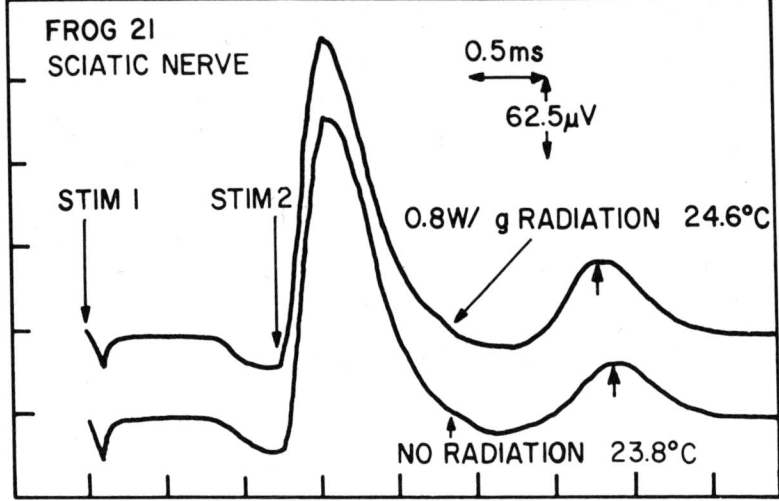

Figure 19 Change in temperature and resultant change in 2nd response of frog sciatic nerve during 2450 MHz microwave irradiation.

The basic measurements are illustrated in Figure 20. These graphs are averages made with a computer of average transients. Stimulus width of 100 - 300 μsec with sufficient strength to bring in both the preganglionic B fiber and C fiber mediated responses was used. Microwave radiation was applied at the indicated levels for periods of 1 min with 1 min between irradiation allowing for control measurements. It can be seen clearly that there is no significant difference between the control and exposure period latencies. A slight temperature increase in the range of $0.1°C$ or more was measured in the solution just exiting the waveguide when the absorbed power densities were over 100 W/kg (50 mW/cm^2 incident). However, the effect of this temperature elevation was well within the confines of the behavior of the ganglia when the temperature of the circulating fluid was increased by an equivalent amount in the absence of microwave radiation. The temperature dependences of the response latencies were found to be 0.5 msec/°C for the low threshold short latency responses (B fiber mediated) and 1.5 msec/°C for the high threshold long latency responses (C fiber mediated). This suggests, within a temperature controlled environment where there is no rise due to absorption of CW microwave radiation, transmission latencies of a mammalian synaptic system are not affected by microwaves.

V. CONCLUSIONS

The behavior of nervous tissues under microwave irradiation has been investigated using both in vivo and in vitro preparations. The microwave frequencies extend from UHF to X band, and the exposure facilities include near zone reflectors, horns and apertures, and waveguide irradiation chambers. With considerable attention directed to dosimetry and instrumentation, it has been demonstrated that the effects of acute exposure to CW microwaves on the electrophysiological properties of the nervous system preparations considered in this study are thermal in nature, that is, all observable phenomena are accompanied by a concurrent temperature elevation of the involved tissues and can be replicated in its entirety by conduction heating methods. Specifically, it has been found that the conduction and transmission latencies and amplitudes of both evoked potentials in the CNS of intact cats, and in isolated nerves of frogs and cats and ganglia of rabbits, are affected in a manner very similar to that of localized conduction heat. Moreover, the threshold of occurrence of these latency changes appears to fall between 2.5 - 5.0 W/kg of absorbed power density in the involved tissues which is about one-quarter to one-half of the normal metabolic rate of brain tissues. This would correspond to 5 - 10 mW/cm^2 incident upon the head of a cat or an animal with an equivalent head diameter. However, the incident power density required to produce an equivalent figure of

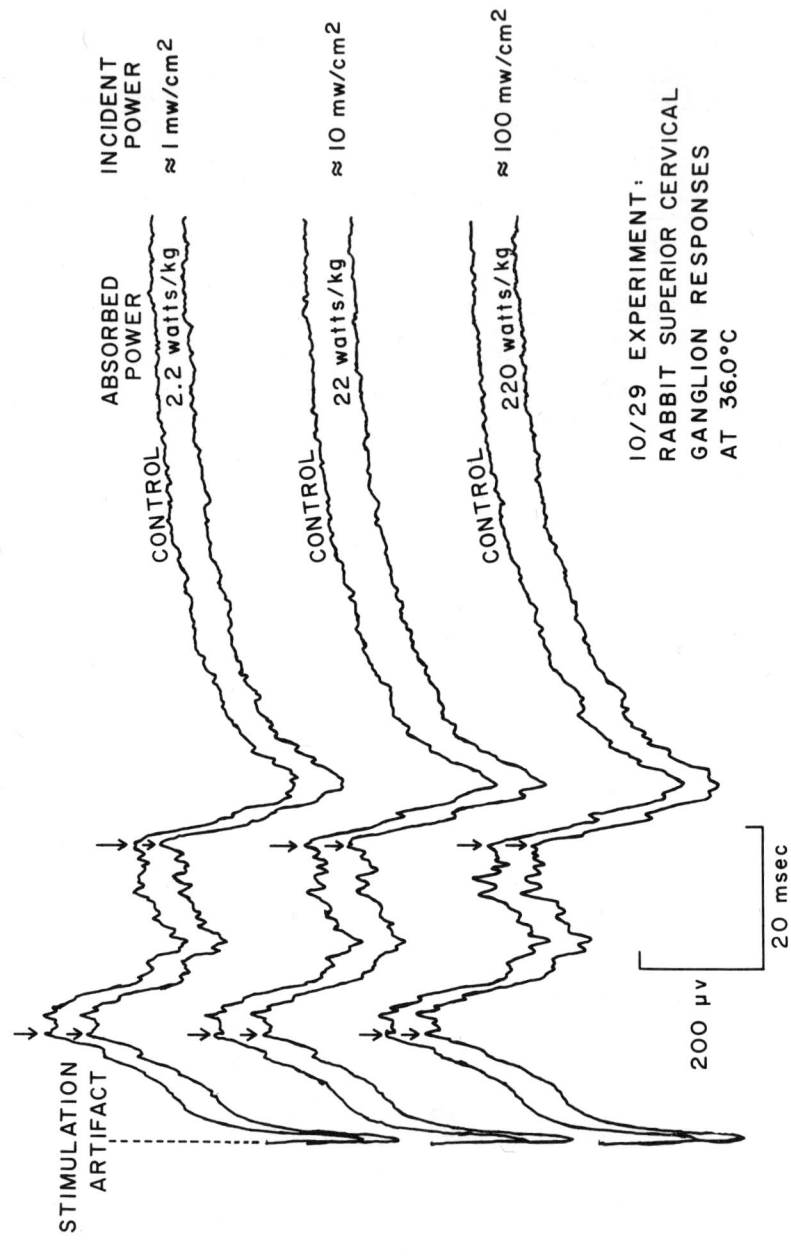

Figure 20 Evoked responses of extracellular recording of isolated rabbit superior cervical ganglion exposed to 2450 MHz CW radiation.

absorbed power in the head of man is between 10 to 25 mW/cm^2. The corresponding incident power required on the head of a small animal or bird would be 0.5 to 1.0 mW/cm^2, based on volume equivalent tissue spheres.

The induced auditory phenomena in man and animals exposed to pulsed microwaves of high peak intensity and low average power has been authenticated. However, the precise mechanism(s) of interaction still remains to be elucidated. Electrophysiological recordings on both intact animals and animals with cochlear disablement suggest that the site of pulsed microwave interaction with mammalian auditory systems is a peripheral one in that the microwave auditory sensation is perceived in a manner similar to that of conventional acoustic stimuli. Recent theoretical and experimental studies seem to indicate that the auditory effect may actually be the result of acoustic energy release from rapid thermal expansion due to power absorption in the gross structure of the head, even though the tissue temperature rises only 5×10^{-6}°C. The full implication of these observations, however, will not become apparent without further investigation.

ACKNOWLEDGMENT

This work was supported in part by Food and Drug Administration, Bureau of Radiological Health Grant No. 00646; in part by the Office of Naval Research Contract No. 00014-67-A-0103-0026; in part by National Science Foundation Grant No. GK-34730; and in part by the Social and Rehabilitation Services Grant No. 16-P-56818/0-11.

The authors wish to thank L. Boyd, S. Hoff, C. Sorensen, F. Harris, and E. Taylor for their assistance of the work reported herein.

REFERENCES

(1) BALDWIN, M., BACH, A. AND LEWIS, S.A.: Effects of radio-frequency energy on primate cerebral activity. Neurology 10 (1960) 178.

(2) BAWIN, S.M. et al.: Effects of modulated UHF fields on the central nervous system. New York Acad. Sci. Conf. on Biological Effects of Non-ionizing Radiation, New York (1974).

(3) BYCHKOV, M.S.: Changes in electrical activity of the cerebral cortex during the influence of an SHF field on animals. Trans. Kirov Military Med. Acad. 73 (1957) 58.

(4) BYCHKOV, M.S.: On the mechanism of the effect of an SHF field. Questions of the biological effect of an SHF electromagnetic field. (Abstracts) Leningrad Izd-vo VMOLA (1962) 6.

(5) BYCHKOV, M.S. et al.: Comparative neurophysiological evaluation of the biological effects of continuous and intermittent action of microwaves. Materials of the 4th All-Union Symp. Moscow (1972) 11.

(6) CHOU, C.K. and GUY, A.W.: Effect of 2450 MHz microwave fields on peripheral nerves. IEEE G-MTT Int'l. Microwave Power Symp., Boulder, Colo. (1973) 318.

(7) FOSTER, R.: Microwave hearing: evidence for thermoacoustic stimulation of the human auditory system by pulsed microwaves. Science. 185 (1974) 256.

(8) FREY, A.H.: Human auditory system response to modulated electromagnetic energy. J. Appl. Physiol. 17 (1962) 689.

(9) FREY, A.H.: Some effects on human subjects of ultrahigh frequency radiation. Amer. J. Med. Elec. 2 (1963) 28.

(10) FREY, A.H.: Brainstem evoked responses associated with low intensity pulsed UHF energy. J. Appl. Physiol. 23 (1967) 984.

(11) FREY, A.H.: Biological function as influenced by low-power modulated RF energy (Invited paper). IEEE Trans. Microwave Theory Tech. 19 (1971) 153.

(12) FREY, A.H. and MESSENGER, R.: Human perception at illumination with pulsed ultrahigh frequency electromagnetic energy. Science 181 (1973) 356.

(13) GOLDSTEIN, L. and SISKO, Z.: A quantitative electroencephalographic study of the acute effects of x-band microwaves in rabbits. Int. Symp. on Biologic Effects and Health Hazards of Microwave Radiation. Warsaw, Poland (1973).

(14) GORDON, Z.V.: Biological effect of microwaves in occupational hygiene. Acad. Med. Sci. of USSR. (1966) NASA TT F-633.

(15) GOURNAY, L.S.: Conversion of electromagnetic to acoustic energy by surface heating. J. Acous. Soc. Amer. 40 (1966) 1322.

(16) GUY, A.W.: Analysis of electromagnetic fields induced in biological tissues by thermographic studies on equivalent phantom models. IEEE Trans. Microwave Theory Tech. 19 (1971) 205.

(17) GUY, A.W., HARRIS, F.A. and HO, H.S.: Quantitation of the effects of microwave radiation on central nervous system function. Proc. 6th Annual Impi Symp. Monterey, Calif. (1971).

(18) GUY, A.W. and KORBEL, S.F.: Dosimetry studies on a UHF cavity exposure chamber for rodents. Proc. 7th Annual IMPI Symp., Ottawa, Canada, (1972) 180.

(19) GUY, A.W., LEHMANN, J.F. and STONEBRIDGE, J.B.: Therapeutic applications of electromagnetic power. Proc. IEEE 62 (1974) 55.

(20) GUY, A.W., et al.: Microwave induced acoustic effects in mammalian auditory systems and physical materials. New York Acad. Sci. Conf. on Biological Effects of Nonionizing Radiation. New York (1974).

(21) JOHNSON, C.C. and GUY, A.W.: Nonionizing electromagnetic wave effects in biological materials and systems (Invited paper). Proc. IEEE 60 (1972) 692.

(22) KAMENSKII, Y.I.: Effect of microwaves on the functional state of the nerve. Biofizika 9 (1964) 758.

(23) KHOLODOV, Y.A.: The effect of electromagnetic and magnetic fields on the central nervous system. Acad. Sci. USSR (1966) NASA TT F-465.

(24) KLIMKOVA-DEUTSCHOVA, E. and ROT, B.: The effect of irradiation on the human EEG. Chekhoslovatskoye Meditsinskoye Obozreniye 9 (1963) 254.

(25) KRITIKOS, H.N. and SCHWAN, H.P.: Hot spots generated in conducting spheres by electromagnetic waves and biological implications. IEEE Tran. Bio-Med. Engng. 19 (1972) 53.

(26) LIBET, B.: Generation of slow inhibitory and excitatory postsynaptic potentials. Fed. Proc. 29 (1970) 1945.

(27) LIVANOV, M.N. et al.: Concerning the question of the effect of an EMF on the bioelectric activity of the rabbit cerebral

cortex. Byulleten' Eksperimental' noy Biologii i Meditsiny 49 (1960) 63.

(28) McAFEE, R.D., BERGER, C. and PIZZOLATO, P.: Neurological effect of 3 cm microwave irradiation. Biological Effects of Microwave Radiation (Peyton, M.F. Ed.). Plenum Press. New York (1961) 251.

(29) MICHAELSON, SOL M.: Human exposure to nonionizing radiant energy - potential hazards (Invited paper). Proc. IEEE 60 (1972) 389.

(30) NIKONOVA, K.V.: Data for a hygienic evaluation of high-frequency EMF (medium and long wave ranges) Masters Dissertation. Moscow (1963).

(31) PARDZHANADZE, SH.K.: The mechanism of the effect of a UHF electrical field on the organism. Collected Works of State Scien. Res. Inst. Health Resorts and Physiotherapy of the Georgian SSR. (1954) 198.

(32) PRESMAN, A. et al.: Apparatus for investigating the excitability of nervemuscle preparations during microwave radiation. Biophys. 6 (1964) 73.

(33) PRESMAN, A.S. (Ed.): Electromagnetic Fields and Life. Plenum Press. New York (1970).

(34) ROTHMEIR, J.: Effect of microwave radiation on the frog sciatic nerve. The Nervous System and Electric Current. Plenum Press (1970) 57.

(35) SCHWAN, H.P.: Electrical properties of tissues and cells. Advan. Bio-Med. Phys. 5 (1957) 147.

(36) SCHWAN, H.P.: Interaction of microwave and radio frequency radiation with biological systems (Invited paper). IEEE Trans. Microwave Theory Tech. 19 (1971) 146.

(37) SHAPIRO, A.R., LUTOMIRSKI, R.F., and YURA, H.T.: Induced fields and heating within a cranial structure irradiated by an electromagnetic plane wave. IEEE Trans. Microwave Theory Tech. 19 (1971) 187.

(38) SINISI, L.: EEG after radar application. EEG and Clin. Neurophysiol. 6 (1954) 535.

(39) TAYLOR, E.M. and ASHLEMAN, B.: Analysis of central nervous system involvement in the microwave auditory effect. Brain Res. (1974) in press.

(40) TAYLOR, E.M. et al.: Some effects of electromagnetic radiation on the brain and spinal cord of cats. New York Acad. Sci. Conf. on Biological Effects of Non-ionizing Radiation, New York (1974).

(41) TOLGSKAYA, M.S. and GORDON, Z.V.: Pathological Effects of Radio Waves. Consultants Bureau. New York-London (1963).

(42) TRENDELENBURG, U.: Some aspects of pharmacology of autonomic ganglion cells. Ergebnisse Der Physiol. 9 (1967) 1.

(43) WEVER, E.G.: Electrical potentials of the cochlea. Physiol. Rev. 46 (1966) 102.

(44) WHITE, R.M.: Generation of elastic waves by transient surface heating. J. Appl. Physiol. 34 (1963) 3559.

-DISCUSSION-

FLOOR - In calculating the power in Watts per kilogram for the nerve work in particular, did you calculate the total power absorbed and then divide by the mass of the tissue?

LIN - This depends on the particular experiment. The absorbed power of the nerve work is calculated using formula (2) from our paper; an attenuation factor of 4α multiplied by the forward power minus the reflected power divided by the cross section area of the waveguide and this was then attenuated by 2α over the distance between the surface of the solution and the location of the nerve. In this case, the distance was one centimeter.

FLOOR - Did you observe an electrophysiological effect in the single light wave pulse, or do you have to have a certain repetition?

LIN - Although the experimental results shown are recordings made with stimuli at one per second, you could induce the hearing phenomenon with a single pulse.

GLASER, Z. - What does 3 Watts per kilogram correspond to roughly in terms of the brain basal rate of heat production?

LIN - The metabolic rate of brain is about 11 Watts per kilogram. However, the average metabolic rate of heat, I think, is about 3 Watts per kilogram. As I have shown using the auditory evoked responses, we could see a change in the latency between 2 to 5 Watts per kilogram. In the sciatic nerve case there is no temperature elevation associated with it.

FLOOR - I have two questions. First, what might be the functional significance of the local potential changes that you observed with continuous exposure, that is, CW exposure? I suppose I am asking whether you have any behavior data to correspond to electrophysiological data. And secondly, would it be possible to obtain parallel data on the significance of the elevation of body temperatures, say, by producing a pyrogen-induced fever?

LIN - Let me take the second question first. The reason that we use the circulating heat exchanger rather than pyrogen is because of two things. First, we have to produce a temperature gradient in the brain similar to that of microwave-induced temperature distribution. Although I did not show you the comparison between these two, the comparison was quite reasonable between these two methods. Since pyrogen is a fever-producing agent, it is going to produce general body heating which is not exactly the same as localized heating. Also, as I have indicated previously, if general body heating is used, the effects are different. Instead of only the second component, the one I indicated with the second arrow, changing, both the first and the second change with temperature elevation. Therefore, in order to mimic the microwave situation, we had to use the heat exchanger. The first question, as far as the functional significance - I presume you are talking about the intact animal experiments. We are actually doing that at this time. We are trying to correlate the relationship between the observed electrophysiologic responses and the behavioral changes, but as you know it is not a very easy task.

LOWY - Do you consider it possible that the rather distinct reduction in latency that you consistently observed might be due to the fact that you started at a temperature of the central nervous system that was abnormally low. You were between 35 and 36 centigrade, which is a temperature below the optimum for recording. By microwave heating you brought the temperature of those tissues closer to normal. That is also borne out by another graph where you showed that if you go over 38 you don't get it anymore. So there, it would only be a confirmation of what you said anyhow, but it would indicate that you started at a very favorable zero point, so to speak.

LIN - Yes. Thank you.

PORTELA - You are working in nerve preparations, and trying to learn about changes in propagation of the action potential, which you are inferring from your paper on latency measurements. Which were your criteria to determine the stimulus intensity necessary to stimulate the A and C fibers? When you are using external stimulating electrodes for periodic excitability analysis there are changes in the external fluid volume between the fibers of

EFFECTS OF ELECTROMAGNETIC FIELDS ON ANIMALS

PORTELA - the trunk which modify the local flow of current that is needed to fire the nerve fibers. Which were your criteria to consider that, because these give you an idea, from the point-of-view of excitability, about the pattern for evoking the train of nerve impulse events, from which later you have analyzed the velocity of propagation, etc.

FLOOR - I think that is a very important point.

LIN - Yes, that is certainly a very important point. Our idea of doing that experiment was rather than to study the neurophysiology of it, to study the effect of microwaves on the particular response.

PORTELA - But the response will depend on the time period of stimulation, necessary for evaluating the excitability and functional conditions of the nerve fibers. In addition, you are talking about a population of cells, each of which has a different threshold. Then, you have to increase the magnitude of stimulus to find out the threshold for populations, not for one single cell. Thus, the question is this: which is the amount of energy you introduced with the stimulus in order to fire the population of nerve fibers?

LIN - The stimulus voltage was about 5 volts. I don't know the current.

PORTELA - You were always using the same voltage?

LIN - Right.

JOHNSON - I want to make two points. First, Dr. Lin describes a certain biological effect in terms of watts per kilogram. I think this is the correct way to document the dosimetry. Tissue absorbed power cannot be related directly to incident radiation intensity in milliwatts per cm^2; or a given incident radiation intensity, the tissue absorbed power may vary 10 to 100:1, depending upon irradiation conditions, animal body orientation with respect to the field vectors, etc. Dr. Lin has convincingly indicated that there is very little effect of microwaves on nerve propagation velocities other than thermal. But I would like to point to an experiment by Dr. James Lords, who has indicated on a turtle heart preparation that instead of causing the expected thermal tachycardia in the turtle heart, moderate levels of microwave exposure cause bradycardia. Lords has shown that the bradycardia is due to a direct microwave-nerve interaction. So in other more sophisticated biological preparations where more of the tissues are left intact, in this case the muscle tissue and nervous tissue with motor end plate, there are apparently some effects which

JOHNSON - cannot be obviously explained in terms of thermal effects.

LIN - Perhaps I ought to point out that one of the reasons that we did the ganglia studies is to try to clarify the effects of microwaves on nervous tissue. As you realize, the thalamic preparation is limited, because it is so complex, the interpretation is difficult. And, the nerve study is also limited because nerve is nothing but a transmission line, in the terminology of electrical engineering, so it is also limited. The ganglia is one thing that has distinct characteristics because it transmits information. This is one of the reasons we looked at that particular preparation to see whether this property is changed. Apparently, there is no effect as far as continuous wave is concerned.

NYBORG - In respect to the mechanisms for the hearing, which you mentioned toward the end of your talk, you favored the one, I believe, that had to do with the production of heat. Now, is there a specific assumption about where that heat is being produced, and just what it is that expands, and just how it communicates the mechanical effect to the ear?

LIN - This question has been studied both theoretically and experimentally. The theory is really based on the simplified case of a plane parallel layer of lossy dielectric with an impinging radiation. In the studies done, both by Dr. White and Gurney, they assumed a third factor, the depth of penetration. Within this depth of penetration, because of the absorption characteristics, heat is generated.

NYBORG - Penetration of what now?

LIN - The microwave field into the tissue, for instance.

FLOOR - Anatomically what part of tissue is it that you are considering?

LIN - This is the theoretical model that people used. Based on this, we can make some generalizations. This layer could be a biological tissue, or in your case, a liquid. In the animal case or in the human case, what could be happening is this: the incident energy impinges on the head, since we all know that at high frequencies the penetration depth is limited. This impinging energy would be absorbed at the surface, either of the bone or of the soft tissue. Since the absorbed energy will be transformed into heat within a very short duration, rapid expansion would occur in these tissues. The resulting acoustic energy may then be transmitted via bone conduction to the inner ear. That is one

LIN - possibility. We have also calculated the pressure available at the surface of the brain for human. Take a 20 microsecond pulse, and the observed threshold energy of 40 microjoules per cm^2; we calculated that the pressure available at the surface of the brain is between .25 to .7 dynes/cm^2, which is incidentially above the hearing threshold by bone conduction.

SENSATION AND PERCEPTION OF MICROWAVE ENERGY*

Sol M. Michaelson

University of Rochester, School of Medicine & Dentistry
Department of Radiation Biology & Biophysics
Rochester, New York 14642

ABSTRACT

Sensing or perception of microwave/radiofrequency energy is accomplished through various mechanisms. In mammals, the main phenomena of sensation or perception are those of thermal sensation and, in selected cases, audition. Thermal sensation is accomplished by stimulation of thermosensitive nerve endings in the skin. Although some investigators believe that "hearing" or audition is evidence of direct nerve stimulation, the most recent data show this phenomenon to be due to electromechanically induced vibrations in tissue and normal reception in the cochlea of the ear.

Introduction

Reports in the literature note that many organisms, including man, can sense or perceive microwave/radiofrequency (rf) energies through various mechanisms. On a physiologic basis sensation is the experiencing of a primary stimulus by means of the functional properties of receptors, afferent nerve fibers, and central sensory systems up to some poorly defined intermediate level of complexity (29, 43). Perception encompasses the total behavioral experience of sensory stimuli, patterns, selectivity, contrasts

---

*This paper was prepared under contract No. FDA 73-30 (PHS, FDA, DHEW) sponsored by the EMR Project Office, BUMED and SURG, Dept. of the Navy, and under contract with the U.S. Atomic Energy Commission at the University of Rochester Atomic Energy Project and has been assigned Report No. UR-3490-550.

and similarities between stimuli, and apprehension of their serial order in time (29).

Studies are described in which microorganisms make specific orienting responses in a microwave field (2, 3). This response, no doubt an example of electromagnetic field-induced orientation of non-spherical particles, may be a manifestation of polarization. Electrical charges appear on the boundaries of the particles, thus creating an electrical dipole and aligning themselves with the electrical field (28). In mammals, especially man, the main phenomena of microwave/rf sensation or perception are those of thermal sensation and, in selected cases, audition.

Cutaneous Thermal Sensation

Sensations of warmth and cold are evoked by radiation exchanges between the skin and the environment (15). Thermosensitive nerve endings can be defined in two different ways: a) by the specific thermal sensation aroused by stimulation of a receptor and b) by the response of a receptor (as indicated by action potentials) to temperature changes (21).

Weber, in 1846, proposed that the rate of change of skin temperature, dT_s/dt, was the effective stimulus for temperature sensation (46). Many investigators subsequently confirmed the fact that under certain circumstances, a direct relationship could be established between dT_s/dt and temperature sensation. Hardy and Oppel (14) and Hendler and Hardy (19) made quantitative measurements of this relationship using infrared and microwaves as sources of radiant energy. Lele et al (27), on the other hand, suggest that sensation is caused by a difference in temperature between two different layers of the skin. This temperature difference is proportional to the temperature gradient. Other investigators suggest that at least three quantities, temperature, rate of change of temperature, and surface area, have to be considered for a description of thermal stimulation (22).

With infrared, the threshold for warmth perception is reached at a warming of the skin at a rate of about .001-.002°C per second in the skin temperature range of 32°-37°C. Because of spatial summation (effect of size of area of stimulation in altering the sensory threshold and the intensity of sensation), threshold and intensity of temperature sensation depend to a large extent on the size of the skin area changing temperature. Similarly, the minimal time of warming the skin before a temperature sensation is elicited depends on the size of the area affected and on the density of the specific temperature receptors in that area.

Oppel and Hardy (31, 32) found that, for the forehead, using an area of 7.5 cm^2, irradiation with infrared at an intensity of 6.7 mW/cm^2 causes a threshold sensation in 3 seconds. This amount of radiation gives a temperature increase of the order of 0.05°C at a few tenths of a millimeter below the skin surface.

TABLE I. Stimulus Intensity and Temperature Increase to Produce a Threshold Warmth Sensation Over 37 cm² Forehead Surface Area*

Exposure Time (sec)	3000 MHz	10,000 MHz		Far Infrared	
	Power Density (mW/cm²)	Power Density (mW/cm²)	Increase in Skin Temp. (°C)	Power Density (mW/cm²)	Increase in Skin Temp. (°C)
1	58.6	21.0	.025	4.2 - 8.4	.035
2	46.0	16.7	.040	4.2	.025
4	33.5	12.6	.060	4.2	---

*data from Hendler (17, 20).

There are several studies which describe the cutaneous perception of microwave energy. Hendler and associates (17, 20) outlined the cutaneous receptor response of man to 10,000 MHz and 3,000 MHz microwaves and far infrared. The forehead was selected for investigation of warmth sensation, since previous studies had shown that the temperature receptors in the skin of the forehead are relatively numerous and evenly distributed, so that it constitutes a low-threshold region of uniform temperature sensitivity (31, 32). Far infrared is almost entirely absorbed in the first 1/10 mm of the skin, and 3000 MHz microwave intensity does not diminish appreciably by absorption in the first 2 mm of the skin (44). Irradiance levels of 10,000 MHz and 3000 MHz microwaves as well as far infrared stimuli producing a threshold sensation of warmth are summarized in Table I.

Calculations of the temperature gradient (dT/dx) occurring within the cutaneous layers for the infrared and microwave stimuli producing threshold sensation revealed no consistency in these layers for both the infrared and microwave cases. For the range of exposure times used, it was found that for both infrared and microwave stimuli which are capable of evoking threshold sensations, a threshold of warmth is elicited when the temperature of a more superficial layer of subcutaneous tissue about 200 μ below the skin surface is increased about 0.01-0.02°C over the temperature of a deeper layer lying about 1000 μ below the surface (17, 20).

Studies in man indicate that when a 40 cm^2 area of the face is exposed to microwaves, thermal sensation can be elicited within 1 second at power densities of 21 mW/cm^2 for 10,000 MHz and 58.6 mW/cm^2 for 3000 MHz. Within 4 seconds the threshold is lowered by approximately 50%, i.e. 12.6 mW/cm^2 (10,000 MHz) and 33.5 mW/cm^2 (3000 MHz) (17). On this basis, if the entire face were to be exposed (assuming uniform temperature sensitivity of the facial skin), the threshold for thermal sensation to 10,000 MHz would be 4-6 mW/cm^2 within 5 seconds or approximately 10 mW/cm^2 for a 0.5 second exposure (13).

Schwan et al (37) found that if a person's forehead is exposed to 74 mW/cm^2 of 3000 MHz microwaves, the reaction time (the time which elapses before the person is aware of the sensation of warmth) varied between 15 and 73 seconds. Warmth perception of 56 mW/cm^2 ranged between 50 seconds and 3 minutes of exposure.

Vendrik and Vos (44) and Eijkman and Vendrik (4) also studied warmth sensation induced by infrared and microwaves in humans. Irradiation of a 13 cm^2 area of the inner forearm with 3000 MHz (pulsed) microwaves resulted in a threshold temperature rise of 0.4-1.0°C, depending on the subject, which is in agreement with the threshold temperature rise obtained with infrared at a depth of about 0.3-0.5 mm (44). For stimulus duration longer than 3-5 seconds, a pronounced influence of the rate of change of temperature was found. The change in the temperature gradient caused by irradiation with microwaves is so small that, during exposure times of not more than about 10 seconds, heat conduction is almost negligible. With a constant intensity of irradiation, therefore, temperature in the skin increases linearly with time. These authors suggest that a threshold sensation is obtained when the temperature of the warmth receptors is increased by a certain amount, ΔT. For durations of the stimulus longer than 3-5 seconds, the rate of change of temperature has a very pronounced influence. This is an adaptation phenomenon. These authors also described effects of "peripheral" and "central" adaptation, and related these to the electrophysiological findings in cats which had been reported by Hensel and Zotterman (22).

It should be pointed out, however, that due to shape factors and non-uniform sensitivity, it is likely that the figures for threshold thermal sensation may be somewhat low, although the practical significance of corrections to these figures is probably small. In this context, it is well to note that the subjects used for threshold sensation experiments may be well trained and particularly attentive to stimuli they could expect. Extraneous sensory stimuli may be removed or kept at some low, constant level. Consequently, these conditions are appreciably different from those to be expected in most practical situations, where "naive" personnel may be exposed to microwaves under very distracting circumstances (18).

The experiments with microwave irradiation indicate that persistent warmth sensations may be experienced even after irradiation has stopped and the skin temperature is rapidly falling. This suggests the existence of an effective thermal gradient between the cutaneous tissue layers (20).

Audition

It has been known for a long time that electrical current passed through the human head by employing various types of electrode systems gives rise to hearing sensations (40). The principles of such acoustic reaction had been described as early as 1800 by Volta (45).

In 1930, Nrunori and Turrisi (30) reported that individuals can perceive electromagnetic energy in the microwave range, attributing this capability to effects on the autonomic nervous system. This response was investigated in detail by Stevens (42), who in 1937 designated the phenomenon "electrophonic effect" to describe hearing caused by any type of electrical stimulation to, or in proximity to, the brain.

In 1949, Sheyvekhman (39) applied 6 meter energy to the head region of humans for 5 minutes by contact. The auditory threshold for tones of 300, 1000, and 4000 cycles per second (cps) was tested before, during, and after irradiation. At 300 cps, the threshold change was within the experimental error of 2 decibels (db); at 1000 cps, the threshold rose 4 to 6 db; and at 4000 cps, the threshold rise was 6 to 10 db. It was noted that recovery time was longer at the higher frequencies. Auditory responses among radar workers were also reported by Barron and Baraff (1) in 1958.

The studies of Stevens and others (5, 23, 24, 42) indicated clearly that the hearing sensations had their origin in electromechanical phenomena generating vibrations in bone or tissue structures outside the cochlea and being perceived through the cochlea in the normal way. Some of the original speculations that the electrophonic effect might be the result of a direct stimulation of auditory nerve activity were soon disproven. Although evidence had been presented to show that it is possible in selected cases to stimulate the eighth nerve directly with sinusoidal current applied with electrodes in the middle ear, such indiscriminate nerve stimulation is perceived only as broad band noise (40).

Frey (6, 7), in a series of papers starting in 1961, reported that the human auditory system can directly detect radiofrequency energy transmitted through air by electromagnetic waves. Several radiofrequencies, transmitted by a pulse modulated microwave transmitter, were used. Subjects, exposed to peak power densities of 200 to 300 mW/cm^2 (wavelength 10-150 cm, pulse duration 1-6 μs, repetition rates 600-2500 pulses/s) and electric field strengths in the order of 15 volts/cm, reported hearing these modulation pulses. This effect apparently occurred instantaneously and at low average power densities (100 $\mu W/cm^2$). The frequency, modulation, and peak power density were considered to be the important variables. The nature of the perceived sound was described as a buzz, ticking, hiss, or knocking, depending upon the transmitter parameters of pulse width and pulse repetition rate. The apparent source of the sounds was localized by the subjects within or immediately behind the head. The microwave sound was heard by individuals with 50 db hearing losses. When the ears of individuals with normal hearing were plugged, the sound became more distinct. Selective shielding of the head revealed the temporal area to be the most sensitive region. The greatest sensitivity was to the frequency range from 300 MHz to 1200 MHz. At first it was thought that a necessary condition for perceiving the sound was the ability to perceive audio frequencies greater than 5000 cps (6). Later, however, Frey revised his conclusions and reported that this was not a relevant variable (7).

In later tests, Frey and Messenger (8) described the radar sounds as "buzzes and hisses." They conducted experiments with human subjects at 1.245 GHz and used pulses with a 50 Hz repetition rate and pulse widths varying between 10 to 75 μs. They found a threshold for the sensation expressed in terms of peak power density viz: 80 mW/cm^2 with about a 75 μs pulse width. They reported that apparent "loudness" increased with peak power density even though average power density was invariant and that changing average power density only did not change the "loudness."

Recent work by Guy et al (11, 12), however, has shown rather conclusively that with pulse widths of the order of 30 μs the auditory threshold is expressed in incident energy density of a pulse - specifically 20-40 μ joules/cm^2 incident at 1-3 GHz. In experiments with cats they obtained evidence which suggests that the effect originates by conversion of microwave energy to acoustic energy at the outer bone muscle interface in the head with bone conduction of the acoustic signal to the ear completing the process.

Sommer and von Gierke (40), in a classic paper in 1964, reported much of their original experimentation and theory. The data reported in these investigations warrant the conclusion that there is no other novel auditory stimulation to explain the hearing sensations excepting mechanical tissue excitation by electrostatic forces associated with the audio frequency modulated DC fields. Calculated threshold data from others also fall within the level necessary to produce bone conduction. There is no evidence of any direct perception of electrical audio signals which would not go via electromechanically induced vibrations in tissue and normal reception in the cochlea.

This conclusion is supported by Harvey and Hamilton (16) who showed that when using an rf carrier, the pressures required to excite the auditory system also fall within the levels required for bone conduction. Perrott and Higgins (34) also found that the electrophonic effect can be readily explained as a microphonic phenomenon. Electrophonic stimulation produced bone vibrations of sufficient amplitude to account for the sound heard by the listener.

Lebovitz (25, 26) has suggested that temperature gradients induced by low intensity (ca. 10 mW/cm^2) microwave heating of the vestibular apparatus could be transduced by the vestibular apparatus resulting in a neuronal output that could mediate a variety of cardiorespiratory responses as well as evoked potentials in portions of the CNS. Non-uniform joule absorption of microwaves by the intravestibular fluid can establish thermal gradients within the labyrinth, with attendant convective torque in the semicircular canals. This can be related to an equivalent (i.e., virtual) angular acceleration, as a function of average incident microwave power density. Cochlear hair cell structures, on the other hand, appear to be of appropriate size and mass to be significantly perturbed via the field-force effects of microwaves. This suggests a class of mechanisms whereby direct auditory perception of pulse modulated microwaves, with low average incident power density, can occur.

Sharp et al (38), in a discussion of the production of sound when short microwave pulses are directed at an absorber, note that bone is known to have piezoelectric properties and the difference of potential resulting from bone deformation has been measured. If either the radiation pressure or electrostrictive forces are sufficient in the irradiated cranium to engender such a potential difference, then there could be an electrically mediated sensation. Also, there could be direct acoustic excitation of the auditory organ via bone conduction of such a vibration or the basilar membrane may directly couple with the microwave energy.

According to von Gierke (9), there is no evidence to indicate that the auditory phenomenon resulting from microwave stimulation is propagated by direct nerve stimulation or any other method that does not involve the cochlea. Reports indicating that dogs with "destroyed" cochlea can perceive an auditory stimulus (35) can be challenged on the basis of possible tactile stimulation to which the dogs respond in the classical conditioning paradigm (9, 34). Recognition of tactile cues for a limited word vocabulary can also explain why "deaf" persons can discriminate speech. In electrophonic stimulation, most of the word tests used to test speech discrimination in "deaf" persons are limited to approximately ten words. It has been shown that with a limited set of words persons can identify the words solely on the basis of the tactile pattern differences associated with each word (9).

Recent studies in Schwan's laboratory provide additional support that the acoustic effect of pulsed microwave fields may well be explained by the forces rhythmically applied to the middle ear structures as the field is turned on and off (36). Since only the strength of the applied field dictates the force generated, the observation by several investigators that the peak power is the important parameter is readily understood.

The mechanism of the auditory effect may be "thermo-acoustic" conversion as with laser-acoustic conversion (10) studied in the 1960's. The observed level of the generated acoustic signal may be sufficient to stimulate the ear after bone conduction. The earlier work by White (47, 48, 49) on the "thermoelastic" or "thermoacoustic" conversion effect, suggests that the effect is produced by the heating of various surfaces as interfaces with pulsed electron beams, microwaves or lasers. White (47, 48, 49) discovered the effect first with electron beams then with laser or microwave beams incident on some surface. He conclusively showed that the mechanism could not be radiation pressure but was very likely an elastic wave initiated by the mechanical stress generated at a surface by thermal expansion during transient heating. In a later paper White (49) made an extensive analysis of conversion of harmonic signals at surfaces without and with mechanical constraint boundary conditions. He found the fractional efficiency of conversion is likely to be very small, $<<<1$ in general, but is nevertheless much greater if the surface is under constraint.

In biological systems it is not likely that the models of White apply directly. If an interface is found where one of the materials expands quite differently than the other, a condition of surface stress may be present and significant conversion efficiency results at low frequency. Of course, there is some loss of the acoustic signal in its transmission to the ear. Still no great conversion efficiency is required in view of the tremendous sensitivity of the ear (41) which is about 10^{-9} µW/cm^2 incident on the ear for the frequency range of around 1000 Hz. Even at 60 Hz the threshold is still about 10^{-3} µW/cm^2 (33).

The importance of the auditory effect as one means of microwave perception needs to be stressed. Although the thresholds for this effect are not yet well determined, it is clear that exposure to hazardous radar fields in the UHF to L-band range of frequency (i.e., >100 mW/cm^2 average) should yield very noticable auditory effects. In a non-scanned radar this average power density would correspond to 100 watts/cm^2 peak with repetition rates in the range of 500-2000 Hz - the sensitive range of the ear. Also the energy density per pulse typically might be 100 µ joules/cm^2 (or possibly >1000 µ joules/cm^2 for some modern radars which use long pulses). On the other hand, if the radar is scanned, much larger pulse energies would be incident on a person if the average power density were around 100 mW/cm^2. In that case for a fraction of a second, an exposed subject should experience a very intense auditory effect. If this is so, then the degree of sensation needs to be better quantified and described because it could be used to determine whether or not people have or have not been exposed to hazardous radar fields at least at UHF to L-band (33).

While there is thus considerable evidence for the perception of pulsed microwave radiation as an auditory sensation, a number of questions regarding the mechanism responsible for the phenomenon remain. Further analysis of the mechanical, electrical, as well as the neural mechanisms involved is necessary (12).

Acknowledgement

The author wishes to express his appreciation to Drs. Henning von Gierke and John M. Osepchuk for providing much of the material for the "Audition" portion of this paper, to Drs. James D. Hardy and Edwin Hendler for their comments on thermal sensation, and to Mrs. Peggy Bush for organizing the bibliographic material and typing the manuscript.

REFERENCES

1. BARRON, C.I. and BARAFF, A.A.: J.A.M.A. 168 (1958) 1194.

2. BLOIS, S.: IRE Trans. Med. Electr. 4 (1956) 35.

3. BROWN, G. and MORRISON, W.: Inst. Radio Engin, Trans. Med. Elect. 4 (1956) 16.

4. EIJKMAN, E. and VENDRIK, A.J.H.: J. Exptl. Psychol. 62 (1961) 403.

5. FLOTTORP, G.: J. Acoust. Soc. Am. 25 (1953) 236.

6. FREY, A.H.: Aerosp. Med. 32 (1961) 1140.

7. FREY, A.H.: Am. J. Med. Elect. 2 (1963) 28.

8. FREY, A.H. and MESSENGER, R.: Science 181 (1973) 356.

9. von GIERKE, H.E.: Personal Communication (1974).

10. GOURNAY, L.S.: J. Acoust. Soc. Am. 40 (1966) 1322.

11. GUY, A.W., CHOU, C.K., LIN, J.C., and CHRISTENSEN, D.: Ann. N.Y. Acad. Sci. (in press).

12. GUY, A.W., TAYLOR, E.M., ASHLEMAN, R., and LIN, J.C.: Microwave interaction with the auditory systems of humans and cats, In Proceedings of the 1973 IEEE G-MTT International Microwave Symposium, Boulder, Colorado, June 1973. IEEE. New York (1973) 321.

13. HARDY, J.D.: Personal Communication (1970).

14. HARDY, J.D. and OPPEL, T.W.: J. Clin. Invest. 16 (1937) 533.

15. HARDY, J.D. and OPPEL, T.W.: J. Clin. Invest. 17 (1938) 771.

16. HARVEY, W.T. and HAMILTON, J.P.: Hearing Sensations in Amplitude Modulated Radio Frequency Fields. M.S. Thesis, Air Force Inst. of Tech., AD 608-889 (1964).

17. HENDLER, E.: Cutaneous receptor response to microwave irradiation, In Thermal Problems in Aerospace Medicine (Hardy, J.D. Ed.). Unwin Ltd. Surrey (1968) 149.

18. HENDLER, E.: Personal Communication (1970).

19. HENDLER, E. and HARDY, J.D.: Trans. IRE Med. Electr. ME-7 (1960) 143.

20. HENDLER, E., HARDY, J.D., and MURGATROYD, D.: Skin heating and temperature sensation produced by infra-red and microwave irradiation, Ch. 21, Temperature Measurement and Control in Science and Industry, Part 3, Biology and Medicine (Hardy, J.D. Ed.) Reinhold. New York (1963) 221.

21. HENSEL, H.: Electrophysiology of thermosensitive nerve endings, In Temperature Measurement and Control in Science and Industry, Part 3, Biology and Medicine (Hardy, J.D. Ed.) Reinhold. New York (1963) 191.

22. HENSEL, H. and ZOTTERMAN, Y.: Acta Physiol. Scand. 23 (1951) 291.

23. JONES, I.A.: Human Detection of UHF Energy. Thesis, Baylor Univ., Texas (1966).

24. JONES, R.C., STEVENS, S.S., and LURIE, M.H.: J. Acoust. Soc. Am. 12 (1940) 281.

25. LEBOVITZ, R.M.: The Sensitivity of Portions of the Human Central Nervous System to "Safe" Levels of Microwave Radiation. The Rand Corp., Santa Monica, Calif., Report R-983-RC (1972).

26. LEBOVITZ, R.M.: Ann. N.Y. Acad. Sci. (in press).

27. LELE, P.P., WEDDELL, G., and WILLIAMS, C.W.: J. Physiol. 126 (1954) 206.

28. MICHAELSON, S.M.: Proc. IEEE 60 (1972) 389.

29. MOUNTCASTLE, V.B. and DARIAN-SMITH, I.: Neural mechanisms in somasthesia, Ch. 62, Medical Physiology. vol. II, 12th Ed. (Mountcastle, V.B. Ed.). C.V. Mosby Co. St. Louis (1968).

30. NRUNORI, N. and TORRISI, S.S.: Am. J. Phys. Ther. 11 (1930) 102.

31. OPPEL, T.W. and HARDY, J.D.: J. Clin. Invest. 16 (1937) 517.

32. OPPEL, T.W. and HARDY, J.D.: J. Clin. Invest. 16 (1937) 525.

33. OSEPCHUK, J.M.: Personal Communication (1974).

34. PERROTT, D.R. and HIGGINS, P.: J. Acoust. Soc. Am. 53 (1973) 1437.

35. PUHARICH, H.K. and LAWRENCE, J.L.: Electro-Stimulation Techniques of Hearing. USAF, RADC-TDR-63 (1963).

36. SCHWAN, H.P.: Principles of interaction of microwave fields at the cellular and molecular level. In *Proceedings of the International Symposium on Biological Effects and Health Hazards of Microwave Radiation*. Warsaw, Poland, October 1973 (in press).

37. SCHWAN, H.P., ANNE, A., and SHER, L.: Heating of Living Tissues. U.S. Naval Air Engineering Ctr., Philadelphia, Pa. Report NAEC-ACEL-534 (1966).

38. SHARP, J.C., GROVE, H.M., and GANDHI, O.P.: IEEE Trans. Microwave Theory and Techniques MTT-22 (1974) 583.

39. SHEYVEKHMAN, B.YE.: Problemy Fiziologicheskoy Akustiki (USSR) 1 (1949) 122.

40. SOMMER, H.C. and von GIERKE, H.E.: Aerosp. Med. 35 (1964) 834.

41. STEPHENS, R.W.B. and BATE, A.E.: In *Acoustics and Vibrational Physics*. St. Martins Press. New York (1966) 188.

42. STEVENS, S.S.: J. Acoust. Soc. Am. 8 (1937) 191.

43. TEUBER, H.L.: Perception. In *Handbook of Physiology* (Magoun, H.W. Ed. Neurophysiology Section), vol. 3. Williams and Wilkins. Baltimore (1960) 1595.

44. VENDRIK, A. and VOS, J.: J. Appl. Physiol. 13 (1958) 435.

45. VOLTA, A.: Roy. Soc. Phil. Trans. 90 (1800) 403.

46. WEBER: Cited in: HENDLER, E.: Cutaneous perception response to microwave irradiation. In *Thermal Problems in Aerospace Medicine* (Hardy, J.D. Ed.). Unwin Ltd. Surrey (1968) 149.

47. WHITE, R.M.: Acoustic detection of electron impact in electron tubes. In *Proceedings of the 4th International Conference on Microwave Tubes*. The Hague. Elsivier. The Hague (1962) 156.

48. WHITE, R.M.: J. Appl. Phys. 34 (1962) 2123.

49. WHITE, R.M.: J. Appl. Phys. 34 (1963) 3559.

SENSATION AND PERCEPTION OF MICROWAVE ENERGY

-DISCUSSION-

VOGELMAN - It should be pointed out that the reason you were talking about pulse modulation is that you really need the power variation. You could do it just as well with sinusoidal modulation if you could create a device which modulates the beam by 90 percent. The only thing you have to do is to have the modulation. If you run CW, you don't have the hearing sensation for the simple reason there is nothing changing, and what you are really hearing is change. I think if you look at the results that have been enumerated to-date, you will find that the electrical engineer's terminology "modulation index" is the "loudness" that is attributed to the sound; and in fact, pulse width can be traded for pulse amplitude to get the same apparent loudness. If you look at the data that was reported most recently in Science (A. H. Frey, Human Perception of Illumination with Pulse Ultrahigh-Frequency Electromagnetic Energy, Science 181: 356-358, 1973) as a function of modulation index you will find that the author (A. Frey) gets apparent increases with both average power, and peak power changes. What he is really getting are the apparent increases due to the product of pulse width and peak power during the pulse.

CARSTENSEN - What is meant about the ability of the body's own sensors to provide warning signals?

MICHAELSON - Maybe audition -- if we knew more about it. It is perhaps a little too esoteric to be dependable. As far as thermal sensation, here again, I wouldn't rely on that if you have to worry about where you are going to feel it. When we worked at Rome (Griffiss Air Force Base) in the big chamber, you could feel the microwaves when walking into the chamber at 165 mW/cm^2.

VOGELMAN - Basically, you can hear radars only if you have no other auditory stimuli. If you are in the presence of ambient noise, you really don't hear it.

LIN - In many of our experiments the cat could perceive the pulsed electromagnetic energy as an auditory sound as determined by electrophysiologic means with the ambient noise levels up to 65 db. This information is included in our paper which was presented.

VOGELMAN - Cats have more efficient auditory sensors than humans. There is no question about it.

LEHMANN - Our staff has listened to microwave beams. A very nasty click is heard, which is very distinct from the ambient noise.

VOGELMAN - If you are scheduled to listen for it, you hear it. What I am saying is, I have been in the presence of radars when I wasn't listening for the radar, and I lost the click in the ambient noise.

LEHMANN - Isn't it a question of intensity?

VOGELMAN - It is a question of recognition.

LEHMANN - It is so nasty. I have never heard such a nasty click in my head before. This is my personal perception, and I think all the rest heard it the same way.

LIN - Once you hear it there is no question about it. It is very distinct. It is different, as Dr. Michaelson pointed out. Most of the time it seems to originate from somewhere behind or within the head. Individual clicks could be heard just as clearly; a number of pulses compressed together. In this case, a chirp is audible.

VOGELMAN - The question I am discussing is not whether you hear it, but whether you can use it as a safety.

MICHAELSON - I would not feel completely secure about it.

VOGELMAN - The problem in that experimental set-up is that you had those kinds of noises present all the time. If you are looking for a distinctive sound, then I think you can hear the radar signal. If you are not looking for it, if there are clicks all around the place -- you can't use it as a safety warning device.

OSEPCHUK - You people are talking about what you heard under power densities that were quite low. What was the average power density?

LIN - The average incident power density required varies with the individual. The highest one ever used was about 1 mW/cm^2.

OSEPCHUK - I didn't know you went that high. Presumably, you didn't go much higher than that also, did you?

LIN - You are right. I didn't.

OSEPCHUK - The point is that if somebody were in the field of 10 mW/cm^2 average or presumably a 100 mW/cm^2 average where you might really get heated, what would the intensity of sound be like, or what would the sensation be like? If it is according to Dr. Lehmann, "a little bit nasty" to listen to right now when

OSEPCHUK - he discerns it, with a 10 or 100 times higher power density, what would the sensation be then?

VOGELMAN - I am still concerned about the warning capability. If somebody is walking into a dangerous area and you shout at them, "Stop!" they just keep walking anyway. I just don't think it is useful as a device for early warning for people.

OSEPCHUK - It is not a question of device for early warning. It is another artifact that should accompany high fields. And, it is very important to know what we are differentiating.

VOGELMAN - I think in high fields out in the open area - in front of radars which I didn't intend to get myself in front of - I have never really been stimulated enough to recognize that I was hearing them. Maybe it's me.

LEHMANN - Do you talk about 100 mW/cm^2 average?

OSEPCHUK - Average, yes.

LEHMANN - Then, you are already about 50 mW/cm^2 away from cataract formation. You are right on the edge.

OSEPCHUK - If you stay forever. In military situations, people occasionally claim that they have stepped out in front of the antenna when it was on.

VOGELMAN - Or it came by when you weren't looking.

OSEPCHUK - OK, and, if they did this, and, of course, as far as cataracts are concerned at those power densities you will have to stay there at least a half-hour. They may have been exposed for only thirty seconds. But in the thirty seconds one should at least perceive a strong auditory sensation if the thresholds are 20 or 30 db lower in energy than the pulses that are being experienced at that situation.

HARDY - An interesting characteristic of the perception of microwaves as heat is that the sensation lasts so long. If one put his hand in front of the microwave horn and then removed it, he will feel a sensation of marked warmth, which persists for a couple of minutes after removal of the hand. In terms of the current theories about temperature sensation, this seems to me a complete mystery. Neither Dr. Lele's thermocouple theory, nor the old gradient hypothesis seem to offer a satisfactory explanation; because the skin temperature is falling, and the gradient is developing in the wrong direction. I think there is a lot more to be done using microwave apparatus in terms of trying

HARDY - to elucidate the mechanisms of thermal perception.

MICHAELSON - That is an interesting point. I always thought it is because of the gradient that is set up in the skin. If you take a rat and expose him the temperature still goes up for quite awhile after you stop the exposure, when using core temperature as the indicator. There is some re-distribution of the temperature. It continues for about 5 minutes or longer. In dogs, on the other hand, you don't see delayed heating; as soon as you stop the exposure, they start cooling. But in the rat, you continue getting this increase.

As far as using thermal perception as a safety factor it is not an unpleasant sensation if it is in the far field and you are working at about 100 mW/cm^2.

BEISCHER - I would like to mention the hearing sensation which has repeatedly been claimed by members of households where 60 Hz is used for utilities. This phenomenon was discussed in 1963 in a meeting at the Brain Research Institute, UCLA.

We have exposed human volunteers to magnetic fields at 50 Hz and field strength of 1 gauss and the subjects reported no hearing sensation. However, Otto H. Schmidt and R. D. Tucker described the perception of moderate strength low frequency magnetic fields by man in an IEEE publication in 1973. I would also like to mention exposure of man to low frequency, high tension electrical fields at the University of Munich. The subjects were systematically exposed to 50 Hz alternating fields at 20,000 volts and reported no auditory or other sensation of the field (R. Hauf and J. Wiesinger. Biological effects of technical electric and electromagnetic VLF fields. Int. J. Biometeor. $\underline{17}$: 213-215, 1973).

I would like to add that to my knowledge no nystagmus was observed in animals or man exposed to radar or low frequency electrical or magnetic fields.

LOWY - Well, nystagmus can, of course, be produced by application of DC.

BEISCHER - Yes.

LOWY - It is equally known that tissues have some rectifying effect. So, it would seem that if you put on a high enough field you can get vestibular responses on application of about 2 mA. With 4 mA, people can hardly stand it. They fall over and so forth. So, certainly what you say makes good sense. You would expect this under certain conditions.

LOWY - I don't know if the question of the possibility of stimulating, say, with audio frequencies - if there are any effects on anything besides the peripheral ear. And, at the present time, there has been accumulated a considerable amount of evidence, that it is so. It is possible nowadays to completely destroy the inner ear by means of toxic drugs; for instance, Kanamycin. Under these conditions it can be shown that direct electrical stimulation is capable of giving an electrophysiological response in certain auditory fibers. In other words, the peripheral auditory nerve, which is stationed next to the haircell, does not degenerate under all conditions when the haircell goes. Ototoxic influences are particularly important in that respect. This is of extreme practical value, because when people have no haircells they are deaf and no hearing aid can help them, of course. However, if it were possible to make use of direct electrical stimulation of auditory nerve fibers, one might try to help them a little in terms of recognizing, first of all, auditory rhythm, but also probably a little more than that. And, there are now several places where these things are being studied experimentally also on humans. And, there can be no doubt that it is possible to stimulate individual parts of the peripheral auditory nerve, and to cause some sound sensation. As a matter of fact to a certain extent there is even evidence that the individual sub-sets of the auditory nerve are tuned in a certain respect. The first reports of this, which came from clinicians, of course, were accepted with very great reluctance and suspicion. But, it seems that when you take very small electrodes and put them into the auditory nerve in a fashion so that a few of them, say, five or six, are in there, they are connected with different sub-sets of auditory nerve fibers. It is possible to give one individual electrical pulse and get some sort of ping sensation. These depend on which nerve fiber you stimulate. In other words, there is such a thing as is claimed by several people, particularly Simmons. It is definitely possible to use electrical stimulation of auditory nerve fibers in the absence of the inner ear, which would exclude any possibility of what is classically called the electrophonic effect. Because electrophonic effect simply means that you have an electromechanical phenomenon leading to vibration of either all or parts of the peripheral ear.

Ultrasound—Biological Effects

ARE CHROMOSOMAL ABERRATIONS RELIABLE INDICATORS

OF ENVIRONMENTAL HAZARDS?

John R.K. Savage

M.R.C. Radiobiology Unit
Harwell, Didcot, Oxon.
England

ABSTRACT

A very large number of agents and treatments are now known to produce structural chromosome aberrations. The types produced appear to be qualitatively (but not quantitatively) identical, and can be accommodated for scoring purposes within the classifications developed for aberrations produced by ionizing radiations.

The majority of the cellular test systems have intrinsic problems which need to be appreciated when planning and interpreting experiments.

Given that a treatment induces aberrations, the question of their significance must be considered at at least three levels; that of the cell, that of the organsim, where significance will vary from tissue to tissue, and that of the organism's progeny. Extrapolation between these levels is often very difficult.

The actual resolution afforded by visually detectable aberrations is extremely poor in molecular terms, so that their absence after a particular treatment cannot be used as a criterion for safety.

INTRODUCTION

The very obvious structural changes which can be induced in chromosomes, and the relative simplicity of the cytological methods which render them observable, have made "chromosomal aberrations" a popular criterion for the evaluation of potential environmental hazards.

Since the bulk of the genetic information which controls the function of the cell is located in the chromosomes it is natural to assume that the integrity of these bodies is vital to the well being not only of the cell, but also of the organism. The validity of this assumption is borne out by a wealth of experimental and clinical data.

If it can be shown that a treatment or substance induces structural changes in chromosomes observable at the microscopic level, it is a fair inference that it is also capable of inducing changes below the visible level, possibly right down to the molecular level. In this case there are grounds for suspecting that the treatment or substance constitutes a genetical hazard.

The failure to induce visible aberrations, however, does not necessarily indicate that a treatment is "safe". As will be shown below, the limits of observable chromosomal alteration are extremely gross measures of damage when considered in molecular terms. Moreover there are complications in the modes of action of some substances such that a particular test system may be incapable of yielding aberrations.

Even if aberrations are produced, there still remains the question of their significance, and this must be considered at at least three levels; that of the cell, of the organism, and of the organism's progeny. It is often very difficult to predict the effects and consequences of a treatment given _in vivo_ from observations made on isolated cells _in vitro_.

Before considering this question of significance, there are certain "hazards" in the experimental test methods which need to be mentioned.

ABERRATION CLASSIFICATION

All agents known to produce true structural changes in chromosomes produce qualitatively similar kinds. There are differences in relative frequencies of the various types, and sometimes differences in localization to specific chromosome regions, but basically the standard classifications (e.g. 4, 12, 17) which have been developed for use in ionizing radiation studies will cover all the types likely to be found. One of the great problems in this field is the propensity of workers to invent their own categories for aberrations, and this can make comparison and collation of data extremely difficult.

Care should be taken to distinguish clearly between chromosome-type changes, where the lesions were produced and interacted before the chromosome region duplicated, and chromatid-type, which are induced predominantly in duplicated or duplicating chromosomes. This distinction (provided it is coupled with a knowledge of whether the cell being scored is in its first, second or later divisions after treatment) can tell us much about the mode of action of the agent. For instance, the majority of chemical agents act during the DNA synthesis phase (or rather, the lesions they produce give rise to structural changes during DNA synthesis), and produce only chromatid-type changes (8). Thus a test system containing only post-synthesis cells will not show any aberrations at the first post-treatment mitosis with such agents, although subsequent divisions may carry heavy damage. Conversely, a test system with cells in pre-synthesis phase at the time of treatment may show chromatid-type changes at first post-treatment division since all cells must pass through DNA synthesis allowing the lesions induced before synthesis to mature into aberrations (18). One of the puzzling features of the initial reports of chromosomal structural changes induced by ultra-sound (9) was the predominance of chromatid-type changes although a pre-synthesis phase population (unstimulated blood lymphocytes) was sonicated. This led some later workers to suggest that sonication had induced secondary toxic substances into the medium and these were then acting as the mutagen (5).

A second important distinction which must be made is that between "breaks" (true discontinuities (12)) and "gaps" (achromatic lesions (12)). This distinction is not always easy to make at metaphase, and the "break" frequency should always be checked at anaphase since "gaps" do not lead to acentric fragments (6, 12). Gaps are of very frequent occurrence, and represent a heterogeneous collection of events, but all the evidence indicates that they are a transitory phenomenon, and do not give rise to aberrations at second or subsequent divisions (6, 21). When this distinction is made, it is usually found that "simple breaks" are a relatively uncommon aberration, and by far the majority of fragments are compound, being derived from interchanges and certain types of intra-change.

THE CELL SYSTEM USED

Usually the cell systems used for tests are chosen for their convenience and versatility rather than for their relevance to hazard assessment, and this raises problems of extrapolation within and between organisms.

The majority of established mammalian cell lines bear very little resemblance to the diploid cells of the original organism, and some at least were derived from tumorous material in the first place. Nearly all of them show karyotypic instability both as regards number and structure. Such instability renders them of doubtful value for aberration work of any kind, for one can never be certain that treatment is not enhancing the already existing mechanisms of aberration formation.

Agents added to the medium, or treatments given in vitro may reach the cells in a way, at a concentration, and frequently in a form most unlikely in vivo. Reactions may occur in the very complex culture media to enhance or mitigate the treatment.

It is often overlooked that the powers to alter substances (and in many cases detoxify them) by the gut, the liver and some other tissues of the intact organism are extremely efficient (22). It follows therefore that parallel in vivo tests should never be omitted.

Culture in vitro is in any case a very unnatural environment. When the vast range of substances and treatments claimed to induce aberrations is considered, and also the fact that the aberration types produced are so very similar, the possibility exists that aberration formation is a rather non-specific end-point and in some cases more indicative of cell "unhappiness" than of mutagenic activity.

THE SIGNIFICANCE OF ABERRATIONS

The above are just a few of the factors which need to be remembered in aberration work. We turn now to consider the question of significance.

Significance to the cell

Are all aberrations lethal? The simple answer is no, but this answer must be qualified, for it depends upon the type of aberration and the type of cell involved.

Clearly a cell that is not going to be called upon to divide can sustain very large amounts of chromosomal damage without apparent functional impairment. This may be deduced by the resistance of such cells to known aberration inducing agents like radiation and also from the fact that unstable aberrations may be found in cells stimulated to divide many years after treatment (21). Dividing cells on the other hand are very susceptible to the mechanical problems and fragment loss occasioned by the asymmetrical structural changes.

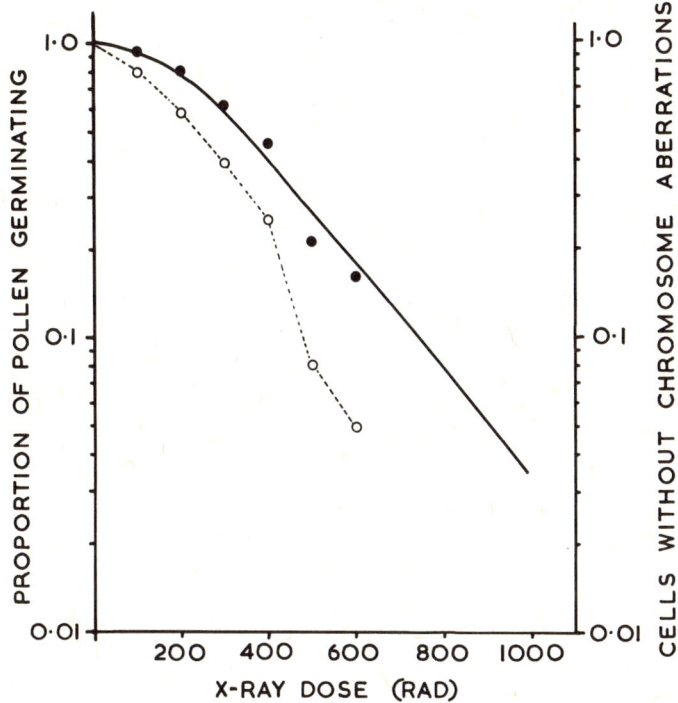

Figure 1. <u>Tradescantia paludosa</u>. Correlation between loss of germinating ability of pollen grains, and cells without chromosomal aberrations at the first pollen grain mitosis. Line; proportion germinating when sown 8 days after X-radiation (data from 11). Closed circles; cells without dicentric or ring chromosome-type aberrations. Open circles; cells without interstitial deletions (data from 14 and SAVAGE, unpublished). The cells can tolerate some loss of interstitial fragments.

The established heteroploid cell cultures referred to above can withstand some segree of fragment loss without impairment of their reproductive activity. In Syrian hamster, truly diploid cells (untransformed) cannot withstand any loss of chromosome type fragments of sufficient magnitude to form a micronucleus, apparently regardless of where that fragment comes from in the diploid karyotype (7). A similar picture emerges from the work of Preston on suppression of germination of pollen in <u>Tradescantia paludosa</u>, where the loss of any major chromosomal fragment (again from anywhere in the haploid karyotype) is sufficient to stop germination. In this system, we can also test the loss of small interstitial fragments ($\sim$ 1μm) and it appears that loss of these (even though any one represents loss of as many as $\sim 5.5 \times 10^8$ nucleotide pairs (15)) does not always involve suppression of germination (Fig. 1).

Both these instances are for chromosome type aberrations. We have no corresponding information for the immediate effects of loss of chromatid fragments, but even if such loss is not lethal after one cell generation, it will certainly become so in second or subsequent ones. In either case, the resulting cell death rapidly removes unbalanced asymmetrical changes from the population (3).

It would seem therefore that a truly diploid cell cannot afford the loss of any major part of its chromosomal material in vitro. This may not necessarily be true for a highly specialised cell in vivo.

Symmetrical changes, (translocations and inversions), unless they are incomplete, do not give rise to fragments or mechanical difficulties at mitosis, and provided there is no concomitant genetic loss in the cell, they are likely to be transmitted intact to subsequent (somatic) cell generations.

Significance for the organism

This must be considered from the short term and long term points of view.

In the short term extensive cell death requires replacement, and whilst that death is in itself an efficient selection process against unbalanced karyotypes, replacement cells may carry considerable rearrangements and some small deficiencies.

Fig. 2 shows the analysis of a fibroblast cell cultured from a skin biopsy from a woman of 78 who has received therapeutic radiation. 95% of the cells which grew out from the biopsy contained one or more visible symmetrical translocations, yet histologically the sectioned skin appeared normal. This cell was one of a clone which had obviously arisen by cell division in vivo. The interesting fact is that this patient received her therapy for acne 60 years ago. The dose is unknown, but it was large for she received weekly treatments for over a year.

These findings suggest that in the short term, these changes have not been significant for the cells have apparently functioned satisfactorily.

Of course this might not have been the case had the cell been

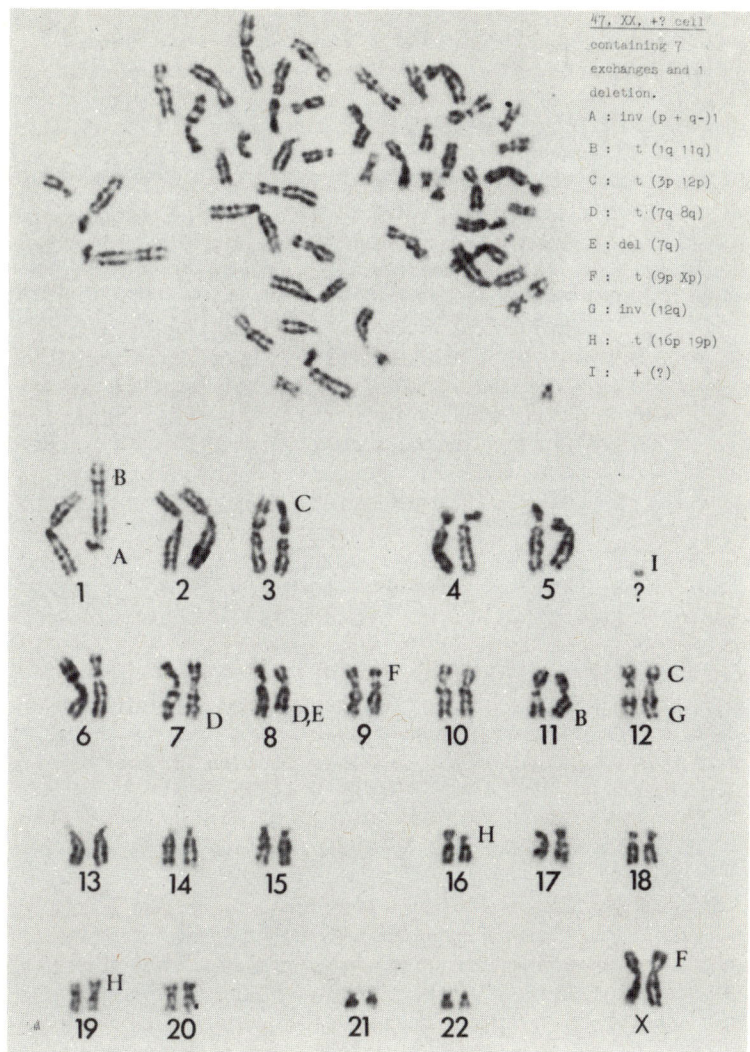

Figure 2. Aberration analysis of a human skin fibroblast cultured from a 75 year old woman who received radiotherapy 60 years prior to the biopsy. There are 7 detectable symmetrical changes, a deletion and an unidentified fragment. This cell was from an in vivo clone (WATSON, BIGGER and SAVAGE, unpublished data).

in another organ, and certainly would not had it been in the germ line.

This cell can be used to illustrate another point, namely

the very limited resolution which chromosomal aberrations can
give us in molecular terms. The chromosomes have been "banded"
by the ASG technique (20), and we can detect the loss of a dark
band (63 In the Harwell nomenclature (2), del(7)(q32) in the
Paris standardization (10)). This represents about the limit
of aberration detection with conventional light microscope techniques. What does this represent in terms of DNA? The diploid
human karyotype has $\sim$ 5.8 pg of DNA (1). If we assume this to
be in B configuration there will be 10 nucleotide pairs per
3.4 nm (13), with an average molecular weight (including shared
phosphates) of 616 per pair. This means a total DNA length of
1.929×10^9 nm (1.93m). The Harwell ASG banding norm has 1304
bands in the chromatids of the diploid female karyotype, and if
we make the further assumption that this DNA is distributed
uniformly between bands (which in reality is most unlikely) the
average length per chromatid band is 1.479×10^6 nanometers
which could correspond to a lot of genetic information. It is
not surprising therefore that we do not find many cells surviving
in skin lacking a band! The main point I wish to stress however
is that there is room for a large amount of significant structural
and genetic change well below the detectable level, so that the
absence of obvious change is no criterion for safety.

From the long term point of view, the hazard is neoplasia,
and although there is no direct evidence that structural chromosome
change is a causal factor, nevertheless many agents known to be
carcinogenic are also able to induce chromosomal aberrations.

Significance for the progeny of the organism

Chromosomal changes in germ line cells present a number of
different problems, both from the points of view of induction and
transmission. The selection pressures against unbalanced karyotypes are much greater than in somatic cells, for there are the
additional stages of embryogenesis which constitute a severe test
of the integrity of the genetic information. Symmetrical changes
which as we have seen, can in some cases function satisfactorily
in somatic cells, are a problem at meiosis and lead to sterility,
or reduced fertility. There is evidence that even apparently
balanced but abnormal karyotypes which manage to get past the
meiotic barrier are still selected against in post meiotic mitoses
leading up to sperm formation (19)

The situation differs in male and female mammals since in
the latter, all potential ova are present at birth, and there
is no mechanism for replacement if permanent damage is done.
Moreover, oocytes are "resting" in an arrested state of early
meiosis (the actual stage differing between mammalian genera)

so that the potential susceptibility to aberration formation from a given agent, may be quite different for the two sexes.

Since these cellular systems are so specialised, it becomes an extremely difficult task to assess the likely effects by extrapolation from an in vitro system. However, we know that some numerical and structural chromosome abnormalities do get through to progeny, and many (though not all) have considerable clinical significance. It is therefore wise to assume that an agent producing damage in somatic cells is likely also to do so in the germ line.

QUANTITATIVE ASPECTS

I have confined discussion in this paper to the qualitative aspects of hazard assessment. The problems with which the experimenter is faced when it is necessary to quantitate the damage, e.g. to produce a dose-response curve, are formidable. This is particularly the case when there is an intrinsic variation in sensitivity between cells in the population, and the agent under test causes delay and perturbs the orderly progression of cells towards mitosis. Those wishing to follow up this aspect will find discussions in some of my earlier papers (15, 16).

CONCLUSIONS

Provided the biology of the test system is understood, and taken into account in the experimental design, the induction of aberrations can be taken as a valid criterion of genetic hazard. The problems of extrapolation and resolution, however, are such that the absence of aberrations cannot be taken as a criterion of safety.

REFERENCES

1   ABRAHAMSON, S., BENDER, M.A., CONGER, A.D. and WOLFF, S.: Uniformity of radiation-induced mutation rates among different species. Nature, New Biol. 245 (1973) 460.

2   BIGGER, T.R.L., SAVAGE, J.R.K. and WATSON, G.E.: A scheme for characterising ASG banding and an illustration of its use in identifying complex chromosomal rearrangements in irradiated human skin. Chromosoma 39 (1972) 297.

3   CARRANO, A.V. and HEDDLE, J.A.: The fate of chromosome aberrations. J. Theoret. Biol. 38 (1973) 289.

4   CATCHESIDE, D.G.: Genetic effects of radiations. Advan. Genet. 2 (1948) 271.

5   COAKLEY, W.T., HUGHES, D.E., SLADE, J.S. and LAURENCE, K.M.: Chromosome aberrations after exposure to ultrasound. Brit. Med. J. 1 (1971) 109.

6   EVANS, H.J.: Chromosome aberrations and target theory. pp. 8-40 Radiation Induced Chromosome Aberrations (WOLFF, S. Ed.). Columbia University Press, New York (1963).

7   GROTE, S.J. and REVELL, S.H.: Correlation of chromosome damage and colony-forming ability in Syrian hamster cells in culture irradiated in G_1. Curr. Top. Radiation Res. Quart. 7 (1972) 303.

8   KILHMAN, B.A.: Actions of Chemicals on Dividing Cells. Prentice Hall, N.J., U.S.A. (1966).

9   MACINTOSH, I.J.C. and DAVEY, D.A.: Relationship between intensity of ultrasound and induction of chromosome aberrations. Brit. J. Radiol. 45 (1972) 320.

10 Paris Conference (1971): Standardization in human cytogenetics. Birth defects: Original article series, VIII, 7. New York: The National Foundation 1972.

11 PRESTON, R.J.: Current research in neutron-irradiation effects on plant chromosomes. pp. 349-371, Symposium on Neutrons in Radiobiology. USAEC Conf.-691106 (1970).

12 REVELL, S.H.: The breakage-and-reunion theory and the exchange theory for chromosomal aberrations induced by ionizing radiations: a short history. Advan. Radiation Biol. 4 (1974) 367.

13 RICH, A.: Molecular structure of the nucleic acids. Rev. Mod. Phys. 31 (1959) 191.

14 SAVAGE, J.R.K.: Non-interaction of radiation-induced chromosome lesions in Tradescantia microspores. I. Fractionated x-ray dose studies. Int. J. Radiation Biol. 11 (1966) 287.

15 SAVAGE, J.R.K.: The use and abuse of chromosomal aberrations as an indicator of genetic damage. Intern. J. Environmental Stud. 1 (1971) 233.

16 SAVAGE, J.R.K. and PAPWORTH, D.G.: The effect of variable G_2 duration upon the interpretation of yield-time curves of radiation-induced chromatid aberrations. J. Theoret. Biol. 38 (1973) 17.

17 SAX, K.: Types and frequencies of chromosomal aberrations induced by x-rays. Cold Spring Harbor Symp. Quant. Biol. 9 (1941) 93.

18 SCOTT, D. and BIGGER, T.R.L.: The induction of chromosomal aberrations by sulphur mustard in marsupial lymphocytes. Chromosomes Today (Darlington, C.D., Lewis, K.R. Eds.) 3 (1972) 162.

19 SEARLE, A.G.: Mutation induction in mice. Advan. Radiation Biol. 4 (1974) 131.

20 SUMNER, A.T., EVANS, H.J. and BUCKLAND, R.A.: New technique for distinguishing between human chromosomes. Nature New Biol. 232 (1971) 31.

21 UNITED NATIONS GENERAL ASSEMBLY, 1969. Report of the United Nations Scientific Committee on the effects of Atomic Radiation, 24th Session, Suppl. 13 (A/7613) New York.

22 WILLIAMS, R.T.: Detoxication Mechanisms. Chapman Hall, London, (1959).

-DISCUSSION-

HILL - Is there any known case where some chemical or physical agent, which is known to be capable of producing mutagenic events, does not give rise to visible chromosomal aberrations?

SAVAGE - Yes. There are a number. Kilhman lists several in his book on chemical mutants ("Actions of Chemicals on Dividing Cells, Prentice Hall, N.J., 1966).

CZERSKI - You were speaking about the selection against chromosomal anomalies. But simple statistics show that about one percent of living beings in the human population shows chromosomal anomalies. So, it may somewhat depend on type of chromosomal anomaly.

SAVAGE - Certainly. Chromosomal anomalies do appear and persist that is a perfectly correct qualification. For example, chromosome 9 apparently can undergo a lot of internal changes, and people can lose the last two or three bands of chromosome X,

SAVAGE - without any apparent anomaly. However, when you actually analyze the number of potential changes that can be produced, and the number that actually have survived, there are very few that manage to survive.

CZERSKI - But there's also another thing, some agents may favor chromosome non-dysjunction or lagging.

SAVAGE - Yes, but that is outside the field of structural aberrations.

CZERSKI - Certainly, but I feel it's quite relevant to the question of whether any environmental factors act on chromosomes. What's your opinion about the chromosomes as a biological dosimeter for X-rays?

SAVAGE - Well, the peripheral blood system can be used and has been used successfully in this country by Brewen and Preston (Brewen, J.G., Preston, R. J., and Littlefield, L. G., Rad. Res. $\underline{49}$: 647-656, 1972), as well as in other places for a reasonable biological dosimeter. But it has tremendous problems: (a) because we know the kinetic situation within the lymphocyte culture is a very problematical one; and (b) it's really only applicable to a whole body exposure. The difficulty of the dilution factors and the other things if you have a partial body exposure, which is what most people have in accidents, renders it almost unusable.

CARSTENSEN - There is some controversy over whether or not ultrasound induces chromosomal abnormalities in animal tissue; in plant tissue ultrasound seems to produce chromosomal aberrations. A number of people have reported this. How do you tell in a plant tissue, like the meristem of a root, whether the chromosomal anomaly is mutation or aberration?

SAVAGE - Well, let's go back to the first point which you made. I think one's got to be very careful about the genuineness of some of these aberrations. The real criterion is do they lead to micronuclei. If micronuclei are present after division, that's the direct evidence that we've had fragments. And, you can guarantee that in nearly all cases the loss will be lethal in truly diploid animal cells. Plants, of course, have very much more DNA than mammalian cells, and consequently there is probably a large amount of genetic redundancy. I say that with my tongue between my teeth. This might be borne out by the fact that the Tradescantia data I presented indicated losses of paired minute chromosome fragments amounting to 8×10^7 base pairs (which is a lot of information) which does not suppress survival, as measured by ability to germinate. The criterion

SAVAGE - I have used as evidence for aberrations is the presence of micronuclei. I have looked at a lot of Vicia cells that were ultrasonicated, and have never seen any micronuclei.

CARSTENSEN - But you have seen aberrations?

SAVAGE - Not true, conventional structural aberrations.

MILLER - But you are looking for and scoring well-spread metaphases. Would you say that there could be other types of aberrations, let's say stickiness, that could be scored?

SAVAGE - Yes, but those types of anomalies are not structural aberrations in the sense that I've defined.

DUNN - Have you looked at plants as well as mammalian cells?

SAVAGE - Yes.

DUNN - You are including plants in that statement?

SAVAGE - Yes. I have not been involved myself in direct experimental work with ultrasound, but I have had a lot of slides sent to me for scoring and comment from various people who have been involved.

NYBORG - There have been reports (Interaction of Ultrasound and Biological Tissue, DHEW Publication (FDA) 73-8008, 9/72; R. M. Schnitzler, pp. 69-72 and R. D. Brock, W. J. Peacock, G. Kossoff and D. Robinson, pp. 83-86) of chromatid breaks in animal cells exposed to ultrasound in solution. Are gaps formed?

SAVAGE - There is no evidence that gaps are of any significance.

NYBORG - Do you have some idea why that might be true?

SAVAGE - Gaps increase in frequency as you get closer to division. Now, the chromosome that we see in metaphase is a highly coiled structure. And, if you treat cells before you examine them with uncoiling agents, you can see this coiled structure of the chromosome. It may well be that a chromosome, having duplicated itself, is then packing up preparatory to division. If the "packing" is interfered with, errors or failures in coiling could lead to the appearance of gaps. But we can follow these cells through to the next generation, when you might expect a great burst of aberrations moulting from the high gap frequency, and they don't come. The gaps just disappear.

LEHMANN - Of those changes you have seen in chromosomes after

LEHMANN - ultrasonic exposure, can you tell us something about the mechanism by which they were produced?

SAVAGE - I can't tell you anything about mechanism or how they are produced even for ionizing radiation! This is a very difficult point. The biophysical evidence suggests that "lesions" in macromolecules of the size of a DNA helix are quite sufficient biophysically to account for aberrations we find. The difficulty arises in going up the many orders of magnitude from changes produced at the molecular level to those we observe at metaphase which appear to involve the whole of the coiled chromosome arm. The interesting thing is that a large number of those that are reported after ultrasound appear primarily to be chromatid-type. And this, I think, may be quite significant, because chromatid-type changes are relatively common spontaneously.

One has to remember that the aberration you see and score is only seen some time -- a long time in biochemical terms -- after the incident which initiated it. We can't see them immediately, not at least with standard microscopic techniques, so we have to guess the intermediate stages.

SUESS - You suggested that these changes will kill the cell. So, if the number of these changes is small relative to the overall number of cells, you shouldn't really worry about them too much. Is this correct?

SAVAGE - No, I think this is a matter of degree. It depends where the cell is. Let's suppose it was in liver. It may well be that a large amount of chromosomal damage might be quite irrelevant to the functioning of my liver. Whereas the same amount, or an even smaller amount, situated in the germ cell line, might be passed on to the progeny and be extremely important.

SUESS - But you have regeneration to some extent and bad cells may be replaced without any damage to the overall function of organs.

SAVAGE - But aberration bearing cells may be called upon for replacement. As I showed you in skin such cells are being called upon to replace the dead cells. Ninety-eight percent of the cells that grew out of that biopsy carried some detectable changes like those that you saw. But as far as the functioning of the fibroblasts was concerned, whatever they do in the skin, this was apparently normal. I think it depends entirely where your anomaly-bearing cell is. And, there's a great difficulty in extrapolating from what you see in the test tube to what will happen if this change occurs in

SAVAGE - the body. That was my main point about the selection that I brought up. Things do get through, but the selection pressures are very high in the sex line.

CZERSKI - The chromosomal pictures obtained from spontaneous abortions, show a whole range of various chromosomal anomalies. We are led to suppose that abortion is somehow connected with chromosomal changes. From a humanitarian point-of-view, it is certainly better to abort than to have a defective child. Nevertheless, on a scale of a population, this is a consideration to be taken into account. So, it all depends on your point-of-view. Yao and Jiles described a rather interesting phenomenon related to exposure to microwave irradiation. Some chromosomal changes appeared suspiciously like changes in coiling, especially in metaphase and metaphase chromosomes. A similar phenomenon was obtained by microwave exposure, this time not of animals, but of human lymphocytes in vitro (Stodolnik Baranska, Biologic effects and health hazards of microwave radiation: proceedings of an international symposium. Polish Medical Publishers, Warsaw, 1974, pp. 189-195. See also Nature 214: 202, 1967). It appears that such changes can be obtained often while in vitro. I am just wondering what your opinion would be about possible consequences in changes of chromosomal coiling.

SAVAGE - Presumables that could easily lead to abnormal disjunction or to chromosomes sticking together, and so on. That, of course, would be lethal in a diploid cell. Plant cells can tolerate numerical variation to some extent; much more so than animal cells. But the normal diploid cell can't tolerate an additional chromosome, other than, say, mongolism and so on. You never get three 1's or three 2's except in some aborts you see.

CZERSKI - Yes, but there are reports of three 17, three 16, and trisomy in the D group. So, I think we have a whole range of anomalies to select from.

LELE - Would any mechanical forces, such as high acceleration rates and shear stresses induce chromosomal anomalies?

SAVAGE - It would produce physical breakage, I am quite sure. Whether the physical breakage is related at all to the aberrations that are produced by chemicals or X-rays is a moot point. But they certainly will produce physical breakage of the chromosomes.

LELE - So that might be one reason why we might find more abnormalities in cells irradiated in suspension than in a monolayer?

SAVAGE - These, of course, can be distinguished though. If you find a cell in which the only type of aberration are pieces broken off, and there are no exchanges or exchange-type aberrations, then I think you would suspect it very much. Because, in reality, when you look very carefully simple "breaks" are (i.e., true terminal deletions) rather rare, certainly very rare in plant cells. Nearly all the fragments that are found are compound fragments that have been derived from exchanges.

NYBORG - In aqueous solutions ultrasonic cavitation leads to rather generous supplies of free radicals. These will probably cause chromosomal changes, don't you think?

SAVAGE - Yes, I suppose so. It depends whether the effect is an indirect or direct one.

ALTMAN - I think indirect mechanisms via radicals are important. Free radicals do act as significant agents, and there is every reason to expect that kind of an action when ultrasonic cavitation occurs, similar to the events in the presence of ionizing radiation.

HILL - I think there is an interesting point here, but it doesn't necessarily follow. In the ionizing radiation case, the free radical is deposited probably very close to the chromosome, whereas in the cavitation situation, the free radical is being formed as far as we know in the cavity and is then being released to the medium where it is subject to chemical interaction with components of the medium. So, it has then got some way to go to before it can produce its action on the chromosome.

ACTION OF ULTRASOUND ON ISOLATED CELLS AND CELL CULTURES

C.R. Hill

Physics Division, Institute of Cancer Research

Royal Marsden Hospital, Sutton, Surrey, U.K.

ABSTRACT

Understanding of the experimental evidence on the effects of ultrasound on living cells calls for an appreciation of the various biophysical mechanisms that may be involved. These include heat, cavitation in its various forms, microstreaming and other possible but unsubstantiated mechanisms. Biological endpoints that have been studied in relation to these mechanisms include cell disintegration, modification of proliferation patterns, genetic mutation, induction of chromosomal aberrations, modification of the cell membrane and of intracellular organelles and modification of cell function and radiosensitivity.

Tentative, concensus conclusions from published experimental data in this field are that : (1) mutation and chromosomal abnormalities have not been consistently demonstrated to result from ultrasonic exposure, (2) for producing cell death and a variety of reversible and irreversible non-genetic cellular modifications, cavitation can be a very effective mechanism, (3) other non-thermal mechanisms may operate but have not been consistently and repeatably demonstrated.

Human beings are more than cell cultures writ large and extrapolators operate in this territory at their own peril.

INTRODUCTION

In this paper the action of ultrasound on living cells is considered from a biophysical point of view. First the physical nature of the ultrasonic field is considered, to the extent that it may be expected to give rise to interactions with living cells. Following this is a discussion of some of the various biological parameters and endpoints that are available for an appropriate to quantitative investigations in this field. Finally the published experimental evidence on effects of ultrasound on cells is reviewed in the light of these physical and biological considerations.

PHYSICAL MECHANISMS

It is useful to try to identify and distinguish between various possible physical mechanisms which may be involved in the causation of biological change following ultrasonic exposure. For present purposes these may be taken as the following :

> Heat
> Cavitation
> Microstreaming
> Others

Of these, heat is specifically excluded from consideration in this paper by the session title under which it is presented.

The term "cavitation" describes a broad group of phenomena, by no means fully understood, which may occur in liquid media under various physical conditions, one of which is exposure to an ultrasonic field. Broadly speaking cavitation entails the creation in the medium of cavities, which may contain gas or vapour, their oscillation in various modes and eventual collapse or elimination (1). Its significance for living cells existing in a medium which can cavitate is twofold. In the first place the cavitation process can lead to local magnification, by several orders of magnitude, of the physical stresses experienced in an ultrasonic field. In particular, very high localised velocity gradients may be set up, with associated shear stresses that are sufficient to rupture biological membranes. Secondly, by a process that is probably associated with ionization taking place within the cavity, an abundant supply of highly reactive chemical free radical species may be released into the liquid. Of the two the former (mechanical) effect is generally of most biological significance as the chemical action tends to be buffered by constituents of the cell-supporting medium (2).

The physical conditions which control the occurrence and magnitude of cavitation activity are not known with great precision but some approximate data are available (3). The various parameters are interdependent but, in the frequency region of interest in medical ultrasonic applications, continuous wave beams can excite cavitation at peak intensities of about 1 W/cm^2 and above (at 1 MHz) with the intensity threshold rising with frequency. Cavitation activity is affected by pulsing, being inhibited by the use of very short pulses (of the order of a few tens of microseconds). It is also related to the nature of the medium and, at least for moderate ultrasonic intensities, appears to be inhibited in gels and intact human tissues. Monitoring and measurement of cavitation activity is an important but imperfectly developed branch of technology (4).

Microstreaming is a phenomenon that may be associated with cavitation but can also be produced more generally in the presence of any non-uniform ultrasonic field, such as that in the vicinity of the tip of a vibrating needle. An interesting field of ultrasound biology has grown up around this possibility for producing controlled shear gradients but it has proved difficult to extrapolate its results to conditions obtaining in intact tissues or suspended cells exposed to macroscopically uniform ultrasonic fields.

There has been considerable speculation as to the possible significance of other conceivable physical mechanisms that could be of biological significance. For sufficiently large molecular structures for example (molecular weight more than 10^6) the accelerative forces associated with even moderate ultrasonic fields will exceed those due to thermal forces (5), suggesting the possibility of some direct effect of stress. However, there is little systematic evidence available concerning the significance of this, or of other possibilities that have been suggested.

BIOLOGICAL PARAMETERS AND ENDPOINTS

In terms of the biological system whose response to ultrasonic exposure can be examined there is a very wide range of possibilities available, many of which have in fact been examined in some manner. However, in terms of the type of end effect that is to be observed, or parameter to be measured, there is less choice and it is useful to consider the possibilities that exist.

Cell death (or survival) is an endpoint that has been used very widely both in ultrasound and ionizing radiation biology but it is important to note a significant difference between the situation in the two cases. Immediately following a dose of x-rays the cell count is generally not significantly reduced and "survival" is measured as the fraction of cells capable of apparently normal proliferation. Here the cell "loss" is a measure of the proportion of cells whose proliferative mechanism is modified, amongst which may be concealed a small fraction of viable mutants. By contrast, following an ultrasonic exposure, cell death generally appears to be an immediate and unambiguous process.

Partly for this reason there is interest with ultrasonic exposures (as indeed with x-rays) in looking for the possibility of more subtle changes that might reflect modification to the proliferative mechanism of the cells. The techniques of cell kinetics are well suited to this purpose.

The mere death of a limited number of cells in a population seldom constitutes a significant hazard for an organism. At least as far as isolated cell biology is concerned the principal endpoint that will be of relevance as an indicator of possible hazard is that of genetic mutation, and this is also the principal guise in which potential carcinogenicity of an environmental agent can be expected to show itself in studies on isolated cells. Genetic studies are technically difficult with mammalian cell systems however and the observation of induction of chromosomal abnormalities, which appear to be associated with, although not quantitatively correlated with, genetic change, provides a useful alternative approach to this problem.

Whatever may eventually be shown to be the particular biophysical modes of action that it induces, ultrasound is primarily a mechanical agent and it is thus natural to look for possible structural changes in cells that have been irradiated. A number of techniques appropriate to such studies exist. Structural modification to a cell membrane may lead to changes in its ionic permeability characteristics and also to changes in electrical properties, which may be detected by means of cell electrophoretic measurements. On an intracellular scale various techniques exist for studying changes to subcellular organelles, including the direct approach of electron microscopy.

Finally, a variety of techniques are available for looking for possible changes, either reversible or irreversible, in functional and general properties of cells consequent on ultrasonic irradiation. Mechanical movement of flagellates and x-ray radiosensitivity are two examples of such specific endpoints.

In the choice of a cell system with which to study a particular biological endpoint a number of different criteria need to be considered. In the first place most ultrasound biology is man-centred and mammalian cell lines will generally be of most relevance. Where, for technical reasons it is necessary to work with microorganisms, the choice of cell types with mammalian-like properties is often possible.

To obtain satisfactory physical conditions of irradiation it is generally arranged to irradiate cells in liquid suspensions. A convenient mammalian system for this purpose is the fluid ascites tumour cell system but this has the disadvantage of being a very heterogeneous cell system and one that cannot be cultured in vitro following an irradiation. For this reason cultures of cells which tolerate suspension culture are often preferred. When an important experimental criterion is that cavitation activity should be eliminated there is an advantage to be gained from using cells that will tolerate, for a limited period, suspension in a gel.

EXPERIMENTAL DATA AND INTERPRETATION

Cell Death

The evidence that ultrasonic exposure can lead to cell death by disintegration is unequivocal and there is no doubt that cavitation is an important mechanism in the process. What remains in some doubt is whether ultrasound can cause cell death, either by disintegration or otherwise, without the mediation of cavitation. There are a number of reports in the literature indicating that this may be so but, in view of the technical difficulty, referred to above, of monitoring cavitation activity, these must be treated with some reservations. Where such investigations have been carried out systematically (3,6) a clear correlation between cell death and cavitation has generally been found.

No clear evidence for delayed death has been reported. In one experiment designed to show up possible changes in the cell proliferation pattern, such as are found following ionizing radiation exposure, no changes could be found following ultrasonic irradiation in which cavitation had been inhibited by use of very short pulse durations (7). An interesting subsidiary finding of this experiment however was that, when cavitation was allowed to occur, cell disintegration occurred preferentially during the mitotic phase of the cell cycle.

Mutation

From the hazard point of view this is the most important endpoint for consideration and it also seems to be one that abounds in technical pitfalls. The evidence in the literature for ultrasonic induction of genetic changes has recently been well reviewed (8) and the conclusion drawn that, with the possible exception of situations where appreciable heat shock could be involved, there is little likelihood that genetic effects will be found to result from exposures to ultrasound at the levels currently used in medical applications. In one recent experimental investigation an attempt was made to test the effect of extremely severe exposure to ultrasound by looking for evidence of genetic change in the survivors of cell populations exposed to cavitating ultrasonic fields (9). Here changes were only found where appreciable heating or free-radical formation occurred.

Chromosomal Aberrations

Considerable effort has been devoted to this line of investigation over the past few years, stimulated in particular by a report of positive effects following rather low intensity (8 mW/cm^2) exposures (10). A number of useful follow-up studies, including some attempts by the original author to repeat his own work (11), have failed to confirm the original findings and, although this may continue to be an interesting and important area of investigation, the current concensus of evidence here is overwhelmingly against the existence of any effect, at least under medical exposure conditions.

Structural Changes

A number of investigators have looked for evidence of structural changes, not necessarily lethal, which may occur in ultrasonically irradiated cells. The cell surface is an obvious site here and Taylor et al. (12), by means of electrophoretic measurements on ascites tumour cells, have shown that changes occur in the electrical charge density, and thus presumably in structure, on the surface of treated cells. In this work there was some indication against cavitation as the effective mechanism but subsequent work using cultured mammalian cells (13) has shown that these changes are correlated with disintegration death of other cells in the exposed population, almost certainly occurring as a result of cavitation. The later work also showed that the changes involved are reversible and non-lethal.

Electron microscopy studies on organelles of treated cells have led to reports of modifications, and particularly of mitochondrial swelling, although quantitative systematic work in this area seems to be lacking. In this connection one group, in work on intact tissues, has reported evidence for disruption of lysosomes and consequent cell destruction by released lysosomal enzymes (14).

Modifications of Function and Response

This is potentially a very large and interesting field of investigation since, in general, the various functions and responses of cells may depend on many factors, any one of which may be open to influence by an agency such as ultrasound. One such response that has been investigated is that of x-ray sensitivity. Here a number of factors in the response, and notably that of the partial-repair phenomenon, would seem to be open to ultrasonic influence. Early reports on animal work suggested that there might be a synergistic effect between x-rays and ultrasound in relation to tumour regression but follow-up work using cultured mammalian cells appears to exclude non-thermal mechanisms in this effect (15).

A striking positive finding on functional change has been reported in the form of an action spectrum for the mechanical flagellation function of rotifers (polynucleated aquatic microorganisms) exposed to ultrasound in a wide range of frequencies (16). In two frequency bands only (around 270 and 510 MHz) movement was inhibited and this change was found to be reversible for pulse durations up to 30s but irreversible for much longer irradiations, acoustic intensities being of the order of 1 mW/cm^2.

CONCLUSIONS

The extrapolation of data obtained on isolated cells to derive conclusions valid for complex organisms is notoriously dangerous and the situation is further complicated in ultrasound biology by the phenomenon of cavitation, which can produce major biological effects and is strongly dependent on the physical characteristics of the irradiated mechanism.

Nevertheless, with this qualification, cellular ultrasound biology can be highly informative. In particular, genetic techniques can often be carried out with greatest precision and economy at the cellular level and there seems to be no particular reason for suspecting that any conceivable ultrasonically promoted genetic changes which could occur in intact tissue should not be revealed in appropriate investigations on isolated cells. Cavitation induced cell disintegration is a well established and common consequence of ultrasonic irradiation but detailed knowledge is still lacking on the physical parameters involved in its control, and particularly on its significance in *in vivo* human irradiation. Evidence for non-thermal, non-cavitational mechanisms of action is still sparse and unsystematic but this could prove to be an exciting field for future research.

The literature of ultrasound cell biology is large and sometimes inconsistent and it is impossible to do justice to all aspects of the subject in a review of necessarily limited extent.

REFERENCES

1. WEBSTER, E.: Cavitation. Ultrasonics 1 (1963) 39.

2. CLARKE, P.R. and HILL, C.R.: J.acoust.Soc.Amer. 47 (1970) 649.

3. HILL, C.R.: J.acoust.Soc.Amer. 52 (1972) 667.

4. HILL, C.R.: Detection of cavitation, in Interaction of Ultrasound and Biological Tissues p.199-200 DHEW Publication (FDA) 73-8008 (1973).

5. CLARKE, P.R.: Studies of the biological effects of ultrasound and of synergism between ultrasound and x-rays. PhD Thesis, University of London, England (1969).

6. COAKLEY, W.T., HAMPTON, D. and DUNN, F.: J.acoust. Soc.Amer. 50 (1971) 1546.

7. CLARKE, P.R. and HILL, C.R. Expl.Cell Res. 58 (1969) 443.

8. THACKER, J.: Curr.Topics in Radn. Res.Quarterly 8 (1973) 235.

9. THACKER, J.: Brit.J. Radiol. 47 (1974) 130.

10. Brit. J. Radiol. 45 (five related papers) (1972) 319-342.

11. MACINTOSH, I.J.C.: (personal communication).

12. TAYLOR, K.J.W. and NEWMAN, D.L.: Phys. Med. Biol. 17 (1972) 270.

13. JOSHI, G.P., HILL, C.R. and FORRESTER, J.A.: Ultrasound in Med. and Biol. 1 (1973) 45.

14. TAYLOR, K.J.W. and POND, J.: Primary sites of ultrasonic damage on cell systems. Interactions of Ultrasound and Biological Tissues. p.87-92, DHEW Publication (FDA) 73-8008 (1973).

15. CLARKE, P.R., HILL, C.R. and ADAMS, K.: Brit. J. Radiol. 43 (1970) 97.

16. DUNN, F. and HAWLEY, S.A. Ultrahigh frequency acoustic waves in liquids. Ultrasonic Energy (Kelly, E. Ed.) p.66-76. (Univ. of Illinois Press) (1965).

-DISCUSSION-

ALTMAN - How important are the formation of radicals and ionization processes in the action of ultrasound, particularly with regard to any impact on the medium like water to form radicals of water molecules such as occurs in the case of ionizing radiation?

HILL - I think one can't be certain on this, but the general picture is that if you compare, for example, an electron generated from an X-ray beam, and a cavitation bubble, the process of free radical production from the X-ray beam in the vicinity of the electron is going to take place very close to the DNA or the chromosome or whatever you are interested in as a target.

In the cavitation situation, it is going to take place inside the cavity. And, it is only going to reach the "target"

HILL - by some subsequent diffusion. I would guess if the cavity is too close to the target structure, the target structure is going to be disintegrated anyhow. So there is probably a threshold diffusion distance to work on. That is sort of a theoretical answer. A practical experimental answer is that we have looked for this, I am sure we have not done a perfect experiment, but we have convinced ourselves that the mechanical effect of the cavitation is the predominant one, and that it is very hard indeed, if at all possible, to pick up free radical action.

MILLER, M. - We recently published in the British Journal of Radiology (47: 122-129, 1974) an article dealing with the biological effects of 2 MHz ultrasound on the growth and cytology of Vicia faba roots. We observed a variety of perturbations: (1) a reduced growth rate following sonication, (2) reduced mitotic index which might or might not recover 24 hours post-sonication, (3) chromosomal aberrations of a type which are not normally reported and (4) in another article (Radiation Botany 14: 201-206, 1974) we have shown there is cell shattering in the region of the elongation zone (that zone behind the meristem) but not in the meristem.

I want to mention something about the scoring of the chromosomal aberrations. The "standard" screening process is to determine with a low magnification lens where the cells are on the preparation. Once the cells are found scanning of cells is done with high dry lens (450X). At this point, one is scanning the slide looking for something that one can score. Since normally metaphase chromosomes are scored for aberrations one normally looks for well-spread, flat metaphases. When a metaphase cell is found a decision to score or not to score it is made. If a decision to score it is made one switches from high dry ($\sim$450X) to the oil immersion lens ($\sim$1000X). The point I want to make is that one automatically selects cells that are well-spread and easily scored.

Now, the types of aberrations that we have observed and reported from our ultrasound experiments in no way contradict the negative results that Dr. Hill has just presented. We do not see the normal breaks, deletions, exchanges -- i.e., the types of structural aberrations that Dr. Savage illustrated here this morning. But by scanning the slide under oil immersion, we have seen these rather peculiar mitotic figures which we have described as "bridged phases,""bridged metaphases" and "agglomerations."

Were we to have scanned our slides with a high dry lens we would have passed them by. So, as a word of caution, I don't think the question of whether or not chromosomal anomalies result

MILLER, M.- from exposure of cells to ultrasound is really settled because there may well be other types of structural arrangements that were simply not looked for.

LEHMANN - We found exactly what you described in onion roots associated with localized small lesions which were most severe in the center. In the periphery of the lesion the chromosomal changes were very similar or identical with the changes you described. We also found that if we repeated the experiment under pressure adequate to prevent cavitation, we found neither lesions nor chromosomal anomalies provided that the temperature rise was controlled.

SAVAGE - I don't think it is strictly true to say that scoring of chromosomes is as you have described it. If a cytologist scored like that he is going to be biased, because in any aberration scoring the largest damage is going to be in the worst spread cells. Therefore, you try and score everything. But it is correct to say that you do choose metaphases.

The types of aberrations you have described and illustrated are nearly always to be found in prophase or in dividing cells. They look as if the cells were in the process of division when the damage was inflicted. And, this gives us an indication of an immediate effect of ultrasound.

The irradiation of prophases and metaphases by ionizing radiation, unless you use a very high dose, doesn't produce any aberrations; it does produce stickiness, but it doesn't produce structural aberrations, which lead to fragments and loss. That is the only distinction I am making. What you found is genuine, I agree. I don't want to belittle it.

LELE - In the proceedings of the New York Academy of Sciences there was an article on the effects of microwave on dividing synchronized cells. In addition to retardation of growth in G_1 phase, they also found some sort of chromosomal aberration similar to what Dr. Miller described.

NYBORG - In respect to the MacIntosh and Davey work (e.g., I.J.C. MacIntosh and D. A. Davey, Brit. Med. J. 4: 92-93, 1970) I am sure there are lots of people who hope that that has all been laid to rest, but I am not completely sure it has been. MacIntosh has stated recently that he believes, under conditions where the first positive results were recorded, that there was cavitation occurring.

Two kinds of cavitation are worth distinguishing from each other; the transient type which is very violent and which would lead to destruction of the cells. And the stable type, a more

NYBORG - gentle kind, which is the more controlled variety.

Now, it seems to me the possibility is still present that in the original experiments of MacIntosh and Davey, by some accident, the stable type of cavitation was present and caused effects on the cells; the latter survived because the transient cavitation was absent. When transient cavitation is present it might simply kill all the cells so that no effects would be seen if they did occur.

Now, there is a possible relationship of this work to the results here at Rochester with the plant tissue. In the plant tissue there is intercellular gas. And, at least at intermediate levels the pulsation of that gas could be thought as a form of stable cavitation. So here is a possible common denominator for the effects in cells; namely, stable cavitation.

HILL - I am not convinced that this idea of a clear distinction between stable cavitation and transient cavitation is applicable at high frequencies, such as medical ultrasonic frequency. Certainly our evidence is that at any point where we can get any indication of cavitation at all, we can immediately start getting evidence of free radical effects and cell disruption.

If you end up a series of experimental observations on a cell culture system, and you fail either with ultrasound or perhaps microwaves to find any evidence for a genetic effect, how far can one say that this is a good reason for thinking that the intact human organism will not be subject to genetic changes? Are there any biologists who are brave enough to jump into that one?

STEWART - Yes. Dr. Kremkau has reported a reduction in mitotic index in partially hepatectomized rats at diagnostic levels. Does not this finding have some implications with regard to chromosomal aberrations or long-term genetic effects?

KREMKAU - We have not looked at chromosomes so we don't know whether there are any associated chromosome abnormalities. But what has been observed and reported is a reduction in mitotic index in rat liver which has been exposed to diagnostic ultrasound. A very general question that is hard to answer is, what does an effect in rat liver tell us about human tissues?

SAVAGE - Could I just add a word of caution here about the use of mitotic index? Mitotic index, on its own, doesn't tell us anything; it is simply the relative time spent in division, to the time spent in interphase. You can get a falling mitotic index with an increase or decrease in a cell rate (Evans, H.J.

SAVAGE - and Savage, J.R.K., Exp. Cell. Res. 18: 51-61, 1959).
And no change if things are increased or decreased proportionately. So one should always accompany mitotic index scores with some measurements of rate of entry into or exit from mitosis, i.e., some independent measure of cell rate.

KREMAKU - This mitotic index observation is made at one specific point in time following a surgical stimulation of the rat liver to go into mitosis. This could be interpreted in several says as an overall decrease in mitotic activity or as a shift in mitotic peak, (either speeded up or slowed down) to give a reduction at some point. It will take a lot of animals and effort to answer those questions. All we know is that we have a reduction at one point in time.

MILLER, M. - I wonder, Dr. Savage, if you'd be willing to discuss before the confreres the data results from your colleagues at Harwell who have been studying primary cell culture?

SAVAGE - This is from some work by Roger Cox at our unit (Int. J. Radiat. Biol. 26: 193-196, 1974) who has been trying to develop genetic systems using primary cell cultures. So they have been culturing untransformed cells. Once a culture method was established they did a survival curve and they didn't get any shoulder. It was a perfect exponential survival curve. So they repeated this, repeated it in several different ways and still didn't get a shoulder. They then did experiments involving fractionation to look for an Elkind effect; again no recovery. At this point they came to us and got some of our untransformed skin line and did the same again and they got no shoulder. So this, of course, is all against conventional radiation biology.

Starting with a new untransformed diploid line growing out from a biopsy they obtained survival curves at different cell generation times as the cultures grew on, and obtained no shoulder up to about the fortieth passage. Then, suddenly, a shoulder began to appear. The cells were still untransformed, but an examination showed that chromosome instability was beginning to appear. This time corresponds also to a noticeable slowing down in growth rate of the cultures, and also to the time found by others when transformation to give an established cell line occurs. So this does suggest that one has to be very careful, when you think about the cell survival curves from established cultures. As I showed this morning such cultures can survive quite a lot of fragment loss, which would be equivalent to sublethal damage.

Non-Thermal Effects of Ultrasound

On Intact Animal Tissues

K.J.W. Taylor[1] and M. Dyson[2]

[1]Royal Marsden Hospital, Sutton, Surrey, and

[2]Guy's Hospital Medical School, London

ABSTRACT

Investigations on the effects of ultrasound on intact animal tissues are complicated by mutual interaction between the ultrasound beam and the tissue. Most of the reported biological effects may be ascribed to either heat or cavitation. There remain a number of phenomena that are apparently due to other mechanisms although it is particularly difficult to completely exclude all possibility of cavitation. These effects include destructive changes in the central nervous system resulting in 'focal' brain lesions, and spinal cord injury resulting in paraplegia. Pulsed insonation of both liver and spinal cord produces vascular damage and this occurs more easily in the presence of hypoxia. There is evidence of dose accumulation in the production of these changes.

Stimulation of tissue regeneration by low intensity insonation rationalises a therapeutic use of ultrasound. There is evidence that this is a non-thermal effect. Occurrence of red cell stasis has relevance to the use of the energy form for measuring blood velocity since standing waves may easily occur in vivo.

Finally, the effects of ultrasound on developing chick embryos are reported. Continuous insonation with Doppler devices did not affect their subsequent development. When delivery of energy was pulsed, relatively long pulses (20 us) at a high p.r.f. (5000 s^{-1}) produced perverted development if applied during the early stages of organogenesis. An intensity threshold for this effect was found between 10 and 25 W cm^{-2}. Embryos at later stages of development were unaffected even by intensities as high as 100 W cm^{-2}. The results indicate a large margin of safety for the dose-parameters used in current applications.

Introduction

The use of intact animal tissues as experimental models greatly increases the complexity of the experimental method compared with the use of single-cell suspensions. However, it does permit the study of the reaction of the organism in terms of subsequent repair or regeneration. Intact animal tissues are highly heterogeneous, so one must consider not only the effects of ultrasonic energy on the tissue but also the effect of the tissue on the ultrasound beam. For example, in many situations in vivo, it appears probable that a standing wave component may be formed and this greatly modifies the biological effects of the beam. Thus, mutual interaction occurs within the tissue substance and the dose parameters as measured in vitro may have little relation to those existing within tissues. For these reasons, there is a need to develop small hydrophones and capacitor transducers to monitor the field parameters within the tissues even though the invasive technique involved with such detectors would further modify the field.

However, we are most concerned with the safety of diagnostic ultrasound devices and, in clinical practice, these are used on intact animal tissues and most frequently, on embryonic targets the cells of which are actively dividing and differentiating. Thus, results of animal experimentation under complex conditions are relevant to the clinical situation although the biophysical conditions are such that these experiments give little indication of mechanisms of interaction.

Before the relevant literature is reviewed, it is necessary to briefly consider the output parameters of diagnostic devices. These are classified as sonar, Doppler and holographic. The output parameters of such devices are shown in Table 1. The range in frequencies employed largely devolves from requirements for different applications but the variations of peak intensity depends more on the quality of the amplification system. It should be noted that there are two families of Doppler devices, one for obstetrical use in which the peak intensity is less than 50 mW cm^{-2} and the other for circulatory studies in which intensities up to 500 mW cm^{-2} may be used.

Cavitation is first considered as the best documented non-thermal effect of ultrasound. The reported intensity threshold for occurrence of cavitation in degassed water has decreased as methods of detection have improved. Fry and Dunn in 1962 (10) quote an intensity threshold of 1000 W cm^{-2} at 1 MHz while later authors reported a threshold of only 1.5 W cm^{-2} (15, 16). Subtler criteria were used for detecting cavitation including the degradation of DNA and detection of subharmonics, as well as the starch-iodine test. In gassy liquids and in the presence of other nuclei, the threshold would be correspondingly less--about 500 mW cm^{-2} at a frequency of 1 MHz. However, there is no

TABLE I.

Dose Parameters Used in
Varying Applications of Diagnostic Ultrasound

Sonar

    Frequency 1-6 MHz (up to 20 MHz for ophthalmology)
    Peak intensity 1-96 W cm^{-2}
    Pulse length <1 µs
    Pulse repetition rate (p.r.f.) 300-1500 s

Doppler for obstetric use

    Frequency about 2 MHz
    Peak intensity <50 mW cm^{-2}

Doppler for vascular studies

    Frequency 5-10 MHz
    Peak intensity <500 mW cm^{-2}

Holographic

    Marked variations between various machines.

evidence that this phenomenon occurs with the dose parameters used in diagnostic ultrasound devices. This may devolve both from the greater density of solid tissues and from the effect of blood flow in removing minute air bubbles from the sound field before interaction or oscillation occurs. There is also the important effect of pulse length: the threshold for cavitation is greatly affected by reducing the length of the pulse from 10 to 1 ms while no evidence of cavitation was found with pulses of only 20 µs duration (17).

Thus in the sonar regime, the peak intensities are above the cavitation threshold but the pulses are too short to allow occurrence of cavitation. In the Doppler applications, the peak intensities are generally too low although the intensities of the most powerful devices approach the cavitation threshold. No data exists on the possibility of cavitation using relatively long pulses at low repetition rates as are used in holographic techniques.

Non-cavitational Effects

1. Production of focal lesions in brain.

There is well documented evidence for a non-thermal damage mechanism producing focal lesions in brain tissue (5). Indeed this is one of the few described effects that has been confirmed by other workers (18,1). Pond (18, 19) described a range of intensities between 650 and 1500 W cm^{-2} (at a frequency of 3 MHz) in which the same temperature cycle produced by non-sonic means did not simulate the effect of a single pulse of ultrasound. Dunn and Fry (5) produced evidence that this mechanism was effective above a threshold of 150 W cm^{-2} but this intensity is still well above that required for any current diagnostic application.

2. Production of paraplegia and haemorrhagic injury to spinal cord.

Production of paraplegia was described following insonation of frog spinal cord at a frequency of 1 MHz (12) and there was good evidence that this was due to a non-thermal, non-cavitational mechanism (11). The role of heat was eliminated by pre-cooling the mice to 2° C before insonation, and cavitation was suppressed by application of increased ambient pressure. Suppression of both factors did not affect the ultrasonic dose required for production of paraplegia. These results were extended by Taylor and Pond (28). These authors investigated the effects of repetitive pulses of ultrasound on the spinal cord of rats.

The predominant biological effect appeared to be vascular damage as evidenced by occurrence of haemorrhage. Experiments on peripheral nerve showed that axons were intensely resistant to the effects of insonation (29). This finding was in agreement with that of other workers (21). Thus, the onset of paraplegia appeared to be mainly mediated by vascular damage and, in further experiments, the occurrence of haemorrhage was used as a visible criterion of damage sustained. This experimental model was used to study the effects of varying frequencies, pulse lengths, peak intensities and duty cycles. The important conclusions were the apparent safety of frequencies in excess of 5 MHz and the demonstration of dose-accumulation (23, 28, 29, 30). Increasing the intervals between adjacent pulses did not affect the integrated dose-time required for injury. However, this mechanism appeared to be less effective with shorter pulses (100 μs) and, with the duty cycles employed in diagnostic applications, many hours of continuous use would be required before a toxic dose could be accumulated. Since in practice there is only a transient exposure of any organ during examination, total exposure times are never likely to exceed one second.

In further experiments, it was also noted that hypoxia increased tissue sensitivity to damage by ultrasound (23, 28). For example, an integrated dose time of 24 s was required for occurrence of damage at a frequency of 1 MHz but this was reduced to 15 s when the arterial partial pressure of oxygen was reduced to 50 mm mercury. The fetus is highly hypoxic but the data derived from mature tissue cannot necessarily be extrapolated to the fetus. When a small series of mouse embryos were insonated in utero with a paralysing dose, they were not paraplegic at birth. This suggests the possibility that the potential hazard to the mother could be greater than that to the fetus. This is an important concept which is seldom considered.

3. Centrilobular necrosis of liver.

Bell (2) reported that insonation at a frequency of 1 MHz resulted in patchy areas of necrosis with evidence of damage to the smaller vessels. Curtis (3, 4) extended these studies and noted that pulsed ultrasound resulted in necrosis which occurred selectively around the central vein. This finding was confirmed by Taylor and Pond (27) who also investigated the effects of varying frequencies. As in the work on spinal cord, the lowest frequencies investigated (0.5 and 1 MHz) were most injurious producing vascular damage with subsequent centrilobular necrosis. This was evidence for a non-thermal mechanism since heating increases with frequency. There was further evidence against a thermal mechanism since thermal damage to liver results in perilobular necrosis (24). The selective action of ultrasound on the cells nearest the central vein first suggested the possibility of a synergistic effect of hypoxia since these cells are known to be the most hypoxic. The predominance of vascular damage appeared similar to the effects reported on spinal cord and showed a similar variation with frequency. There is good evidence that this is a non-thermal effect but, for the reasons given in the introduction, it is more difficult to completely exclude the possible occurrence of cavitation. However as in spinal cord work, the low intensities used in Doppler and the short pulses used in sonar makes it highly unlikely that this damage mechanism would be operative in clinical applications.

4. Stimulation of rate of tissue regeneration.

Stimulation of the rate of tissue regeneration by insonation was reported by Dyson, Pond, Joseph and Warwick (6). They insonated healing surgical defects in rabbits' ears and found a 32% increase in the rate of regeneration which was paralled by an increased uptake of tritiated thymidine indicating increased synthesis of nucleic acids. The optimum parameters of treatment were: frequency 3 MHz, peak intensity 500 mW cm^{-2} with delivery pulsed 2 ms on and 8 ms off for 5 min three times a week. This is an interesting effect since the dose parameters are so close to those employed in some Doppler devices. It appears to be important that a relatively transient exposure can produce a prolonged stimulation in the rate of tissue regeneration. A similar stimulation of regenerative activity has been reported by Galitsky and Levina (13) who found that the production of granulation tissue was improved after insonation. In the work by Dyson et al (6, 7), there was evidence against a mechanism mediated by heat alone since the variation of intensity, pulse length and duty cycle to produce the same bulk heating, did not produce the same biological effect.

5. Red Cell Stasis

Complete red cell stasis in circulating blood may occur during insonation in a standing wave field (8). Chick embryos were exposed to a continuous beam of ultrasound and it was noted that the red cells formed aggregates at distances equivalent to a half wave length of the incident beam (Fig. 1). The threshold

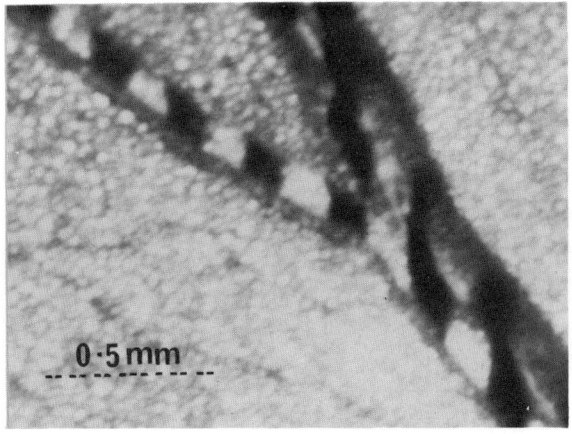

Fig. 1. Blood cell stasis induced in vessels of the area vasculosa of a chick embryo. The centres of adjacent cell aggregates are half a wavelength apart (approximately 0.25 mm at 3 MHz).

for this phenomenon was only 500 mW cm^{-2}. More recent work has shown endothelial damage in some of the blood vessels in which stasis occurred; the plasma membrane on the luminal aspect of the cells was particularly prone to damage (9).

6. Abnormal embryonic development.

There have been reports of maldevelopment of Drosophila after insonation (20) and more recently in mice after exposure to a diagnostic Doppler device with a nominal output of 40 mW (22). Experiments were undertaken to expose embryos to a simulated sonar and Doppler regime, to note occurrence of any abnormalities and if found, to search for intensity thresholds.

Chick embryos provide excellent experimental models for teratological studies. They can be grown <u>in vitro</u> for several days so that they can be serially observed during the process of organogenesis. At the conclusion of the experiment, a 4-day embryo is small enough to be adequately examined as a whole mount and this prevents the need for serial sectioning. This allows large numbers of embryos to be examined.

After 18 hours incubation, chick embryos at the head process stage, corresponding to the third week of human gestation, were

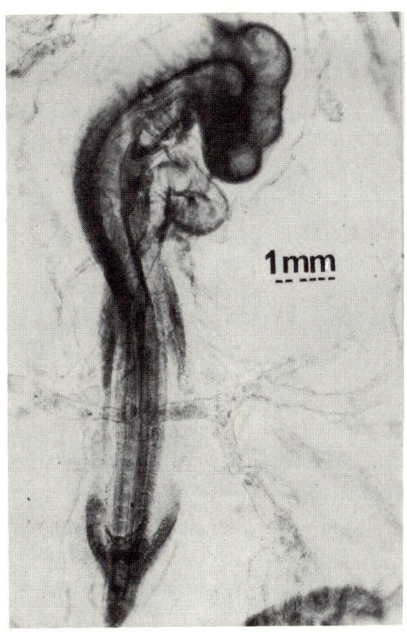 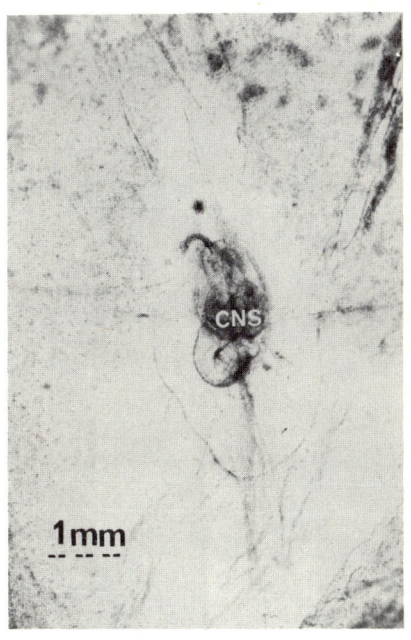

Fig.2a. Control embryo after 66 hours incubation.

Fig.2b. 66-hour embryo insonated at 18 hours. The central nervous system (CNS) is grossly deformed.

subjected to insonation or a sham insonation procedure and then returned to the incubator and allowed to develop for a further 48 hours. The resultant embryos were dissected off the yolk, fixed, mounted and stained. There was an elaborate staging procedure to assess the degree of development and perverted organogenesis (26). The most relevant results are given although many more series were performed with different treatment parameters.

When the treatment parameters were as follows, treatment resulted in abnormal development: frequency 1 MHz, peak intensity 40 W cm with delivery pulsed 20 μs on and 180 μs off for 5 min. Abnormalities were also found after treatment at 25 W cm^{-2} (Fig. 2), but reducing the peak intensity to 10 W cm^{-2} no longer produced abnormalities. This showed peak intensity threshold between approximately 10 and 25 W cm^{-2}. Because of the brevity of the pulses, cavitation should not have occurred. However, it should be recalled that the pulse length is more than an order of magnitude greater than is used in sonar techniques and the pulse repetition rate ten times greater. The experiments were repeated using embryos after 42 hours incubation (corresponding to about 6 weeks human development). At this stage, when organogenesis was complete, the embryos were totally resistant to insonation even though the peak intensities were increased to 100 W cm^{-2} (Fig. 3).

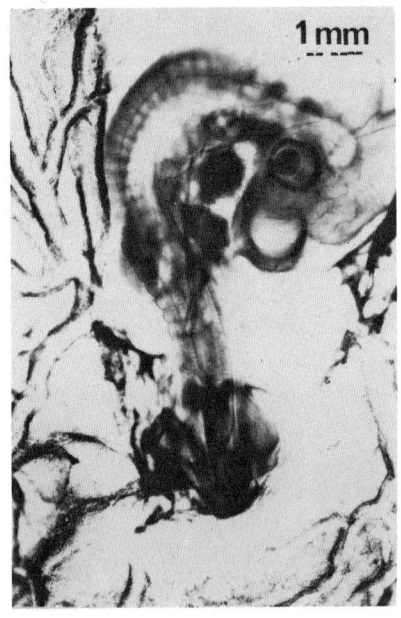

Fig. 3a. Control embryo after 90 hours incubation.

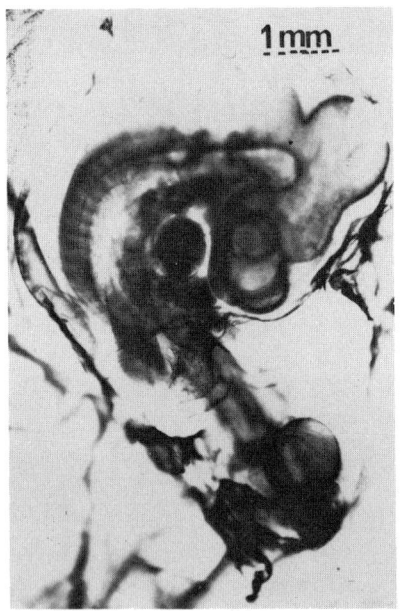

Fig. 3b. 90-hour embryo insonated at 42 hours. Development has proceeded normally.

The conclusions from this study were that abnormal development of embryos may be induced by ultrasonic irradiation and the threshold intensity for this effect is between 10 and 25 W cm^{-2}. The strong possibility of a standing wave component makes these figures approximate. Sensitivity to ultrasonic damage is a transient stage in organogenesis and at an early stage of development the embryos become intensely resistant to insonation (25). The results do suggest a large margin of safety for diagnostic applications but that the lowest intensities should be used if adequate to obtain good data, and that caution should be observed when ultrasound is used in the early stages of pregnancy.

In a further series of experiments, 44 hour embryos were exposed to Doppler devices continuously for 24 hours. The dose-parameters were: frequency 2 MHz, intensity 100 mW (electrical power) with continuous beam. There was no difference between the treated and control groups.

In conclusion, there are many biological effects of ultrasound which are not easily explicable in terms of thermal or cavitational effects. However, because of the complexity of the in vivo experiments, it is difficult to completely exclude all possibility of these mechanisms. Bio-effect data are still grossly insufficient and largely unconfirmed but the results available indicate the safety of present diagnostic applications.

Acknowledgments

We would like to thank Kevin Fitzpatrick for excellent photography.

References

1. BASAURI, L. and LELE, P.P.: A simple method for production of trackless focal lesions with focussed ultrasound. J. Physiol. 160 (1962) 513.

2. BELL, E.: The action of ultrasound on the mouse liver. J. Cell Comp. Physiol. 50 (1957) 83.

3. CURTIS, J.C.: Hepatic injury produced by intense ultrasound. Sci. Proc. 7th Ann. Mtg. A.I.U.M. (1963) 30.

4. CURTIS, J.C.: Action of intense ultrasound on intact mouse liver. In "Ultrasonic Energy" ed. Elizabeth Kelly, Urbana, Ill. (1965) 85.

5. DUNN, F. and FRY, F.J.: Ultrasonic threshold doses for the mammalian central nervous system. Trans. Biomed. Eng., BME 18 (1971) 253.

6. DYSON, M., POND, J.B., JOSEPH, J. and WARWICK, R.: The stimulation of tissue regeneration by means of ultrasound. Clin. Sci. 35 (1968) 273.

7. DYSON, M., POND, J.B., JOSEPH, J. and WARWICK, R.: Stimulation of tissue regeneration by pulsed, plane wave ultrasound. IEEE Trans. Sonics & Ultrasonics, Su-17 (1970) 133.

8. DYSON, M., POND, J.B. and WOODWARD, B.: Flow of red blood cells stopped by ultrasound. Nature (Lond) 232 (1971) 572.

9. DYSON, M., POND, J.B., WOODWARD, B. and BROADBENT, J. The production of blood cell stasis and endothelial damage in the blood vessels of chick embryos treated with ultrasound in a stationary wave field. Ultrasound in Med. Biol. 1 (1974) 133.

10. FRY, W.J. and DUNN, F.: In "Physical Techniques in Biological Research" ed. W.L. Nastuk, Acad. Press Inc., New York, 4 (1962) 261.

11. FRY, W.J., TUCKER, D., FRY, F.J. and WULFF, V.J.: Physical factors involved in ultrasonically induced changes in living systems. II. Amplitude duration relations and the effect of hydrostatic pressure for nerve tissue. J. Acoust. Soc. Amer. 23 (1951) 364.

12. FRY, W.J., WULFF, W.J., TUCKER, D. and FRY, F.J.: Physical factors involved in ultrasonically induced changes in living systems. I. Identification of non-temperature effects. J. Acoust. Soc. Amer. 22 (1950) 867.

13. GALITSKY, A.B. and LEVINA, S.I.: Vascular origin of trophic ulcers and application of ultrasound as pre-operative treatment to plastic surgery. Acta Chirug. Plast. 6 (1964) 271.

14. HILL, C.R.: Acoustic intensity measurements on ultrasonic

diagnostic devices. In "Ultrasonographia Medica" ed. J. Bock and K. Ossoinig, Vienna Acad. Med. (1970) 21.

15. HILL, C.R.: Ultrasonic exposure thresholds for changes in cells and tissues. J. Acoust. Soc. Amer. 52 (1972) 667.

16. HILL, C.R., CLARKE, P.R., CROWE, M.R. and HAMMICK, J.W.: Biophysical effects of cavitation in a 1 MHz ultrasonic beam. Ultrasonics for Industry Conference Papers (1969) 26.

17. HILL, C.R. and JOSHI, G.P.: The significance of cavitation in interpreting the biological effects of ultrasound. Proc. Conference UbioMed-70, Polish Acad. Sci. (1970).

18. POND, J.B.: A study of the biological action of focussed mechanical waves (focussed ultrasound). Ph.D. Thesis (1968) Univ. of London.

19. POND, J.B.: The role of heat in the production of ultrasonic focal lesions. J. Acoust. Soc. Amer. 47 (1970) 1607.

20. SELMAN, G.G. and COUNCE, S.J.: Abnormal embryonic development in Drosophila induced by ultrasonic treatment. Nature (Lond) 172 (1953) 503.

21. SHEALEY, C.N. and HENNEMAN, E.: Revisible effects of ultrasound on spinal reflex. Arch. Neurol. (Chicago) 6 (1962) 374.

22. SHOJI, R., MOMMA, E., SHIMIZU, T. and MATSUDA, S.: An experimental study on the effect of low intensity ultrasound on developing mouse embryos. J. Fac. Sci. Hokkaido Univ. Series VI, Zool. 18 (1971) 51.

23. TAYLOR, K.J.W.: Ultrasonic damage to spinal cord and the synergistic effect of hypoxia. J. Path. 102 (1970) 41.

24. TAYLOR, K.J.W. and CONNOLLY, C.C.: Differing hepatic lesions caused by the same dose of ultrasound. J. Path. 98 (1969) 291.

25. TAYLOR, K.J.W. and DYSON, M.: Possible hazards of diagnostic ultrasound. Brit. J. Hosp. Med. 8 (1972) 571.

26. TAYLOR, K.J.W. and DYSON, M.: Toxicity studies on the interaction of ultrasound on embryonic and adult tissues. Proc. 2nd World Congress on Ultrasonics in Medicine, 1973. Excerpta Med. (In Press).

27. TAYLOR, K.J.W. and POND, J.B.: The effects of ultrasound of varying frequencies on rat liver. J. Path. 100 (1970) 287.

28. TAYLOR, K.J.W. and POND, J.B.: A study of the production of haemorrhagic injury and paraplegia in rat spinal cord by pulsed ultrasound of low megaHertz frequencies in the context of the safety for clinical usage. Brit. J. Radiol 45 (1972a) 343.

29. TAYLOR, K.J.W. and POND, J.B.: Primary sites of ultrasonic damage on cell systems. In "Interaction of Ultrasound and Biological Tissues" ed. J.M. Reid and M.R. Sikov, DHEW Publication (FDA) 73-8008 BRH/DBE 73-1 (1972b) 87.

30. TAYLOR, K.J.W. and POND, J.B.: Experimental ultrasonic injury and safety limits in its use. Acta Radiol. $\underline{13}$ (1972c) 743.

-DISCUSSION-

DUNN - A real question that we are going to have to face has to do with the fact that cavitation seems to be a rather popular mechanism to invoke, or at least to consider. And when you begin thinking about it, much that we know about cavitation comes from investigators who devote all their attention to studying cavitation. They have studied this at considerably lower frequencies than you are talking about today. They have also studied it largely in, or almost exclusively in, water. But in many cases very carefully prepared water. Some have used tap water, but in many cases the water has been very nicely prepared and even the container has been very nicley prepared. And they, of course, have their own ideas about what constitutes cavitation nuclei.

I think a question we should begin thinking about is what in fact constitutes cavitation nuclei in tissue? Whatever it is, is it very profuse? Is it widely distributed in tissue or is it a rather uncommon thing? I am not necessarily asking for an answer to these questions, but I think these are questions of importance that we are going to have to face rather soon.

TAYLOR - Yes. I think one of the most surprising things is the very low intensities at which cavitation occurs in tissues.

DUNN - Do you think you get cavitation in that stasis demonstration?

TAYLOR - I don't think so. There is evidence of some fairly minor damage on the luminal aspect of the endothelium (Dyson, M., Pond, J. and Broadbent, J., 1973, The effect of ultrasound at levels inducing blood cell stasis on the ultrastructure of blood vessels in a chick embryo, In: Abstracts, 2nd World Congress in Ultrasonics in Medicine, Rotterdam, 1973. Page 27 ICS 277, Exerpta Medica, Amsterdam), but I think this is too mild to result from such a catastrophic event as cavitation.

DUNN - The intensity as I wrote it down was 8 $W \cdot cm^{-2}$.

TAYLOR - Yes, In Dr. Dyson's results on blood stasis, a threshold

TAYLOR – intensity of 0.5 – 8 $W \cdot cm^{-2}$ was observed depending upon the vessel size and the velocity within it (Dyson, M. and Pond, J. 1973. The effects of ultrasound on circulation. Physiotherapy 59: 284). Such intensities are well within the clinical therapeutic range. Professor Lehmann stated this morning that he quite happily puts $4 \cdot W\ cm^{-2}$ into patients fairly continuously so that blood cell stasis is a strong possibility in the clinical situation unless the transducer is moved.

NYBORG – May I comment on that, indulging in speculation to some extent, but starting out by remarking that a number of people have been intrigued by this phenomenon. The stasis effect looks like formations that have been seen in physical situations. It is well known that droplets will collect in half wave length striations, in standing waves for example. But as far as I know, attempts so far to explain the process really quantitatively have failed; the aggregation occurs more quickly than is predicted by physical theory based on the usual radiation pressure acting on objects the size of red cells. Suppose there were present very small gas bubbles; this phenomenon wouldn't necessarily be called cavitation by most people. It certainly isn't transient cavitation. These gas bubbles, when they pulsate, would attract particles to themselves, but might be invisible. And, they might migrate quite rapidly toward the pressure maxima of a standing wave field. So you could imagine -- I said this was speculation -- that there were little bubbles present somehow in the blood, that these attracted the erythrocytes to themselves, and covered themselves up so they were invisible, a rather sly method. And, then they migrated, carrying along a load of erythrocytes to the pressure maxima.

DUNN – What do you believe constitutes the nuclei? Is it undissolved gases or might it, for example, be interfaces separating structures having different impedances?

TAYLOR – There are plenty of things knocking around in virtually any water, tap water if you like, that act as nuclei.

DUNN – Well, you expect the nuclei in tissues to be the same as the nuclei in even tap water. Nuclei in tap water are generally undeveloped gasses.

HILL – The evidence we have for cavitation comes from a variety of different types of observations. Some of them are observations of mechanical changes produced by the cavitation -- others acoustical observations. The evidence from our work

HILL - on cavitation is that there does seem to be a fairly sharp threshold.

NYBORG - I would say that the threshold, especially for a stable cavitation, will depend on whether or not the bubbles are already present to be acted upon. If they are not, the threshold that's observed will characteristically be the threshold for the growth of the bubbles. But, if the solution is prepared with bubbles already in it, either accidentally or deliberately, then the threshold may be much lower.

Now, as an example, (again you may not call this cavitation but it is the effect of a pulsating bubble), Rooney carried out an experiment in which he obtained destruction of erythrocytes using a single gas bubble which he provided and placed into a very small container. There was no sub-harmonic involved; the usual indices for cavitation could not apply. And, the threshold was much lower than that usually taken to characterize cavitation thresholds, such as 3/10ths of a watt per centimeter, one atmosphere pressure amplitude.

If one prepares a medium so that it contains one or more bubbles of suitable size, this kind of a cavitation will occur even in a gel, even in viscous media. But in a viscous medium the growth of a bubble will take place very slowly, and hence, I believe, the characteristic belief that the threshold for cavitation is high in viscous media and in gels.

PHYSICAL CONSEQUENCES OF ULTRASOUND IN PLANT TISSUES AND OTHER BIO-SYSTEMS

W. L. Nyborg, D. L. Miller and A. Gershoy

Departments of Physics and Botany, University of Vermont

Cook Physical Science Building, Burlington, Vermont 05401

ABSTRACT

With plant tissue, as well as with selected animal cells in suspensions and tissues, it has been possible to make direct observations by optical microscopy during irradiation with ultrasound. This visual inspection reveals a number of kinds of mechanical action by which ultrasound produces changes in bio-systems. In general, cell membranes are set into vibration. These vibrations are especially vigorous near gas-filled intercellular spaces, which abound in plant tissues. Because of these complex oscillations, cytoplasmic structures behave as if acted on by time-independent forces and torques; also intracellular and extra-cellular fluids are caused to flow. Much of the mechanical action is understandable on the basis of theory for nonlinear acoustics. Second-order approximations to the solutions lead to the phenomena of acoustic streaming, radiation force (including aggregation tendencies) and sonic torque. Bio-systems present unique situations for applying this physical theory, as yet mostly on a qualitative basis. Examples of semi-quantitative agreement exist, and we can expect an increasing ability to make quantitative predictions as systematic experimental and theoretical studies proceed.

Introduction

In this paper we address ourselves to the question of how well we understand biophysical principles involved when ultrasound causes changes in biological systems. In experiments for gaining insight, or for testing hypotheses, plant tissues have proven useful; it is primarily for this reason that emphasis is given here to observations on plant rather than animal test objects. Also our emphasis is on phenomena which are not primarily thermal in origin; heating is considered elsewhere in this Conference. We consider specific means by which sound generates fields of forces, stress and torque in biological media, and thus causes displacement, flow, deformation and structural change to occur. While these responses are nonthermal in origin, they may be strongly influenced by temperature.

In discussing mechanical effects of ultrasound it is usual to distinguish between those associated with cavitation and those which are not. Here we have a problem in semantics. The term "cavitation" is sometimes reserved for a rather violent activity in which bubbles or cavities collapse with supersonic speed, generating shock waves and pulses of high temperature. For this Flynn (9) has suggested the phrase <u>transient</u> <u>cavitation</u>. But in connection with the more subtle manifestations of ultrasound we are concerned with a gentler kind of activity, called <u>stable cavitation</u> by Flynn, associated with pulsations of gas volumes which are maintained over a period of time. Stable cavitation has been given less attention than it deserves; this is partly so because it is more difficult to detect than is transient cavitation.

Among nonthermal and noncavitational bio-effects of ultrasound are those which can be explained in terms of acoustic streaming and radiation forces. Acoustic streaming consists of time-independent circulation set up by sound in a fluid; especially important here are the viscous stresses associated with small-scale eddying motions which are set up under some circumstances. Radiation forces are time-independent forces which act on particles, membranes and other inhomogeneities in a sound field. Acoustic streaming and radiation forces are examples of a broad class of phenomena which arise because the basic equations of acoustics are nonlinear. The various manifestations of cavitation are also a result of nonlinearity. In general it seems clear that adequate understanding of ultrasonic bio-effects can only come about through development and application of the subject of nonlinear acoustics.

Cavitation

There are hundreds of reported investigations in which ultrasound is applied to solutions or suspensions. (7). Among these there are very few in which chemical or biological effects are shown to occur in the absence of cavitation. Chemical effects are believed in most cases to be the result of activity by free radicals generated during pulses of high temperature in transient cavitation. Degradation of polymers and disruption of cells appear to arise from mechanical action, specifically from hydrodynamic stresses in fields of high velocity gradients near the bubbles. The cavitation is initiated by "nuclei" filled with gas or vapor. Bubbles grow from these nuclei in an ultrasonic field of sufficient amplitude. A complex dynamical situation ensues, involving bubble pulsation, fragmentation and coalescence. Cavitation activity in solutions is strongly dependent on the distribution of pre-existing nuclei, and on factors governing growth and dynamics which are difficult to control. As a result experiments with cavitation in solutions are not always reproducible from one laboratory to another; the literature contains numerous examples of disagreement between investigators.

Pritchard, Hughes and Peacocke (29) used improved conditions for sonating DNA solutions (at 20 kHz) by providing a uniform array of air bubbles to initiate the action, and by maintaining sonic amplitudes low enough to avoid transient cavitation. Rooney (30,31) developed a method for providing complete control by using a single air bubble, employing very low amplitudes, and irradiating only small volumes of about 0.2 ml. In Rooney's experiment the bubble was of resonant size for the frequency used (20 kHz), namely, about 250 µm in diameter. Damage to red cells in suspension was found to occur when the bubble radial oscillation amplitude exceeded about 20 µm. The cell disruption was attributed to hydrodynamic shearing stress in the small-scale acoustic streaming field, or microstreaming field, set up near the oscillating bubble. Calculations based on the theory for acoustic streaming yielded a value of about 5000 d/cm^2 for the shearing stress required to cause hemolysis. Later results from purely hydrodynamical methods are consistent with this result when differences in conditions are taken into account.

Since bulk animal tissues are typically opaque it is understandable that there has been no report of cavitation activity in such tissue, which is based on direct visual observation during the sonication process. Indirect evidence for cavitation in animal tissues has been reported under some conditions (2, 10, 20, 33) but we shall not take up this subject here.

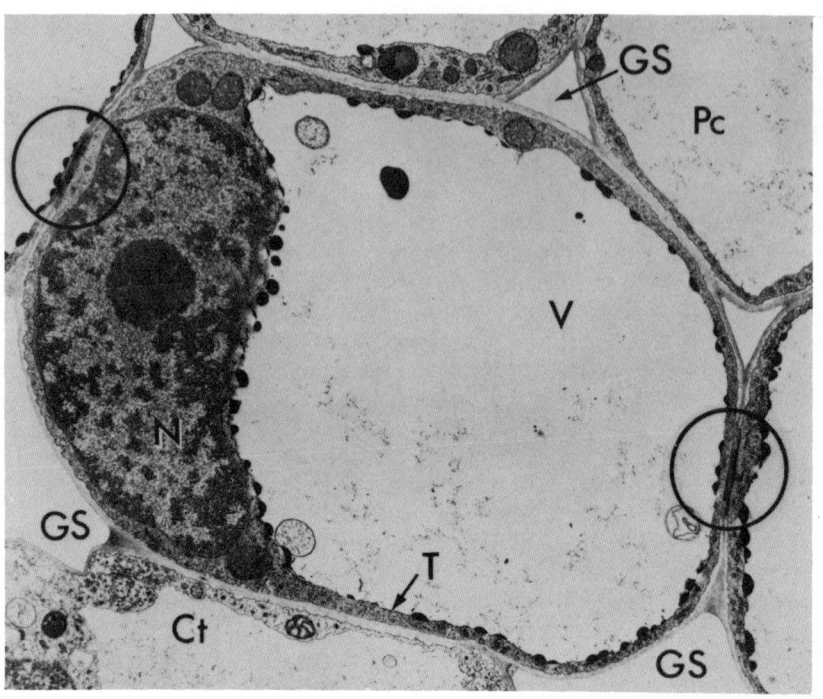

Fig. 1 Electron microscope photograph, from Ledbetter and Porter (19), Fig. 5.4.1a . Here characteristic triangular gas spaces GS (intercellular spaces) are shown in a section. The cell shown here is an endodermal cell of a root in contact with parenchyma cells, Pc and Ct. The large central vacuole V is shown, also the cytoplasmic membrane (tonoplast) T. The two enclosed circular areas show Casparian strips; these are uninterrupted by the gas spaces. Magnification is X 8600 .

Evidence that cavitation occurs in plant tissue was obtained by Lehmann, Herrick and Krusen in 1954 (21) by means of experiments designed to test a cavitation hypothesis. In thin translucent plant tissue it is readily possible to make direct observations with optical microscopy during sonation. It is known that intercellular spaces in plant leaves, stems and roots are filled with gas. See Fig. 1. By observing thin plant leaves or tissue sections microscopically during sonation much intracellular activity can be seen associated with gas pockets. This was first noticed by Harvey, Harvey and Loomis (16). At moderate amplitudes the gaseous bodies persist over a relatively long period of time. This activity is an example of "stable cavitation".

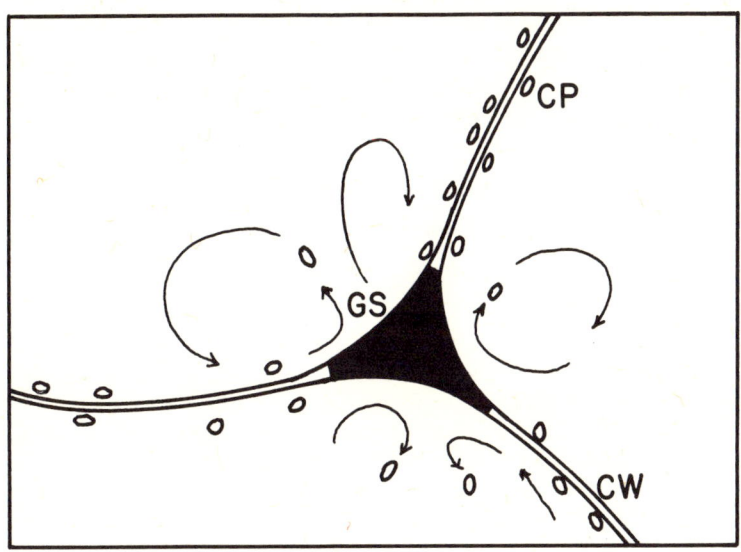

Fig. 2 Diagrammatic tracing of parenchyma cells in contact, taken from a cinematic photograph of sonated tissue in a fleshy leaf of Senecio sp. Magnification about X 1000. Here the gas-filled intercellular spaces GS are characteristically larger than in Fig. 1 and extend into narrow intercellular channels. The acoustic streaming pattern is shown diagrammatically by arrows. Dislodged chloroplasts CP serve as indicator particles. The walls of the intercellular spaces are wetted by thin water films. CW: cell wall.

We have recently done experiments at two frequencies: about 85 kHz and 1 MHz. At the lower frequency a microvibrator is used in an arrangement described earlier. (5,8,11,25,36). With a micromanipulator the rounded tip of a Mason horn (the microvibrator) is brought into contact with part of a cell boundary, while under view on the stage of a microscope. This tip, a few tens of microns in radius, then drives the cell boundary at the horn frequency, about 85 kHz. At the higher frequency we used a method first described by Harvey and associates (16); in this method the biological specimen of interest is irradiated at 1 MHz by a quartz disk-transducer which also serves as supporting microscope slide. Electroding is omitted in the center of the transducer so that the conventional optics of the microscope can be used. Motion pictures were taken to record significant phenomena.

At both frequencies the ultrasound causes considerable movement of intracellular particles, and other activity. Motion is particularly pronounced near gas-filled intercellular spaces. See Fig. 2. At 1 MHz it is found that when the amplitude is sufficiently high the intercellular space gradually becomes depleted of gas. In the process of "shrinking" of the gaseous spaces, bubbles in the vicinity of resonant size, about 7 μm at 1 MHz, are sometimes formed and are then sites of much visible activity.

Such a small bubble is unstable in the absence of sound; it quickly disappears via diffusion unless the sound field is maintained. By pulsing the ultrasound on and off in one-second intervals, it has been possible to demonstrate situations where the bubble shrinks during the "off" interval, then grows to its former size when the sound comes "on" again. The growth is evidently an example of "rectified diffusion" described initially by Harvey, et. al., (15); subsequently the subject has been treated theoretically by a number of authors (6,9,17). The bubbles do not seem to grow beyond resonant size, and may generate a subharmonic signal and/or surface waves. These observations on shrinking and growth have some significance in reference to procedures where tissue is sonated, then prepared histologically. Clearly the absence of bubbles in histological sections is unsatisfactory proof that bubbles were not present during sonation. Active bubbles in the micron scale which might have been present would have shrunk to invisible size within a second after sonation ceased. Examples of activities associated with gas in plant tissues will be demonstrated by a motion picture accompanying this talk.

Theory from Nonlinear Acoustics for Mechanical Bio-effects

A variety of movements and intracellular changes have been reported by investigators from 1927 to the present. While these findings may not have been anticipated from basic principles, it now seems possible to find physical explanations for much of what has been observed. As yet the physical theory is not capable of precise quantitative application. Nevertheless it has proved very useful as a qualitative (sometimes semi-quantitative) guide. In this theory approximate solutions are obtained to the basic nonlinear equations of acoustics (23). These are usually arrived at by a method of successive approximations in which it is assumed that each of the field quantities can be represented as a converging series of functions. Thus for the pressure p we write

$$p = p_o + p_1 + p_2 + \ldots . \tag{1}$$

Here p_o is the zero-order contribution, representing the pressure in the absence of sound. The first-order contribution p_1 is the usual quantity obtained from linear acoustics. We shall assume here that the field arises from a source which vibrates sinusoidally in time with angular frequency ω; then the pressure p_1 also varies sinusoidally in time with the same angular frequency. If a typical source amplitude is A, then the amplitude of p_1 at any point is proportional to A. It is found that the second-order contribution p_2 consists of two parts. One part varies sinusoidally in time, but with (second harmonic) frequency 2ω and amplitude proportional to A^2 (the square of the source amplitude). The other part is independent of time and thus represents an addition (positive or negative) to the static pressure; its magnitude at any point in space is proportional to A^2.

The fluid velocity $\underline{u}$ is also represented as a series of functions. Assuming there is no flow in the absence of sound we take $\underline{u}_o$ to be zero and obtain

$$\underline{u} = \underline{u}_1 + \underline{u}_2 + \ldots \quad . \tag{2}$$

By analogy to p_1 we interpret $\underline{u}_1$ as a contribution which is obtained from linear acoustics for the problem of interest; it represents velocity which varies sinusoidally in time with angular frequency ω and amplitude proportional to A. Also $\underline{u}_2$ is found to have both a second harmonic and a time-independent component; for each of these a typical magnitude is proportional to A^2.

When each of these series is substituted into the dynamical equation, product-terms are obtained, some of which are independent of A, some proportional to A, others to A^2, etc. By grouping these the equation can be written

$$[\quad]_o + [\quad]_1 A + [\quad]_2 A^2 + \ldots = 0 . \tag{3}$$

This equation is valid for all values of A and it therefore follows that each of the brackets $[\ ]_o$, $[\ ]_1$, etc. is separately equal to zero. In this way, and by time-averaging, it has been shown that the following equation applies, to a Newtonian fluid with shear viscous coefficient η in a steady sound field:

$$\eta \nabla^2 \underline{u}_2 - \nabla p_2 + \underline{F} = 0 \tag{4a}$$

where

$$\underline{F} = -\rho_o < (\underline{u}_1 \cdot \nabla)\underline{u}_1 + \underline{u}_1 (\nabla \cdot \underline{u}_1) > ; \tag{4b}$$

here ρ_o is the density of the fluid (in the absence of sound); the symbols $\underline{u}_2$ and p_2 here represent only the time-independent parts of the second-order contribution to $\underline{u}$ and p (the second harmonic part having been averaged out). Equations (4) are a second-order time-averaged approximation to the dynamical equation and are basic to discussions of acoustic streaming, radiation pressure and allied topics. The expression for $\underline{F}$ is rather complex, but we note the important facts that $\underline{F}$ depends only on $\underline{u}_1$ and that (because of the time-averaging operation) it is a function only of spatial coordinates. Once the first-order velocity field $\underline{u}_1$ is known $\underline{F}$ can be calculated. It is also worth noting that Eq. (4a) has the same form as the governing equation for slow viscous flow caused by a force field of external origin; the quantity $\underline{F}$ (with units of force/volume) is the exact analog of a driving force field and we shall refer to it as an "effective force".

In using Eqs. (4) the formal procedure is to first find the first-order time-dependent velocity field $\underline{u}_1$, by solving the linearized acoustical equations. Then the time-independent function $F(x,y,z)$ is determined from Eq. (4b). Finally solutions of Eq. (4a) are sought, yielding the time-independent quantities $\underline{u}_2$ and p_2. The quantity $\underline{u}_2$ at any given point gives (to second-order approximation) the time-averaged velocity at that point. We frequently also wish to know a quantity $\underline{U}$ which gives the average velocity of a specified portion of the medium (as it moves); $\underline{U}$ is obtained by adding to $\underline{u}_2$ a velocity transform (involving $\underline{u}_1$) which we shall not discuss here. (35,23). Either $\underline{u}_2$ or $\underline{U}$ is the acoustic streaming velocity and represents time-independent circulation or eddying set up by a sound field. The quantity p_2 represents an addition to the static pressure.

Standing Waves in a Membrane

A common observation in sonated plant cells is the occurrence of eddying motions in the vacuole as marked out by particles free to move, or in partially disrupted cytoplasm. (5,8,11,16,25). Similar motions have been observed in nuclei of marine eggs, indicated by movements of the nucleolus. (32,36). These eddying motions or circulations are evidently examples of acoustic streaming. In general the streaming field is complex and difficult to analyze with any precision. But we gain insight on the phenomenon by taking up a simplified situation where the acoustic streaming problem can be solved.

For this purpose we imagine a model cell whose boundaries are planes which are normally at $z = 0$ and $z = h$. In the presence of a sound field the boundary at $z = 0$ is set into standing wave oscillations; suppose the displacement of this boundary in the z direction is given by the real part of

$$\zeta = -iB \cos kx \, e^{i\omega t} . \tag{5}$$

Then if viscosity is ignored the x-component u_1 and z-component w_1 of the first-order velocity $\underline{u}_1$ in the fluid near the membrane are given by

$$u_1 = B\omega \sin kx \, e^{-kz} e^{i\omega t}$$

$$w_1 = B\omega \cos kx \, e^{-kz} e^{i\omega t} \tag{6}$$

According to these equations the path of a fluid particle near the boundary is an ellipse. When $kx = 0, \frac{1}{2}\pi, \pi$, etc. the ellipse becomes a straight line. When $kx = \pi/4, 3\pi/4$, etc., the ellipse becomes a circle. The expression for u_1 in Eqs. (6) does not satisfy the nonslip boundary condition which we should expect to apply. To meet a nonslip condition it is necessary to take viscosity into account; the first-order solution then yields for u_1

$$u_1 = B\omega \sin kx \, e^{-kz}(1 - e^{-mz}) e^{i\omega t}, \tag{7}$$

where $m = (1+i)\beta$ and β is given by

$$\beta^2 = \omega/2\nu \tag{8}$$

where ν is the kinematic viscosity coefficient η/ρ_o. The reciprocal of β has dimensions of length and is typically of the order of microns for water-like fluids at ultrasonic frequencies. The expression for u_1 in Eq. (7) is an approximate one, valid when β^{-1} is small compared to the amplitude B and the wavelength $2\pi/k$. High gradients of the velocity u_1 typically exist near the boundary $z = 0$, in a boundary layer of thickness characterized by β^{-1}. From Eq. (7) and a counterpart for w_1 we can evaluate the effective force distribution $F(x,z)$ according to Eq. (4b). For the x component of F we obtain

$$F_x = (1/4) \rho_o k\omega^2 B^2 \sin 2kx \, (nC + nS + 2C - e^{-2n}), \tag{9}$$

where $\eta = \beta z$ and

$$C = e^{-n} \cos n ; \quad S = e^{-n} \sin n . \tag{10}$$

The variation of F_x with height above the boundary is shown in Fig. 3. We see that F_x is relatively large near the boundary, changes sign when $\beta z = 2$, and again when $\beta z = 5$. Its magnitude decreases rapidly with distance from the boundary, and F_x is hardly significant for $\beta z > 8$.

One may similarly obtain an expression for F_z, but this component is often negligible. Knowing the components of $\underline{F}$ one may now proceed to Eqs. (4) and obtain expressions for the streaming velocity $\underline{u_2}$ and U. For the x-component u_2 of the velocity $\underline{u_2}$ it is found that near the boundary

$$u_2 = -2k\omega B^2 \sin 2kx \, [1 - 8S + 2nC - 2nS - e^{-2n}], \tag{11}$$

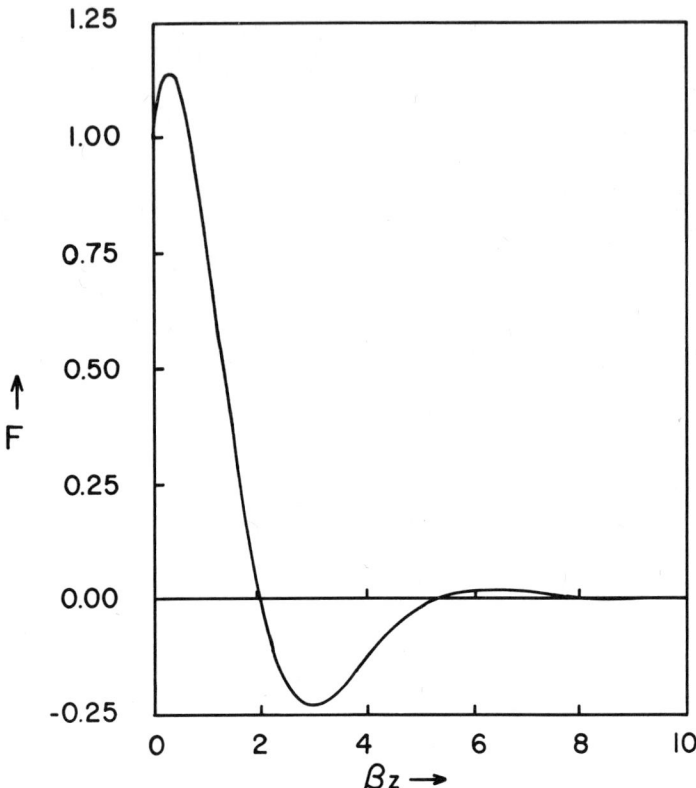

Fig. 3 Plot of "effective force" F generated by membrane vibration, based on Eq. (9). F is in arbitrary units. βz measures distance from the membrane in units of the boundary layer thickness β^{-1}.

where S is as defined in Eq. (10). These expressions are valid only for distances z from the boundary that do not much exceed the boundary layer thickness β_2^{-1}. A typical magnitude for the velocity is seen to be $\omega k B^2$. The velocity gradient near the boundary is given in order of magnitude by $\beta \omega k B^2$ and the shearing stress by $\eta \beta \omega k B^2$.

By proceeding further a solution was found which is valid in the general interior of the "cell"; from this were obtained the streamlines which are plotted in Fig. 4. These plots are valid outside the boundary layers. An array of circulations alternating in direction are spaced one-fourth wavelength apart along the membrane. The direction is such that near each displacement maximum of the membrane vibration pattern the flow is away from the boundary.

Comparing these predictions with observations it must be remarked that regular arrays of the kind shown are seldom seen. In intact, highly vacuolate plant cells, the central vacuole is traversed by a system of channels which connect with parietal cytoplasmic layers (channels). These internal channels complicate the acoustic streaming motions. An approximation to free movement in the cell obtains when structures have been rather thoroughly destroyed or "homogenized" by some means (such as previous sonic irradiation at relatively high amplitude). Eddying patterns with varying degrees of resemblance to Fig. 4 are then seen in cells with such homogenized contents. Examples will be shown in a film accompanying this talk. The theory for Fig. 4 was based on the assumption of regular standing waves extending along the entire cell boundary, a condition that is seldom approximated.

In plant cells with intercellular air spaces the sonically produced eddying occurs primarily near those spaces. See, for example, Fig. 2. Examples of this microstreaming will be shown in the motion picture film which accompanies this talk. This motion is evidently generated because the air-spaces pulsate in the sound field, and thus set the bounding membranes into vibration.

Sonic Torque

When free streaming motions are seen of the type exemplified by Fig. 4 the cell has been badly damaged and may not recover. In considering thresholds for biological effects of ultrasound we are perhaps more interested in phenomena which occur at lower amplitudes. A common observation at low levels is a tendency for small intracellular bodies to rotate when the cell boundary is set into vibration. When the vibration is of very low amplitude and applied only for a short time the body may rotate through, say, an angle

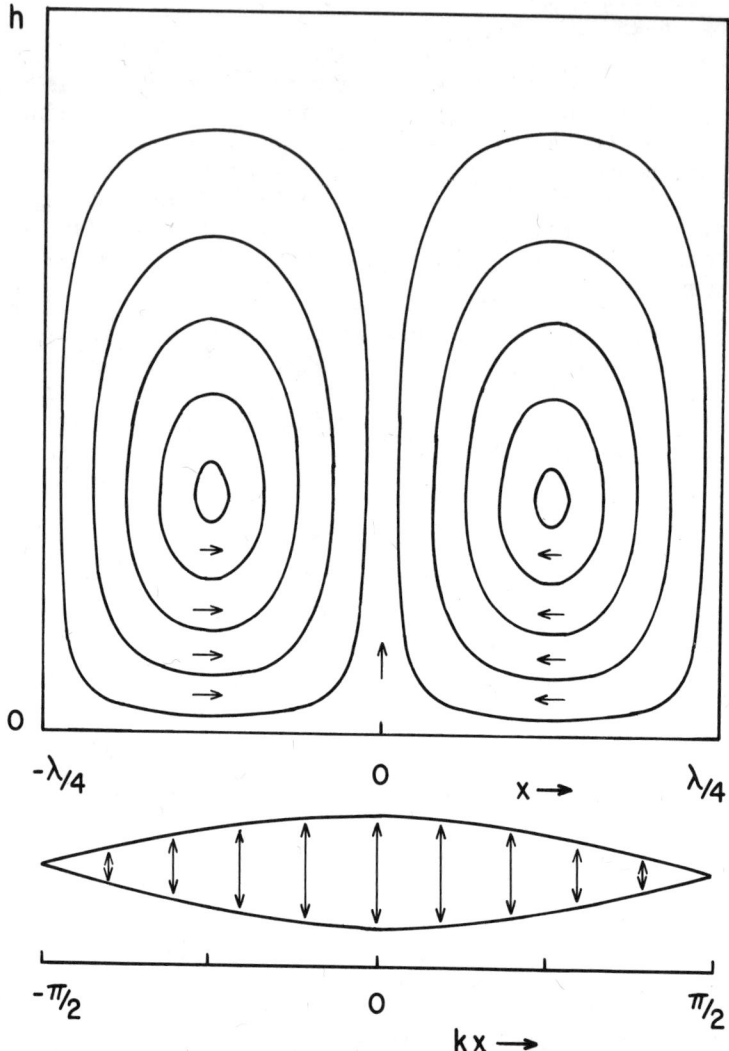

Fig. 4 Computer-generated streamlines approximating those expected near a vibrating membrane.

of 90°, then return to its original orientation when the sound is removed. This tendency to "unwind" is an indication of structure in the cytoplasm to which the body is attached. But when the sound is applied at higher amplitudes or for a longer time the body may rotate continuously in a given direction with no tendency to unwind when the sound is removed. Apparently whatever structural elements exist normally to restrain rotation of the body have then been stretched and torn by the sonic torque.

In explanation of these rotations we refer to theory showing that any volume element in a sonicated medium may be subject to time-independent torque. (22,23). Specifically the sonic torque L on a spherical element about its center is given by

$$\underline{L} = (I/2\rho_o)|\nabla \times \underline{F}| , \qquad (12)$$

where $\underline{F}$ is the "effective force" defined in Eq. (4b), and I is the moment of inertia of the sphere about its center. While Eq. (12) was derived for an element of homogeneous fluid it is also applicable to a spherical inhomogeneity suspended in fluid (such as a small cytoplasmic body) provided that the specific acoustic impedance $\rho_o c$ (where c is the velocity of sound) is similar for fluid and inhomogeneity. This condition is satisfied rather well in biological situations and we therefore can use Eq. (12) for considering effects of sonic torque on spherical intracellular particles. If the sonic torque on a spherical cytoplasmic body causes it to rotate with angular velocity $\dot\theta$ in a Newtonian liquid of shear viscosity coefficient η the drag torque will be $8\pi\eta R^3 \dot\theta$. (18). In steady state rotation we therefore have (since $I/2\rho_o$ is $4\pi R^5/15$)

$$\dot\theta = (R^2/30\eta) |\nabla \times \underline{F}| . \qquad (13)$$

There is a variety of acoustical situations in which sonic torque is generated. Consider for example the boundary layer generated near a vibrating membrane as considered earlier. In Eq. (9) and Fig. 3 we see that F_x varies with distance z from the membrane. Since F_z is relatively small the principal contribution to $\nabla \times \underline{F}$ is a component along the z direction whose magnitude is given by $\partial F_x/\partial z$. From the nature of the plot in Fig. 3 we see that for a small body in the boundary layer near a vibrating membrane the magnitude and direction of torque on the body will depend on its position.

Such torque distributions seem to be important biologically. In many plant cells there is a thin layer of cytoplasm in a parietal layer of 5-10 μm thickness, contiguous to the cell boundary. When this is vibrated a boundary layer is set up within the cytoplasm, assuming its specific acoustic impedance is not greatly different from that of the vacuolar water. An $\bar{F}$ field is then established in the cytoplasm, with an associated distribution of sonic torque. When the torque acts on inhomogeneities such as chloroplasts or other bodies in the cytoplasm they may rotate, but are subject to restoring torque exerted by surrounding cytoplasm. Only if a yield value of the torque is exceeded will the inhomogeneity "break loose" from its constraints and spin freely.

A particle need not be in a boundary layer to be subject to sonic torque. An object in any field where the local particle velocity has perpendicular components differing in phase by angles other than integral multiples of π will be subject to unidirectional torque. In other words such torque exists at points where particles of the fluid execute open elliptical orbits during each sonic cycle. This situation exists, for example, in the neighborhood of a vibrating membrane (outside the boundary layer region) as was pointed out in connection with Eqs. (6).

Still another situation giving rise to sonic torque exists when two independent ultrasonic beams, or a transmitted and reflected beam, intersect. In such an overlap region the sonic torque on a suspended particle is proportional to α/η, the ratio of the acoustic absorption coefficient in the particle to the viscosity of the surrounding liquid. This situation might arise in ultrasonic scanning procedures.

Steady rotation of particles in a sound field has been observed by a number of investigators. Harvey and associates (14) reported on rotation of intracellular particles in many cells and Schmitt (32) noted the rotation of nucleoli in marine-egg nuclei. Since then, others have confirmed and extended the observations. (5,8,11,25,36). Particularly significant biologically are the findings of Dyer (4) that when the boundary of moss protonema is vibrated during the process of cell division the incipient crosswall is set into steady rotation about an axis perpendicular to the protonemal axis. See Fig. 5. After sonication he found that the cell division proceeded in a number of instances, but that the daughter cells and their progeny were abnormal, the abnormality persisting through a series of cultures.

It is clear that sonic rotation is a common occurrence, and that it sometimes has biological significance. But so far little or no systematic study has been made of the sonic torque, nor of its biological implications. And as yet it has not been possible to compare experimental findings with theory.

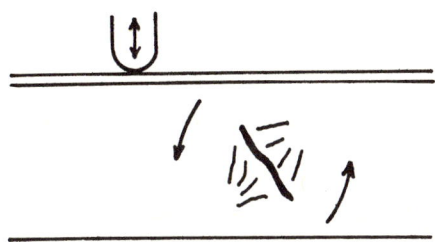

Fig. 5 Rotation of incipient wall in moss protonema as observed by Dyer (4).

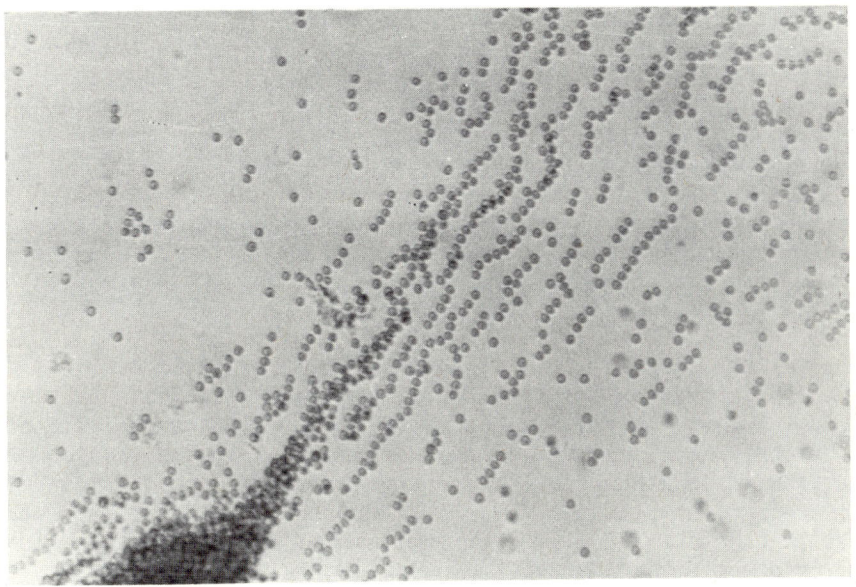

Fig. 6 Pearl-chain formation in a 1 MHz ultrasonic field; the particles are sphered erythrocytes.

Radiation Pressure and Forces

Other visually observed events in cells, tissues and suspensions appear to be the result of time-independent forces or pressures set up in a sound field. We shall have space here only to refer to a few examples.

(a) Particles migrate in standing wave fields to form masses one-half wavelength apart, as observed for erythrocytes in blood vessels by Pond, Woodward and Dyson (28).

(b) Particles move toward a small sound source, such as a portion of cell membrane set into locallized vibration by a microvibrator or pulsating microbubble. (5,11,25,36).

(c) Particles attract or repel each other in a sound field. An example is shown in Fig. 6. Here sphered erythrocytes on the face of a 1 MHz quartz transducer queue up in lines resembling the "pearl chains" observed in 1958 by Herrick (16a) in electromagnetic fields; for recent results see Sher, Kresch and Schwan (1970). These chains are separated from each other by a spacing which is much less than the half-wavelength spacing observed in (a).

All the phenomena referred to in items (a), (b) and (c) are consequences of radiation force. A convenient expression for the radiation force on a particle has been given by Gor'kov (12). For the force F on a small body of volume v and density ρ surrounded by a liquid of density ρ_o his expression is

$$F = vB \, \nabla \bar{T} - v(1 - \gamma) \, \nabla \bar{V} \; ; \qquad (14)$$

here $\bar{T}$ is the time-averaged kinetic energy density and $\bar{V}$ the time-averaged potential energy density in the sound field; γ is the ratio of the compressibility of the body to that of the surrounding fluid and B is given by

$$B = 3(\rho - \rho_o)/(2\rho + \rho_o) \; . \qquad (15)$$

Gor'kov's expression is valid fairly generally, but not for plane travelling waves or fields of similar type. Expressions related to Eq. (14) have been arrived at and discussed by Nyborg (24) and by Crum (3). To use Eq. (14) for a particle of known properties it is sufficient to have available a first-order solution for the sound field in question; from this information $\nabla \bar{T}$ and $\nabla \bar{V}$ can be evaluated in a straightforward manner.

A special application was made to the interaction between a pair of particles suspended in liquid and subject to sound. (26). The sign and magnitude of the interparticle force was found to depend on the angle θ between the local direction of oscillation in the sound field, and a line connecting the particles. For $\theta = 0$ the force is repulsive and for $\theta = 90°$ it is attractive. This result probably goes far in explaining the pearl chains of Fig. 6.

Particles aggregate sideways to form the chains; repulsion between the chains causes the separation seen.

(d) Fluid moves in narrow cytoplasmic channels as a result of gradients in the static pressure contribution p_2 (see Eq. (1)). Theory for this fluid motion was put forth recently by Gershoy and Nyborg (11) to explain swelling of cytoplasmic layers contiguous to vibrating cell boundaries in plant cells. As they point out the theory is analogous to that used successfully by Piercy and Lamb (27) to explain motion in a capillary attached as a side arm to a vessel in which an ultrasonic beam is propagated. This swelling phenomenon is very important since it leads to rupture of the vacuolar membrane. Cytoplasmic material is thus exposed to the vacuolar environment and suffers irreversible damage.

Conclusion

A number of phenomena have been observed in biological cells, especially in plant cells, which can be accounted for in terms of nonlinear acoustics. In a number of instances we can be rather certain of our understanding qualitatively and, to some extent, semi-quantitatively. But much remains to be done to bring the subject to an adequate quantitative state.

It is clear that intercellular gas is an important factor in making plant tissue sensitive to ultrasound. At the lower sound levels these gas pockets execute relatively high pulsation amplitudes and are very effective in setting adjoining cell boundaries into vibration. Vibration of cell membranes leads, in turn, to time-independent forces and torques on cytoplasmic bodies, as described above.

Significant reductions have recently been reported in growth rates and mitotic activity in plant roots irradiated with 1-2 MHz ultrasound at levels as low as 1 W/cm^2. (1,13,34). Gregory, et. al. (13) state that the effect is nonthermal. Perhaps some of the nonlinear effects we have described may be relevant here.

REFERENCES

1 BLEANEY, B. I. and OLIVER, R.: British Journal of Radiology 45 (1972) 358.

2 CURTIS, J.C.: Effects of ultrasound on tissues and organs - a review, Session 3:2, Interaction of Ultrasound and Biological Tissues (J. M. Reid and M. R. Sikov, Eds.). DHEW Publication (FDA) 73-8008 (1972).

3 CRUM, L. A.: J. Acoust. Soc. Amer. 50 (1971) 157.

4 DYER, H. J.: J. Acoust. Soc. Amer. 37 (1965) 1195A.

5 DYER, H. J. and NYBORG, W. L.: IRE Trans. Med. Electron. ME-7 (1960) 163.

6 ELLER, A. and FLYNN, H. G.: J. Acoust. Soc. Amer. 37 (1965) 493.

7 EL'PINER, I. E.: Ultrasound: Physical, Chemical and Biological Effects, Consultants Bureau, New York (1964).

8 EL'PINER, I.E., FAIKIN, I.M., and BASURMANOVA, O. K.: Fed. Proc. 25 (1966) T716.

9 FLYNN, H. G.: Physics of acoustic cavitation in liquids, Ch. 9, Physical Acoustics, Vol. I, Part B (W. P. Mason, Ed.). Academic Press. New York (1964).

10 FRY, F. J., KOSSOFF, G., EGGLETON, R.C., and DUNN, F.: J. Acoust. Soc. Amer. 18 (1970) 1413.

11 GERSHOY, A. and NYBORG, W. L.: J. Acoust. Soc. Amer. 54 (1973) 1356.

12 GOR'KOV, L.P.: Soviet Physics - Doklady 6 (1962) 773. [Transl. from Dok. Aka. Nauk. SSSR 140 (1961) 88].

13 GREGORY, W. D., MILLER, M.W., CARSTENSEN, E.L., CATALDO, F.L. and REDDY, M.M.: British Journal of Radiology 47 (1974) 122.

14 HARVEY, E.N.: Biol. Bull. 59 (1930) 306.

15 HARVEY, E.N., BARNES, D.K., McELROY, W.D., WHITELEY, A. H., PEASE, D.C., and COOPER, K.W.: J. Cellular Comp. Physiol. 24 (1944) 1.

16 HARVEY, E.N., HARVEY, E.B. and LOOMIS, A.L.: Biol. Bull. 55 (1928) 459.

16a HERRICK, J. F.: Pearl-chain formation, Proc. 2nd Tri-Service Conf. on Biol. Effects of Microwave Energy. Sept. 1958.

17 HSIEH, D.Y. and PLESSET, M.S.: J. Acoust Soc. Amer. 33 (1961) 206.

18 LAMB, H: Section 334 in Hydrodynamics. Dover Publications. New York (1945).

19 LEDBETTER, M. C. and PORTER, K.R.: Introduction to the Fine Structure of Plant Cells, Springer-Verlag, New York 1970.

20 LEHMANN, J. F. and HERRICK, J.F.: Arch. Phys. Med. Rehab. 34 (1953) 86.

21 LEHMANN, J.F., HERRICK, J.F. and KRUSEN, F.H.: Arch. Phys. Med. Rehab. 35 (1954) 141.

22 NYBORG, W.L.: J. Acoust. Soc. Amer. 25 (1953) 938.

23 NYBORG, W.L.: Acoustic Streaming, Ch. 11, Physical Acoustics, Vol. II, Part B (W. P. Mason, Ed.). Academic Press, New York (1965).

24 NYBORG, W. L.: J. Acoust. Soc. Amer. 42 (1967) 947.

25 NYBORG, W. L. and DYER, H. J.: pp 391-396 in Proceedings of the Second International Conference on Medical Electronics, Paris, 24-27 June 1959. Illiffe and Sons Ltd. London (1960).

26 NYBORG, W. L. and GERSHOY, A.: Microsonation of cells under near-threshold conditions, Proceedings of the 2nd World Congress on Ultrasonics in Medicine, Rotterdam, 1973. Excerpta Medica. Amsterdam.

27 PIERCY, J. E. and LAMB, J.: Proc. Roy. Soc. A226 (1954) 43.

28 POND, J. B., WOODWARD, B., and DYSON, M.: Phys. Med. Biol. 16 (1971) 521.

29 PRITCHARD, N. J., HUGHES, D.E. and PEACOCKE, A. R.: Biopolymers 4 (1966) 259.

30 ROONEY, J. A.: Science 169 (1970) 869.

31 ROONEY, J. A.: J. Acoust. Soc. Amer. 52 (1972) 1718.

32 SCHMITT, F. O.: Protoplasma 7 (1929) 332.

33 SENAPATI, N., LELE, P. P. and CAULFIELD, J. B.: J. Acoust. Soc. Amer. 55S (1974) S6 (A).

34 SHEPSTONE, B. J. and HERING, E. R.: British Journal of Radiology 45 (1972) 786.

34a SHER, L. D., KRESCH, E. and SCHWAN, H. P.: Biophysic. J. 10 (1970) 970.

35 WESTERVELT, P. J.: J. Acoust. Soc. Amer. 25 (1953) 60, 799, 951.

36 WILSON, W. L., WIERCINSKI, F. J., NYBORG, W. L., SCHNITZLER, R. M. and SICHEL, F. J.: J. Acoust. Soc. Amer. 40 (1966) 1363.

―DISCUSSION―

CARSTENSEN - Would you explain the mechanism of pearl chain formation again?

NYBORG - Consider two particles, such as erythrocytes, near each other in a sound field where the oscillatory fluid motion is, say, in the North-South (NS) direction. Let AB be the direction of a line connecting the two erythrocytes. Then the theory says that if AB is perpendicular to NS the cells will attract each other, while if AB is parallel to NS they will repel. This qualitative explanation seems to be enough in itself to explain when the chains form in a direction perpendicular to NS, separated from each other along the NS direction.

HILL - What do you know about the nature of the sound field: to what extent is this a standing wave situation? How far does this relate to a progressive wave situation, and how far can your results be extrapolated?

NYBORG - The answer to the first part of your question is yes; it is a standing wave: not a simple standing wave but a complicated one. The sample is like a liquid disc, about a centimeter in diameter, equal to several 0.15 centimeter wavelengths. At one face it is bounded by the transducer, at the other by a cover slip, and at its edge by air.

In respect to the observations with plant tissues containing gas, I really wouldn't expect that the nature of the field would make very much difference in respect to the action of a gas bubble. The bubble should respond to the pressure

NYBORG – amplitude in its own vicinity, without regard to whether the field is a standing wave or a progressive wave.

VOGELMAN – I noticed in the film that the cells move toward the source, line up perpendicular to the source, then move off to the left, that is left from my point-of-view. The chains are moving in one direction with respect to the source. Can you explain why that should be?

NYBORG – I am not sure, but I think it is just accidental, as a result of some asymmetry in the field.

HILL – One thing occurs to me. In that experiment where you get the captive bubble with a resonant size at 1 MHz, and you show it growing and contracting, this is very intriguing indeed. When you have got it to that stage, can you then start reducing the sound intensity to the point where you can no longer make it grow?

NYBORG – That would be very interesting to know. An experiment to try that is more or less on the program.

HILL – This would, in a sense, give some answer to the question of the role of cavitation threshold.

NYBORG – I do think in the latter part of the film where this shrinking and growing occurs, that we probably are close to the threshold for transient cavitation. Mr. Miller finds that the detector which is provided by the upper cover slide does pick up some sub-harmonic under these conditions. That doesn't prove it's transient cavitation, but it might be. The detector picks up the sub-harmonic when the intensity level has reached about 2 W/cm, and the bubble is going through this cyclic shrinking-growing process.

CARSTENSEN – You made a very brief reference to Dyer's mutagenic effect in moss. Is that, in your opinion, a reasonably clear example of a mutagenic effect of ultrasound? And, do you know of any other?

NYBORG – It is certainly an inherited change. He still has cultures which you probably can inspect yourself if you wish.

CARSTENSEN – It is clear though that it occurred because of the ultrasound and wasn't just a spontaneous mutation.

NYBORG – Yes. He has compared the frequencies with those of spontaneous mutations and finds the frequency of sonically produced changes to be much greater.

CARSTENSEN - But now, from what you have just shown, almost identical effects occur at one megacycle. So the fact that it was originally discovered using an 85 kiloHertz vibrator is probably not really relevant then.

NYBORG - I think perhaps not. We have not been very confident about making that extrapolation previous to these recent findings, however. But now, in view of the obvious significance of gas in plant tissues, it seems to me probable, rather than simply speculative, that there will be such effects. I would certainly regard sonically generated intracellular movement as a strong candidate as a mechanism for the interesting findings here at Rochester, and the previous ones by Oliver and Bleaney on reduction of growth in root tips.

MICKEY - If I understood your description of that, it was a partially reversible effect on the cell plate which is the structure between the two nuclei that ordinarily will result in cytokinesis or separation of the two cells. Did you say that this was partially reversible?

NYBORG - I am not sure which phenomenon you are referring to.

MICKEY - I am talking about the protonema where the cytoplasm of the cell failed to complete division and therefore you had two nuclei still remaining in the same cell. This is a different phenomenon from mutation as such. This is an effect upon the cell plate. Now, you can cause this by a number of agents, both physical and chemical. For example, you can put a little caffeine in the cell and the cell plate will not form, thus resulting in a binucleate cell. And, when the caffeine concentration is reduced, sometimes this cytokinesis can occur again and separate these nuclei into separate cells. But sometimes it doesn't, so you maintain a binucleate condition. And also, occasionally these two nuclei will fuse into a single nucleus to produce a tetraploid or a cell with four sets of chromosomes. So I suspect that you are not talking about a mutation as such; you are talking about a physical phenomenon of cell plate formation which affects whether the cell will divide into two separate cells or whether it will be a single cell with two nuclei.

NYBORG - There is a tentative hypothesis here of what we would call a mechanical effect. It is sometimes found as the result of the whirling of the incipient wall, that both of the nuclei got in the same daughter cell with none in the other.

MICKEY - We have to be careful about using the word mutation. It is a broad term, and it includes not only gene mutations but also chromosomal aberrations. But it does not include structural features such as the cell wall or spindle formations. I wouldn't call that a mutation.

NYBORG - Dyer himself has been hesitant to call it a mutation; but it is an inherited effect.

MICKEY - Cells can retain the condition through a number of cell divisions.

NYBORG - A very large number of cell divisions, yes. His hypothesis is that an unequal division of chromosomes was produced by the original sonic action which continued through subsequent generations.

Medical Applications

THE USE OF NON-IONIZING RADIATION FOR THERAPEUTIC HEATING

Justus F. Lehmann and C. Gerald Warren

Department of Rehabilitation Medicine
University of Washington School of Medicine
Seattle, Washington 98195

Various forms of non-ionizing radiation are used for the therapeutic application of heat. As part of this symposium, a brief overview will be given on the use of therapeutic heat in the form of shortwave, microwave and ultrasonic radiation. Goals in therapeutic heat application are as follows: (1) to increase extensibility of collagen tissue; (2) to decrease "joint stiffness"; (3) to produce pain relief; (4) to relieve muscle spasm; (5) to assist in resolution of inflammatory infiltrates, edema and exudates; and (6) to increase blood flow.

One of the more recently studied effects is the increased extensibility of collagenous tissues when heated. Limitation of joint range is frequently a result of the deposition of collagenous tissues in the form of scars in the synovium of joints, in the joint capsule and in other tissues. Limited range of motion due to tightness of joint capsules resulting from immobilization can also be attributed to a large degree to this type of tissue. Such contractures are greatly disabling. If it is possible to increase the extensibility of the collagenous tissue by heating, this effect can be used in conjunction with physical therapy, in the form of stretching and range of motion exercises, to correct limitation of the joint.

Many patients with rheumatoid arthritis or other collagen diseases complain of stiffness, especially in the morning. This is not only disabling, but can also cause a great amount of discomfort. Such stiffness can be greatly alleviated by the application of heat.

Pain is also a common complaint with musculoskeletal diseases.

It can be relieved by elevating the pain threshold through therapeutic heat application. This is accomplished by temperature elevation at the site of the free nerve endings as well as at the peripheral nerve, without noticeably affecting sensory or motor function.

Pain can also be alleviated indirectly by relieving muscle spasms which cause it. Such spasms are often secondary to some underlying pathology and can produce ischemia in the muscle which then becomes painful. Muscle spasm is reduced directly by an action on the spindle mechanism, causing the activity of the gamma fiber system to be reduced. It is also possible that there is some reflex effect through temperature receptors in the superficial tissues. Even though heat can relieve muscle spasms and subsequently the ischemic pain, the underlying pathology may not be affected. However, the patient does benefit greatly from this therapeutic application. An example is the muscle spasm resulting from an acute ruptured disc impinging on the nerve root.

Finally, heat is also applied to improve the body's defense mechanisms. Mild degrees of temperature elevation produce an increase in blood flow and a moderate increase in vascularity. Higher temperature levels will produce the hallmarks of an inflammatory reaction, which can be produced in varying degrees depending upon the temperature elevation.

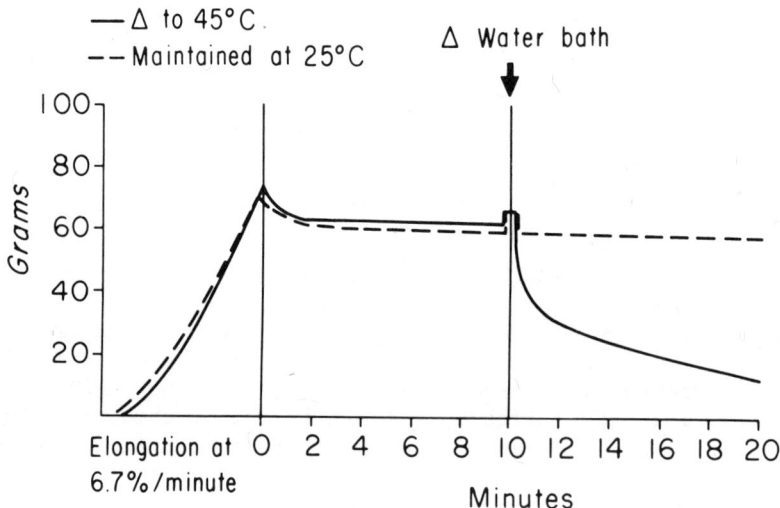

Figure 1. The effect of temperature on tendon extensibility. Ref. 1

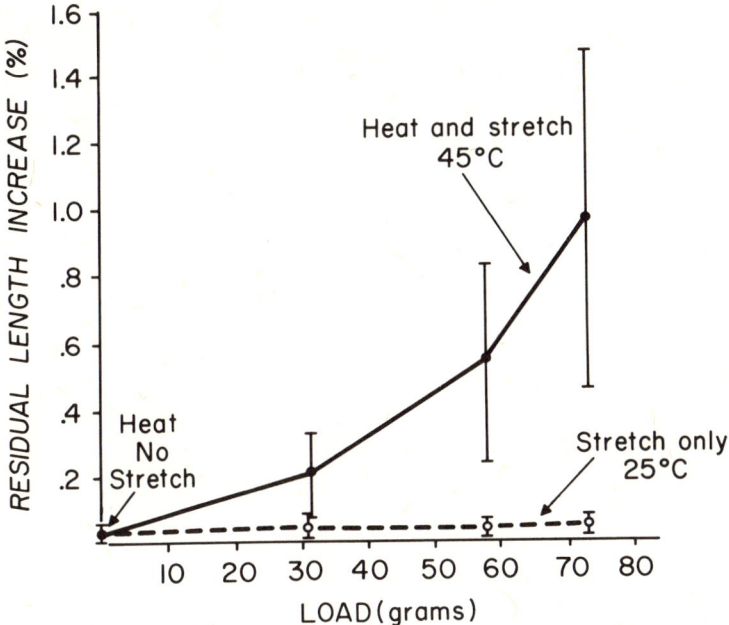

Figure 2. Residual tendon length as measured after loading at the indicated levels of 45°C and 25°C baths. Ref. 2

More recent investigative results support this concept of therapeutic heat application. The viscoelastic properties of tendons and of other collagenous tissues have been studied at therapeutic temperatures. At normal body temperature, the elastic properties are dominant and the tissue shows only a minimal amount of creep. Figure 1 shows, as an example, the behavior of tendons when they are elongated to the point where they develop a corresponding tension. If this length is maintained, the tendon will show a slight decay in tension. If the temperature is changed from normal tissue temperature to a therapeutic temperature of 45°C, the tension decays significantly in the treated sample, whereas the control at normal tissue temperature essentially maintains its tension. If tendons are heated and stressed (Figure 2), residual elongation is greater than if they are stressed, but not heated. If the tendons are subjected to the same loads and maintained at normal tissue temperatures, little residual elongation is produced. After a therapeutic application of heat and tension, however, a permanent elongation can be achieved. The amount of elongation achieved in the laboratory, when compared with elongation produced in clinical practice as measured by increase in joint range, seems to be of the same order of magnitude. In clinical studies done some time ago in which only heat was applied, no increased range of motion was demonstrated because the tissue was not simultaneously stressed in conjunction with therapeutic heat application. This is demonstrated by comparison of the heated and non-heated tendon with no load applied, Figure 2.

Joint stiffness has been studied more recently by Backland and Tiselius. The viscoelastic properties of the joint were measured by moving the metacarpophalangeal joint of the index finger passively through the full range of extension and flexion with the patient relaxed. The range of motion was plotted against the resistance offered to motion, measured by strain gauges applied to the crank. The slope of the resulting hysteresis loop indicates stiffness. These studies also show that the measured stiffness, which correlated highly with pain, was reduced by the application of heat.

This discussion of the physiological effects of heat application in a therapeutic setting indicates that it may be a very valuable adjunct to restoration of function, however, it is not a cure for disease. New information and apparatus resulting from research over the past years have made it possible to apply heat effectively to almost any part of the body for therapeutic purposes. Dosimetry, in the form of feedback of actual temperatures in the tissues during therapeutic application, is not available in most cases. Therefore, one has to rely heavily on the sensation of pain, which is an adequate warning signal that tolerance levels have been exceeded. Therefore heat should not be applied, or it should be applied only with special precautions in areas which are anaesthesic or lack pain sensation, or if the patient is obtunded. Based on experience, it is quite safe to exceed the pain threshold of a patient for a brief moment, if it is not done consistently.

Malignancies should not be treated, since depending upon temperature elevation, the growth rate may potentially be increased even though the tumor can be destroyed at higher temperatures. In addition, an increase in the rate of development of metastases can be anticipated because of the increase in vascularity and blood flow.

Since vascularity and blood flow both increase with the application of heat, it is obvious that any hemorrhagic diasthesis may be aggravated.

Treatment should not be applied to areas of ischemia, since the elevation of tissue temperature increases metabolic activity, potentially producing a situation where the reduced blood flow could not keep up with the demand.

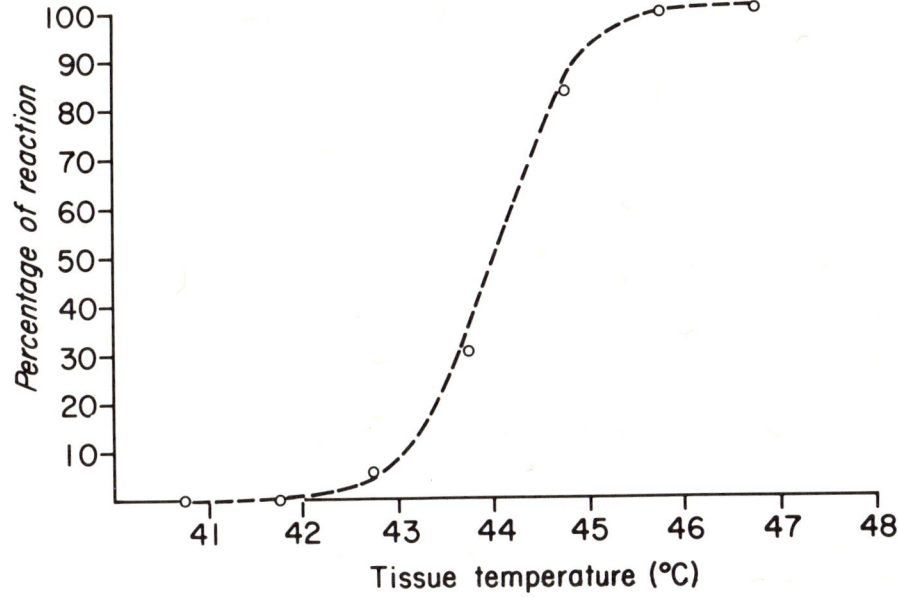

Figure 3. Dependence of hyperemia on tissue temperature. Ref. 3

Heat in general should not be applied to the reproductive organs, the developing fetus or the eye, if temperature elevations to therapeutic levels are anticipated.

The extent and degree of the biologic reaction to therapeutic heat application depends first on the level of tissue temperatures (Figure 3).

Second in importance is the duration of the tissue temperature elevation. Some reflex responses are also dependent on the rate of temperature elevation and the area of tissue exposed to heat. In Figure 3 it is apparent that the therapeutic range of temperature is a narrow one, the upper part of the curve being associated with destructive phenomena. Thus it is imperative to use adequate techniques of application to ensure that the desired temperature range is reached in the tissues to be treated. This is especially important when vigorous therapeutic effects are to be obtained, such as when an effect on collagenous tissue is desired. The temperatures required to produce good results are close to the maximally tolerated temperature levels. Therefore when the site of the contracted tissues has been determined, it is necessary to select a form of energy which will be propagated and absorbed in such a way that it will selectively raise the temperature in the area to be treated.

In this manner, the peak temperature can be brought to the maximally tolerated level without excessively elevating the temperatures in other areas.

The various heating modalities each produce highest temperatures at different sites in the organism, based on the form of energy and the technique of application. In selecting a modality for a given purpose, it is important first to determine which area should be heated and then to select the modality and technique of application which will selectively raise the temperature in that desired area.

An understanding of the way such temperature distributions are produced by the various modalities is essential as a basis for therapy. First, the pattern of relative heat, the amount of energy converted into heat at any given depth of the tissues, is important. Second, the thermal properties of the tissues will determine the resulting temperature rise. If time is sufficient, conductivity of the tissues may modify the temperature distribution. In addition, physiologic factors also greatly modify the ultimate temperature distribution. If blood flow is selectively triggered in one area, increased cooling of the heated area occurs. In most applications, the initial temperature distribution of the organism is modified by superimposing over it the therapeutic temperature distribution produced by the application of one of these modalities. Usually the initial temperature distribution is favorable for deep heating, because tissues have lower temperatures at the surface and higher temperatures closer to the core. While predictions of heating can be made from the pattern of relative heat and from the thermal properties of the tissues, the ultimate information on the resultant temperature distribution can be obtained only in vivo, and ultimately in man.

The following examples show temperature distributions which are produced by the typical application of various forms of non-ionizing radiation. Figure 4 shows the temperature distribution which results from the application of shortwave diathermy with a special induction coil applicator to human volunteers. The highest temperatures are usually produced in superficial musculature or at the interface between subcutaneous fat and musculature. If a capacitor applicator were used, the highest temperatures would be produced more often in the deep subcutaneous tissues and to a lesser extent in the superficial musculature.

The technique of application used may markedly modify the results. If an internal electrode (vaginal or rectal) is used with a large belt electrode over the abdomen, a high concentration of energy is produced around the internal electrode, and the temperature is selectively raised in the pelvic organs. This selective heating of the pelvic organs is only possible with this technique

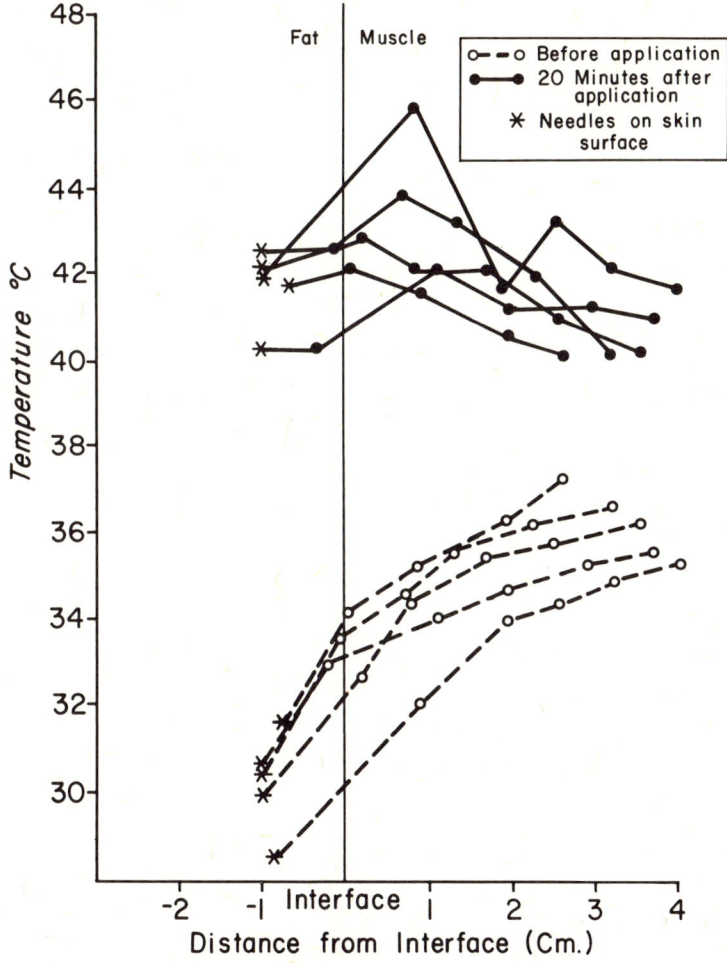

Figure 4. Temperature distribution in five human thighs before and at the completion of 20 minutes of exposure to shortwave with 2 cm air space between applicator and skin surface. Ref. 4

of shortwave diathermy application. It cannot be duplicated by any other technique of application or by other forms of energy such as microwave or ultrasound. In this case it is possible to measure tissue temperature actively during application, since the internal metal electrode readily assumes the surrounding tissue temperature, and its temperature can be measured with an alcohol thermometer. Because of the fact that the blood flow in the pelvic organs can be greatly increased, it is possible that the demand on the heart may be noticeably increased, and therefore a vigorous application of this type may be contraindicated in cases of chronic heart failure. Similarly, this is the only modality which can significantly increase the temperature in the pregnant uterus, and therefore pregnancy represents a specific contraindication.

When shortwave diathermy is applied, metal implants in the tissues represent a definite hazard. If implants are located in an area where a substantial amount of energy is available, the current may be concentrated in the metal or in the surrounding tissues, and may produce selective burns. Thus metal implants represent a contraindication to shortwave diathermy application and is one example of a specific contraindication for a specific modality.

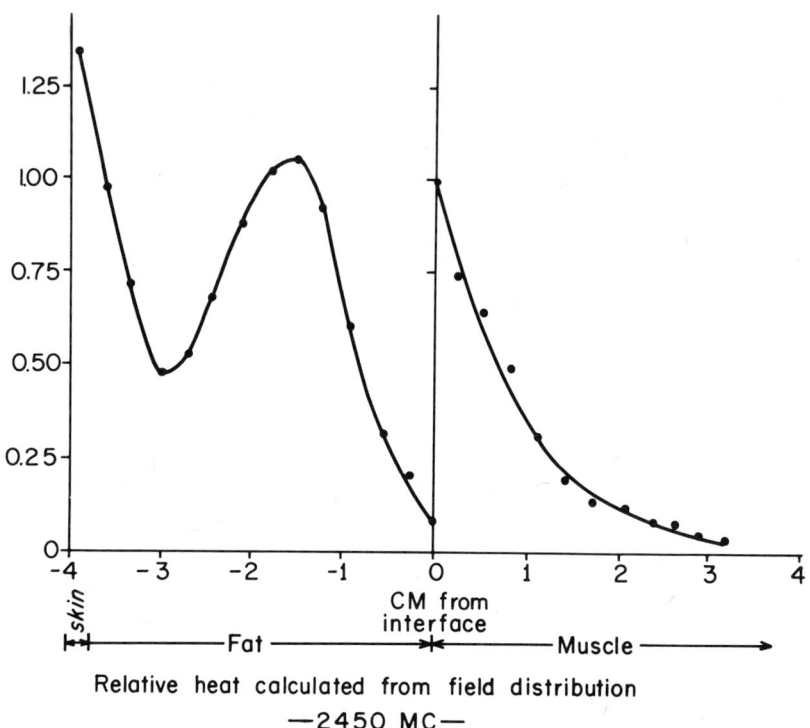

Figure 5. Pattern of relative heating calculated from field distribution at a frequency of 2450 MHz. Ref. 5

Figure 5 shows the pattern of relative heat produced by the application of microwaves at 2,450 MHz with a commercial applicator. At this high frequency, too much of the energy is converted into heat in the subcutaneous fat, too little in the musculature, and

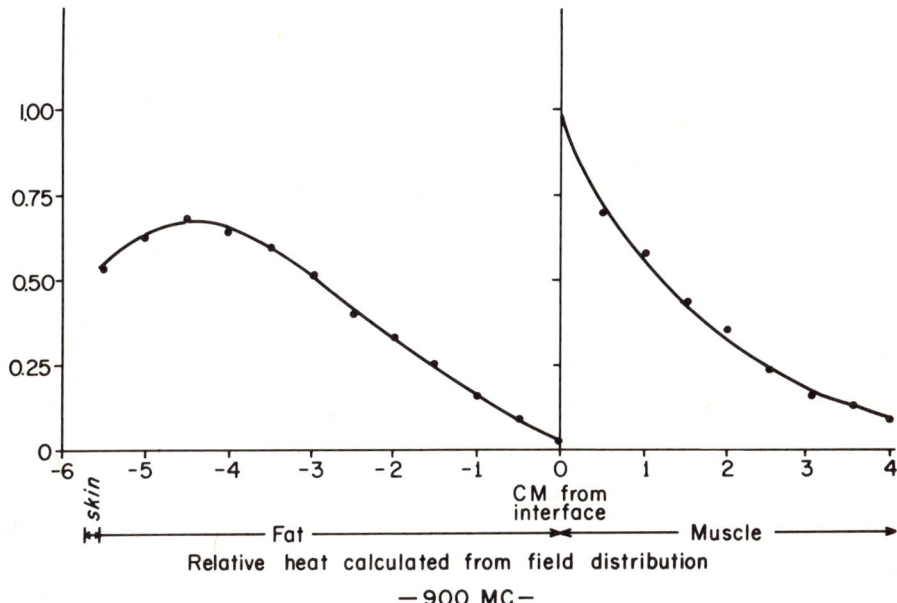

Figure 6. Pattern of relative heating calculated from field distribution at a frequency of 900 MHz. Ref. 6

thus the depth of penetration into the musculature is relatively poor. This compares with Figure 6, which shows the pattern of relative heat produced by lower frequencies such as 900 MHz or below. At these frequencies, relatively little energy is absorbed in subcutaneous fat tissue of an average thickness of two centimeters. The depth of heat penetration and the heating in musculature is much greater. The temperature resulting from the application of microwaves at 2,450 MHz is very similar to that obtained by various methods of shortwave diathermy, which produce the highest temperatures in deep subcutaneous tissue or in superficial musculature. Therefore, it would be advantageous to use the lower frequency of 915 MHz, which is available for medical use in the United States. A special, direct contact applicator which has been developed by Dr. A. Guy and associates, allows cooling and temperature control of the surface when in contact with the skin. Figure 7 shows the resultant temperature distribution in human volunteers, which demonstrates that this modality selectively raises the temperature throughout the musculature from the subcutaneous muscle interface to the periosteum of the bone. The application of microwaves at higher frequencies cannot produce this result. The direct contact applicator has major advantages, for it produces less stray radiation when compared with standard applicators of therapeutic

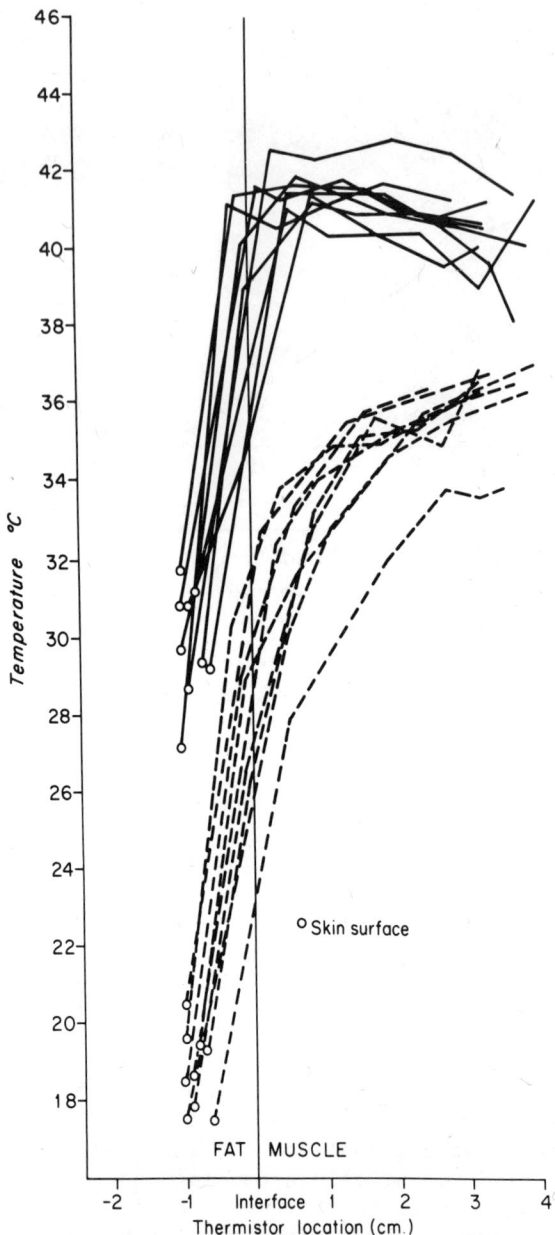

Figure 7. Temperature distribution in all volunteers with ≤ 1 cm of subcutaneous fat before (----) and 20 minutes after (——) microwave application. Ref. 7

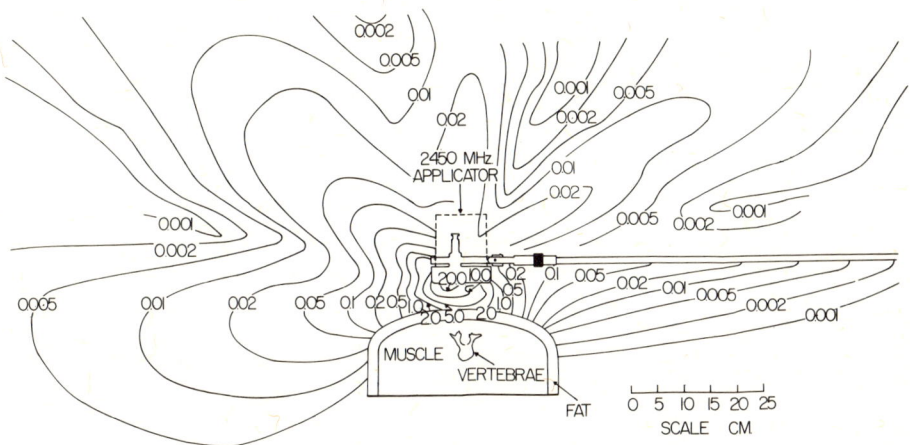

Figure 8. Leakage radiation (mW/cm^2) from phantom model of human back exposed to 2450 MHz C director (1-W input). Ref. 8

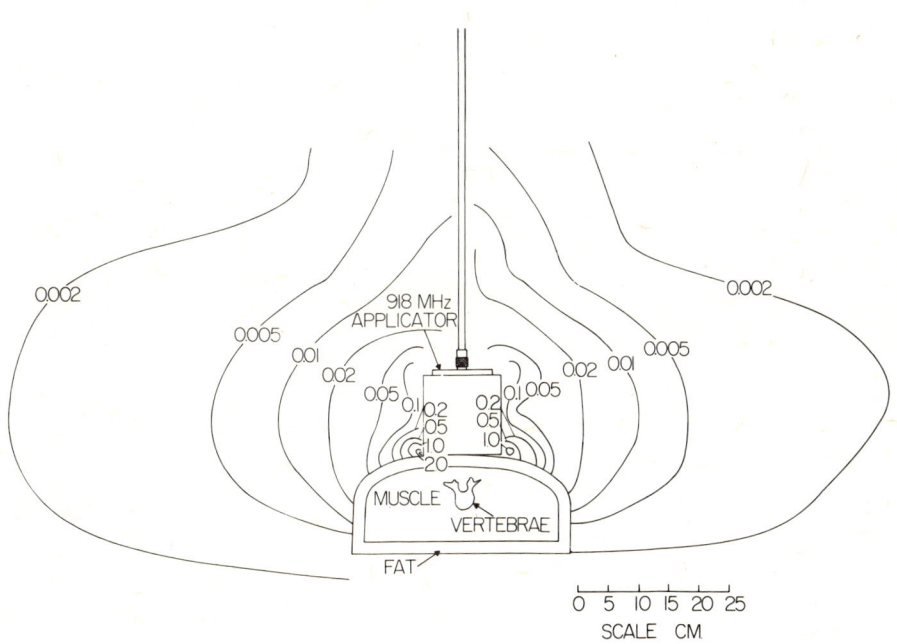

Figure 9. Leakage radiation (mW/cm^2) from phantom model of human back exposed to 918 MHz 13 cm by 13 cm square aperture TE_{10} mode source (1-W input). Ref. 9

machines at 2,450 MHz (Figures 8 and 9). It is highly desirable that the amount of stray radiation be reduced since it is easier to prevent inadvertent exposure of parts of the body outside the therapeutic area. Specifically, the eyes and testicles should not be exposed. Reduction of stray radiation is also desirable since the possibility of inadvertent exposure to the therapist is reduced. As an example of a specific contraindication to microwave application, the eye can be used. Figure 10 shows clearly that radiation exceeding 150 milliwatts per square centimeter incident on the eye may produce cataracts. Other studies by Dr. A. Guy and associates have shown that cataract formation is probably a temperature-related phenomena. Metal implants are another specific contraindication in microwave therapy.

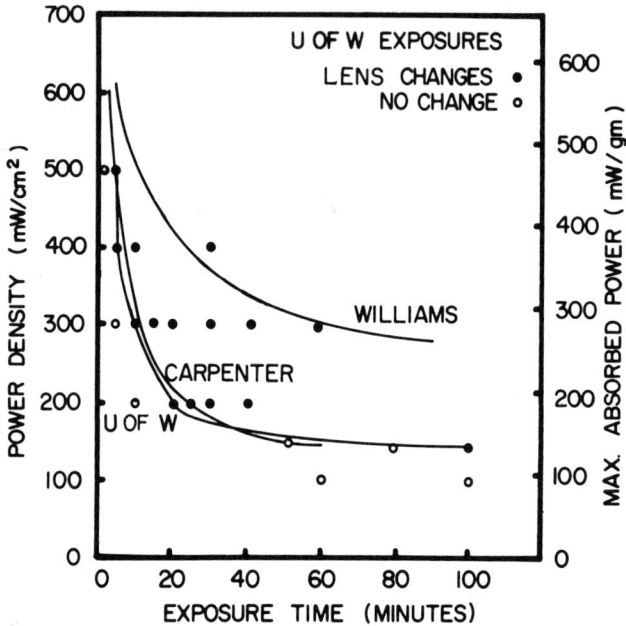

Figure 10. Time and power density threshold for induction of cataractous change in rabbits by single exposure to near zone 2.45 GHz radiation. Ref. 10

The application of ultrasonic energy at a frequency of 0.8 MHz or 1.0 MHz has a calculated pattern of relative heat which shows a very good penetration (Figure 11). The penetration is so good that relatively little ultrasound is absorbed in soft tissue, and therefore this modality is not effective for heating tissue in general.

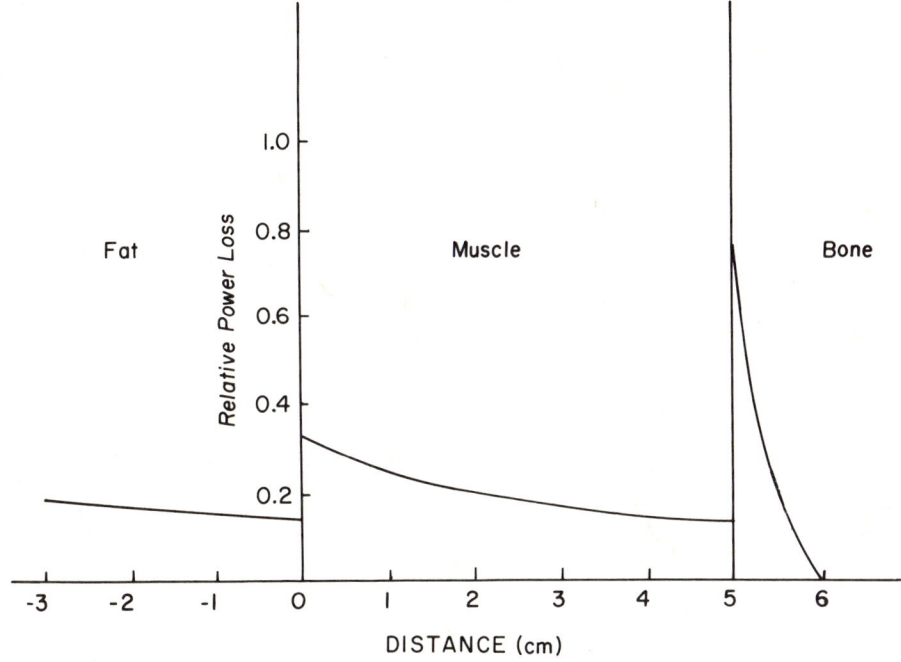

Figure 11. Pattern of relative heating calculated from field distribution at a frequency of 1.0 MHz.

However the temperature can be selectively raised in certain sites in the tissues. For instance joints are selectively heated because bone absorbs eight times more energy than soft tissue, reflection and scattering occurs at the bone interface due to mismatch of acoustic impedance, and the conversion of longitudinal waves into rapidly attenuated shear waves may contribute to rise in temperature at this site. Other areas which can be selectively heated include myofacial interfaces and large nerve trunks. As an example, Figure 12 shows the temperature distribution in a human after ultrasound exposure selectively raised the temperature in front of the joint where capsular contractures occur. As another example, Figure 13 shows the increase in temperature to a therapeutic level in the hip joint of a pig. This heating pattern cannot be reproduced by any other modality, such as shortwaves or microwaves, even though first degree burns are produced in superficial tissues. This experiment again points out that each modality selectively heats certain areas of the body, and that substitution of one modality for another is not possible.

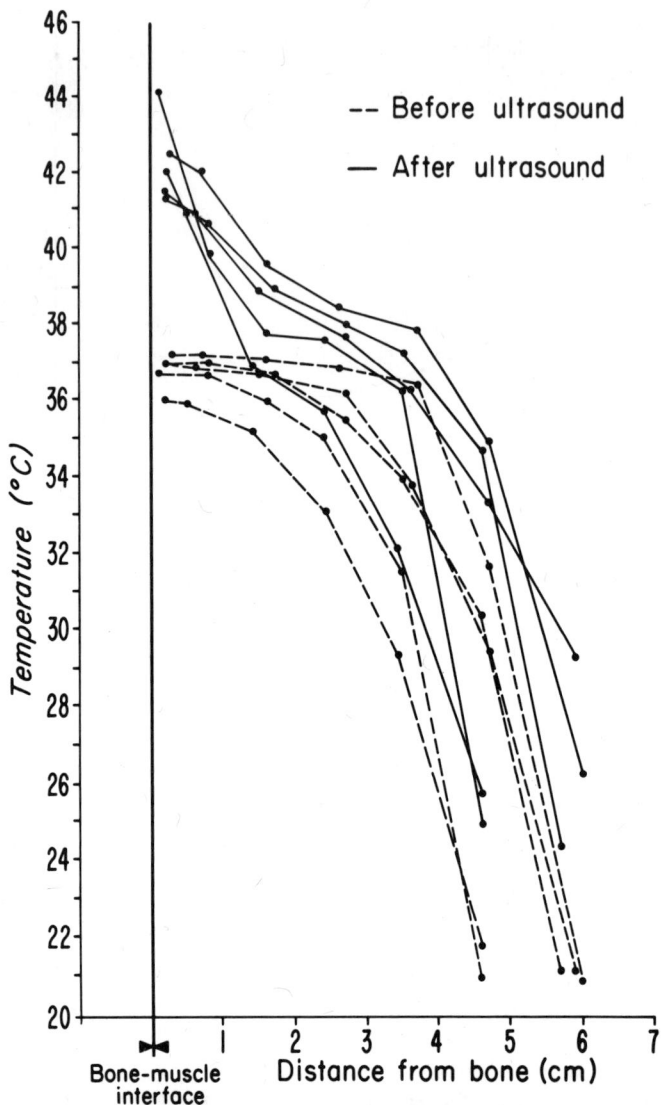

Figure 12. Comparison of temperature distributions in five human thighs before and after exposure to ultrasound, using a mineral oil coupling medium at 18°C. Ref. 11

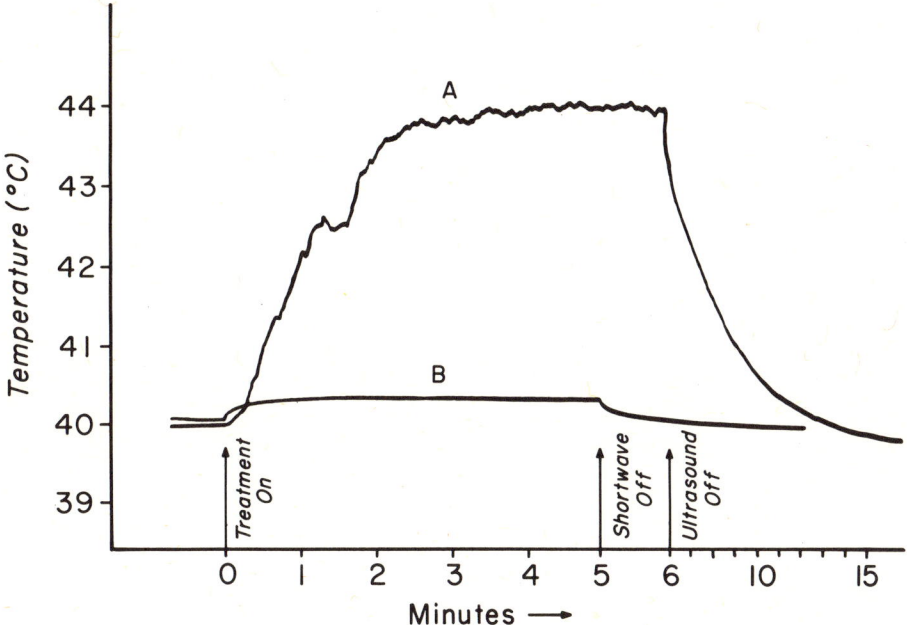

Figure 13. Change in temperature inside hip joint during exposure to A, ultrasound, and B, shortwave. Ref. 12

Controlled clinical studies confirm the specificity of each modality. Such a study was done on a group of patients with para-arthritis at the Mayo Clinic. It was found, as Table 1 shows, that microwave treatment, which heats the subcutaneous tissues and superficial musculature, was not as effective as the use of ultrasound, which selectively raises the temperature of the contracted capsular tissues. Both modalities were used with a standard stretch and exercise program. Within five days of treatment, the gain in range of motion was significantly different between these two groups of patients. Similarly, a well-controlled clinical study on patients with hip fractures showed (Table 2) that ultrasound application produced better results than the application of infrared heating when a standard exercise program was applied to both. The differences in range of motion were significant after five days of application.

The utility of non-ionizing radiation, such as microwave, ultrasound and shortwave diathermy, is most commonly based on the heating effect these modalities produce. Nevertheless non-thermal effects could potentially be valuable or detrimental. So far, review of the literature shows that while many non-thermal effects are proven to exist, there is far less evidence that they also occur under the

Table 1. Gain in Range of Motion after Ultrasonic and Microwave Treatment (Ref. 13)

GAIN IN:	AFTER TREATMENT WITH:	
	Ultrasound	Microwaves
Forward flexion	27.4°; ± 2.3°*	16.1°; ± 1.5°
Abduction	32.6°; ± 2.5°	21.2°; ± 2.1°
Rotation	45.4°; ± 2.8°	17.3°; ± 4.0°

*Standard error of the mean.

Table 2. Comparison on Amount of Change in Range of Motion after One Week of Treatment* (Ref. 14)

	WITH ULTRASOUND			WITH INFRARED				
	N	Mean	S.D.	N	Mean	S.D.	t	p
Hip								
Flexion	15	21.67	9.7	15	5.40	11.4	4.057	.01
Extension	15	10.40	8.2	15	−3.20†	7.7	4.503	.01
Abduction	15	6.33	7.4	15	−1.67†	5.6	3.225	.01
Adduction	15	9.67	6.6	15	−1.20†	6.0	4.567	.01
External Rotation	14	12.86	7.9	15	0.20	7.9	4.178	.01
Internal Rotation	14	10.93	7.6	15	−1.60†	8.8	3.965	.01
Knee								
Flexion	15	18.33	14.9	15	10.33	13.3	1.498	.20
Extension	15	3.60	3.9	15	−3.47†	4.8	4.259	.01

* Ultrasound and infrared groups compared using independent samples method.
† When the mean is expressed as a negative value, range of motion has been lost during treatment.

conditions of therapeutic application. If they do exist, there is as yet no evidence that this may lead to desirable therapeutic results. If on the other hand they are detrimental, as the occurrence of cavitation or degassing in ultrasound application, it is essential to avoid the conditions which would produce these phenomena, since they are destructive. Destructive lesions in mammalean tissue have been proven to be the result of cavitation, because it was found that they were prevented by sufficient external application of pressure. The threshold of the pressure required for prevention was related to the ultrasonic intensity, as anticipated. Rearrangement of cellular components of the blood in the bloodstream, as described by Dyson and others, may also pose a problem with stationary ultrasound application. Since therapeutic application is normally performed with a moving applicator, these phenomena are rare and may not occur at all. If they occur, no clear relationship to any detrimental result in patients has been demonstrated. It is highly desirable however, that these phenomena be explored further, not only to determine whether or not special effects exist, but if they exist, how they are produced. Understanding the mechanism will also allow judgements to be made as to whether they may be useful or detrimental for therapy. If detrimental effects are found, the therapeutic conditions should be modified so that these effects do not occur. If proper equipment, frequencies and techniques of application are used, the high concentration of cells in the blood, body fluids and serum impede the occurrence of gaseous cavitation. Cavitation can be produced at low intensities in fluid media such as saline or in the media of the eye; therefore, this represents a special contraindication to ultrasound application. Ultrasound should not be applied to the eye at therapeutic intensities.

In summary, the whole rationale for the use of the various modalities, shortwave, microwave and ultrasound, rests in the fact that they all heat tissue, but each modality produces a different temperature distribution and therefore has specific indications for useage. Non-thermal effects are documented, some of them being identified as detrimental. These must be avoided in therapeutic application. In therapeutic situations it is obvious that the intensity used must be high enough to produce significant if not major biologic changes. Thus, one looks for a trade-off and utilizes the modality if the potential hazard is minimal as compared with the potential benefit. Further studies in the area of non-thermal effects would be desirable .

The research on which this report was based was in part supported by Social and Rehabilitation Service Grant Number 16-P-56818-10.

REFERENCES

1. Lehmann, J.F., et al: Effects of Therapeutic Temperatures on Tendon Extensibility. <u>Arch. Phys. Med.</u>, 51:481-487, 1970.

2. Lehmann, J.F., et al: Effects of Therapeutic Temperatures on Tendon Extensibility. <u>Arch. Phys. Med.</u>, 51:481-487, 1970.

3. Lehmann, J.F.: The Biophysical Basis of Biologic Ultrasonic Reactions with Special Reference to Ultrasonic Therapy. <u>Arch. Phys. Med.</u>, vol. 34, 1953.

4. Lehmann, J.F., et al: Selective Muscle Heating by Shortwave Diathermy with a Helical Coil. <u>Arch. Phys. Med.</u>, March 1969.

5. Lehmann, J.F., et al: Comparison of Relative Heating Patterns Produced in Tissues by Exposure to Microwave Energy at Frequencies of 2,450 and 900 MHz. <u>Arch. Phys. Med.</u>, vol. 34, 1962.

6. Lehmann, J.F., et al: Comparison of Relative Heating Patterns Produced in Tissues by Exposure to Microwave Energy at Frequencies of 2,450 and 900 MHz. <u>Arch. Phys. Med.</u>, vol. 34, 1962.

7. DeLateur, B.J., et al: Muscle Heating in Human Subjects with 915 MHz Microwave Contact Applicator. <u>Arch. Phys. Med.</u>, vol. 49, 1968.

8. Guy, A., et al: Therapeutic Application of Electromagnetic Power. <u>Proc. IEEE</u>, January, 1974.

9. Guy, A., et al: Therapeutic Application of Electromagnetic Power. <u>Proc. IEEE</u>, January, 1974.

10. Guy, A., et al: Effect of 2450 MHz Radiation on the Rabbit Eye, Submitted to the <u>Proc. IEEE</u>, June. 1974.

11. Lehmann, J.F., et al: Selective Heating Effects of Ultrasound in the Human Being, <u>Arch. Phys. Med.</u>, 47:331-339, 1966.

12. Lehmann, J.F., et al: Comparative Study of the Efficiency of Shortwave, Microwave, and Ultrasonic Diathermy in Heating the Hip Joint. <u>Arch. Phys. Med.</u>, vol. 40, 1959.

13. Lehmann, J.F.: Comparison of Ultrasonic and Microwave Diathermy in the Physical Treatment of Periarthritis of the Shoulder. <u>Arch. Phys. Med.</u>, 35:627-634, 1954.

14. Lehmann, J.F., et al: Clinical Evaluation of a New Approach in the Treatment of Contracture Associated with Hip Fracture After Internal Fixation, <u>Arch. Phys. Med.</u>, 42:95-100, 1961.

-DISCUSSION-

ELY - You have given us contraindications for microwave irradiation of the skull because of the brain and the eye. What do you think of other areas on the head, such as sinuses and so forth?

LEHMANN - I would not do it, even though quite frankly there is and have been a large number of patients being exposed this way without any untoward damage. Specifically, many opthalmologists still apply shortwave and sometimes microwaves to the eye. No matter what you think about it, whether it is really dangerous or not, good sense tells you not to do it because there have been several major lawsuits in the country. Even some limited exposure to shortwave or microwave equipment has been enough to result in claims that cataracts which occurred, some many years later, are due to microwaves. And juries are not always scientific.

GLASER, Z. - A claim has been made that a cataract resulted in the human eye on the side of the body opposite to which some diathermy was applied to the shoulder. Would you care to comment on that?

LEHMANN - Well, that is one of the cases I have in mind. From all we know about the field distributions, the intensities were too low to really produce any effect, nor was exposure prolonged or excessive. So, from all we know, on a scientific basis, I don't believe that it happened. It has been claimed that these cataracts look so different that the clinician can tell that it is a microwave cataract. Most opthalmologists do not believe it. As a matter of fact, in Dr. Lin's and Dr. Guy's experiments with rabbits, done conjointly with our Department of Ophthalmology, they found that they cannot see any differences. There could be many other causes producing exactly the same picture.

HILL - I liked your results on measurements of extension of flexibility of joints. I wonder whether you got any corresponding data on objective evidence for increase in vascularity?

LEHMANN - There have been various studies which clearly indicated that the blood flow is increased. There have been studies just measuring the infrared emission and relating that to increase in vascularity of the skin. There have also been different methods

LEHMANN - used. Some have used flow meters. Other studies have been done more recently with isotope clearance. I think Dr. Hardy can probably tell much more about it than I can.

HARDY - Well, I can mention a recent experiment from our laboratory. We implanted heaters in the abdomen of sheep, and heating the surfaces of the heating elements to around 45°C and maintaining this temperature for some hours. Of course, there were some animals in which for one reason or other the heaters were broken and these elements served as controls because we could not heat them. We observed in looking at the encapsulation of the heaters at an autopsy that those around the heaters that had been maintained at 45°C were highly vascularized; the capsules around those heaters which were broken were just a mass of avascular tissue. These experiments have led us to believe this 45°C temperature level somehow does promote a kind of vascularization.

HILL - The thought I had on this was that you also mentioned that one of the contraindications for heat was in regions of malignancy. I know there is background on this. On the other hand, there is another side to this: that some of the deficiencies in cancer treatment result from lack of vascularity, and, I wonder how far you have thought along these lines?

LEHMANN - Sometime ago, I did some experiments on that when I was at the Mayo Clinic. My concept was that the ultrasound is absorbed more in bone than in the subcutaneous and soft tissue. So, you could produce a selective effect on any tumor in bone because of the absorption characteristics. If you raise the temperature, it likely would enhance sensitivity to ionizing radiation. So, we demonstrated a saving of about 50% of the ionizing radiation dose. It was a temperature effect and we could prove it. If we did not allow temperature elevation it didn't occur; if we allowed the temperature to rise it did occur. If we reproduced the temperature, by means other than ultrasound, it occurred to the same degree. The only problem was -- and there was my miscalculation, the malignancy for which it was primarily applied was ostegenic sarcoma. If they are really malignant, they don't have any bone in them. The selectivity is gone. At that point I stopped.

MILLER, M. - When you showed the picture of the cavity in the tissue, and you spoke of that as "cavitation," did you mean that cavitation process caused it or that there was a cavity formed?

LEHMANN - I meant cavitation caused that. A cavity can be caused by anything, you know, depending on how you prepare the histologic preparation. It may be an artifact. But, this type

LEHMANN - of lesion was very characteristic and reproducible thousands of times. You prevent it by applying pressure and therefore you can say it's cavitation. It is gaseous cavitation; it is not true cavitation.

MILLER, M. - It seems as though the location of the cavity was related to places where you might have spaces in the tissue naturally, where cavities would naturally be present and a bubble would have a chance to form and grow and do its thing. In ultrasonicated plant tissue, we have observed a somewhat similar result (vide Miller et al., Rad. Bot. 14: 201-206, 1974). At the very tip of the meristem, where the cells are tightly packed, there do not appear to be these broken cells. But, back behind this region where the cells are elongating and there are intercellular spaces forming, we have observed disruption of cell walls.

LEHMANN - Yes, I think you are very right. I think we did lots of experiments with mammalian cells and found that the higher the concentration of cells, for instance blood cells in a suspension, the less likely it is that you get cavitation. As I pointed out the same is true for viscosity of the suspension medium. Whether you use saline or a more viscous type like blood serum, the serum inhibits cavitation. We also looked at onion roots and found little lesions which were due to cavitation as proven by the pressure correlation.

MILLER - Did you have any thoughts on other types of non-physical effects other than cavitation?

LEHMANN - Yes. Non-thermal. There are quite a number of possible non-thermal effects. For instance streaming. You can observe it in larger fluid volumes in the body. I think that Dr. Nyborg has shown it in his film. You see it even in eosinophilic leucocytes. The granules move. And biologic membranes; we could prove that part of the increased rate of exchange of ions across those membranes was due to a simple stirring effect due to the streaming the sound produced. Of course, in addition, the temperature changed the permeability of the membrane and the rate of diffusion.

TAYLOR - Can I ask a question similar to the first question of Dr. Miller's? Apart from air-filled tissue spaces in plants, air is frequently present in the clinical situation, for example in the gut. Then you have a tissue air interface.

LEHMANN - You have air in the gut. Yes, it is possible.

TAYLOR - This reminds me very much of some experiments I did about

TAYLOR - five years ago in a similar situation except I was shooting through liver (Taylor and Connolly, 1969). On the distal surface of the liver there was a tissue air interface. I put a series of thermocouples through the liver substance. On the near side, nearest the transducer, the temperature rose a couple of degrees centigrade. On the far side, the temperature increased by 25 to 30°C. In other words, at the reflecting interface where one presumably got sheer node conversion and absorption of the transverse waves, a marked temperature effect was produced. I wonder if in fact, you are getting a thermal effect by a similar mechanism in Dr. Lehmann's experiments. Possibly the tissues may even boil. If you increase the pressure, you raise the boiling point which is again in accord with your observations. (Ref. Taylor, K.J.W., Connolly, C.C., 1969. Differing Hepatic Lesions Caused by the Same Dose of Ultrasound. J. Path. $\underline{98}$: 291).

LEHMANN - We have also studied these things. They are temperature-dependent, but they are not really that temperature-dependent. I can, for instance, with the proper ultrasonic intensity get the same lesion keeping the temperature normal by simultaneously cooling.

TAYLOR - The same intensity?

LEHMANN - Yes, similar intensity, same location. We observed effects of cavitation in other tissues for instance in blood vessels. There was a hole right in the wall of the vessel. The hole had a diameter of perhaps 10-15 microns, and the blood leakage occurred from that hole. So, cavitation also occurred elsewhere and not just in the surface of the gut. So, I think cavitation is a real danger. It is likely to occur in fluid media. Therefore, I would not apply ultrasound in therapeutic intensities to the eye.

MICHAELSON - I would like to clarify one point to have the record clear. You didn't mean to say that the standard is being dropped from 10 mW/cm^2 to 1 mW/cm^2?

LEHMANN - No, I didn't mean to say that. If anything, I am on the other side of the fence. Use one example, what happens when people go overboard in this respect. The hammer is a lethal weapon, and if you want to be very sure that no one can do any harm with it, the best thing to do is to make it out of soft rubber so no matter how hard you hit someone over the head, he cannot be injured by it. It would be a poor hammer, though. And this concept applies to therapy as well.

ULTRASOUND IN SURGERY

Padmakar P. Lele
Departments of Mechanical Engineering and
  Nutrition and Food Science
Laboratory of Experimental Medicine, 26-023
Massachusetts Institute of Technology
Cambridge, Massachusetts 02139

ABSTRACT

Since its introduction into medicine some 25 years ago ultrasonics has made great strides in both the diagnostic and therapeutic fields and is being used to an ever increasing extent in clinical medicine. The reasons for this are two-fold; first, it often provides information regarding tissue structure that is not obtainable during the life of the patient by any other diagnostic technique, and second, conventional radiological techniques have proven to be more hazardous than was believed previously. Thus, for instance, using rather simple pulse-echo techniques, in ophthalmology it is possible to measure the dimensions of the various components of the eye more accurately than is possible using light and it is possible to detect conditions such as detachment of the retina even when obscured by a hemorrhage in structures lying in front of it; in cardiology it is possible to diagnose accumulation of fluid in the pericardial sac or the structural and functional state of several valves of the heart; and in obstetrics it is possible to diagnose pregnancy in the presence of fibroid tumors of the uterus. Thus, in fact, virtually no organ is inaccessible to ultrasonic examination. In addition to the pulse-echo techniques, ultrasonic Doppler methods are also used extensively for detection of movement such as pulsations in blood vessels. This enables detection of pregnancy as soon as the fetal heart is accessible to ultrasonic examination; or diagnosis of thrombosis (blockage) of veins deep in the leg.

In all these applications, as far as is known, no irreversible or permanent alterations are produced in the tissues as a result of their irradiation with ultrasound. On the contrary, it is the beam of ultrasound that is modified by its passage through the tissues and the type and extent of this alteration yields information on organ structure. At higher ultrasonic intensities (and dosages) however, tissues cannot completely recover from the effects of ultrasonic irradiation and permanent structural result. Ultrasound at such intensity levels is therefore used for therapeutic purposes such as in the treatment of Ménière's Disease where ultrasonic surgery is the treatment of choice and can alleviate the illness without producing deafness. Another such application is the surgery of the cataract of the eye. For these diagnostic and therapeutic applications ultrasound is used at frequencies of about 1 to 10MHz. At lower frequencies it is used extensively for disruption of cell walls for extraction of thermolabile intracellular ingredients. A discussion of all the current applications, their underlying physical principles, and instrumentation can be found in some of the general texts [1-4] and specialized publications [5-8]. A brief outline is presented here of some of the work currently being conducted in the author's laboratory. We deal first with the development of focused ultrasound as a surgical tool for clinical and research use.

An ideal surgical tool should have the following characteristics: (1) It should destroy the target and nothing but the target; it should not leave a track of destruction or injury as the D.C., Radio-frequency or Cryogenic probes do; (2) The resultant lesions should be (a) reproducible in size and predictable, (b) their size and shape should be finely controllable, (c) they should be non-hemorrhagic, (d) they should be discrete or sharply demarcated from surrounding tissue which should be totally unaffected, and (e) they should be instantaneous in development; (3) There should be no delayed effects such as those produced by X-rays and other ionizing radiations; and (4) With sub-lethal dosages it should be possible to produce transient alterations of function to assist in target localization.

Focused ultrasound meets these requirements of an ideal surgical tool [9:105]. It has the demonstrated ability to destroy preselected targets located deep within tissue without any damage to the tissue in the path or surrounding the lesion [10:484, 11:513, 12:502]. The margins of the necrotised tissue are sharply demarcated from the surrounding normal tissue [10:484, 11:513], the size of the lesion is predictable to a high degree of accuracy and can be finely controlled [11:513, 12:502], the lesions develop almost instantaneously [13, 14] and there are no late effects such as those associated with ionizing radiations [10:484, 15:345]. Most of these apparently unique effects are attributable to the fact that many of the soft tissues are relatively permeable to ultrasonic energy and permit it to be focused deep within the tissues generating locally high intensities. Some other forms of energy such as infrared are almost totally absorbed at or near the surface. Thus, attempts to use a convergent cone of infrared radiation [16] as well as a focused beam of coherent radiation from a ruby laser for placement of lesions 15-20mm deep within the cat brain resulted only in superficial lesions. No deep lesions were produced. High frequency electromagnetic waves such as those used in medical diathermy and microwaves, on the other hand, do penetrate the tissues; but their wavelenghts are too long for surgical purposes. The frequencies of ultrasound currently used are even lower; but as sound travels at much slower speeds than do the electromagnetic waves, the wavelenghts are very much shorter. Focused ultrasound thus holds a unique position among physical agents.

A simple collimated beam of ultrasonic radiation such as that radiation from a plane disc transducer cannot produce trackless, deep focal lesions, the formation of which depends on the creation of a zone of high intensity at and restricted to the target. This can be accomplished by either aiming two or more collimated or weakly focused beams at the target from different directions [18: 261], or by the use of a single beam converging at the target [19: 494]. The former suffers from the inherent disadvantage that the energy content in the side lobes is rather high and asymmetric; the latter is not only simpler in principle, design and operation, but is probably also more reliable and is being successfully used in several laboratories of basic neural sciences without the continual assistance of physicists and electronics technicians. At the Massachusetts General Hospital it has also been used satisfactorily in patients for transdural irradiation of superficial cortical epileptogenic foci in cases of post-traumatic epilepsy, for performing commissural myelotomy of the spinal cord for intractable pain [20], and for anterior horn neuronolysis in a case of multiple sclerosis, in addition to the irradiation of a number of cases of painful subcutaneous neuromata [21:858].

A general description of the technique will be followed by a discussion of problems particular to its surgical use. Clinical indications for placement of deep focal lesions are commonly to be found in the central nervous system (CNS), which is also amenable to stereotaxic procedures [22]. Though the brain will be used as an illustration, the following description is equally applicable to other soft tissues and organs provided that the acoustical properties of the target are not greatly different from those of the surrounding tissues (e.g., areas of calcification).

Equipment and Technique

Ultrasound is generated by the application of radio-frequency electrical power to a piezoelectric transducer. The transducer material can be either natural or cultured quartz or one of the ceramics, for example barium titanate, lead zirconium titanate, etc. Quartz has greater mechanical strength, lower internal friction, resists electrical abuse to a greater extent, and its function is less temperature dependent than that of ceramics. However, it needs comparatively much higher driving voltages. Ceramics are more efficient, need larger currents at lower voltages but tend to lose their piezoelectric properties if mistreated electrically by overheating or by mechanical shock. They are also inferior to quartz for operation at odd-multiple harmonic frequencies. But because of the lower driving voltages required, the complexity, size and cost of the generator are likely to be lower. On these considerations, although quartz is the transducer of choice for the development of ultrasonic techniques, the ceramics probably are to be preferred for routine application.

The transducer may either be a section of a sphere or a flat disc. In the former, the acoustic output converges at the centre, and the focal length, the solid angle of radiation and maximum working distance are unalterable. The output of the latter is a plane wave (a more or less collimated beam) which is focused by suitable lenses.

The lens is made of a plastic, such as polystyrene or rexolite, is of simple design and is fabricated easily and inexpensively. A disc transducer-lens combination offers flexibility in the choice of focal lengths and thus the solid angle of radiation, beam geometry (spherical, cylindrical or compound focusing) and the shape of the focal region (needle shaped or almost spherical [23]). The lens is coupled to the transducer by a thin film of silicone oil and is held concentric with it by a retainer ring (Figure 2). With the irradiation head machined to the tolerances common in the industry the focus of the system, in relation to the radiating surface of the transducer, is as constant for any given focal length as it is for a focused bowl - even though the irradiation head may be dismantled occasionally for the use of a different lens or for renewal of the silicone oil coupling. Up to 40 percent of the acoustic energy emanating from the transducer may be lost in the lens, depending on its material and average thickness. But the focused output is still more than adequate to meet all probable power requirements. Moreover, it will be seen later that the use of excessively high power levels should be avoided as it is related to the occurence of hemorrhages. The intensity distribution (and thus the focal length) at the focus in water is determined experimentally using a fine thermocouple embedded in soft polyethylene [24].

The size of the smallest lesion that can be made is inversely related to the frequency of the ultrasound, and thus higher frequencies would appear to be more suitable. For the placement of deep lesions at high frequencies much more power is needed at the transducer than at lower frequencies since the attenuation in the intensity of sound in its passage through intervening soft tissues is directly related to the frequency. And since very small lesions deep within the brain are clinically not manifest, frequencies 1, 3 and

5 MHz cover most of the actual needs. The lower frequencies are used for deep lesions in large structures and the higher for lesion placement in small structures such as the spinal cord, where the lesion size is critical. A transducer ground finely to a fundamental resonant frequency of 1 MHz in the thickness mode is used to generate all the three frequencies.

The convergent beam of ultrasound emanating from the lens is transmitted on the soft tissues (the exposed dura mater of the brain) through degassed water contained in the applicator cone. Since the acoustic impedances of air and of bone are very much different from that of water and soft tissues even the smallest bubble of air or spicule of bone lying in the path of sound reflects the energy and distorts the beam geometry. Meticulous care is therefore necessary to prevent either air or bone from protruding into the cone of sound. When subjected to an intense ultrasonic field fluids release dissolved gases in the form of small bubbles. To avoid this, water and saline solutions are degassed thoroughly by boiling for about an hour prior to their use in ultrasonic procedures. After cooling to 37°C under partial vacuum they are stored in airtight, rubber hot water bottles and maintained at a temperature of 37°C.

The focal length of the system with any particular lens is temperature dependent. To maintain a constant focal length, in spite of changing target depths and varying ambient temperature, the temperature of the components in the ultrasonic pathway is maintained at 37°C - the average temperature of mammalian tissues in vivo. A graded series of applicator cones with different working distances as well as a telescoping cone [25] permit quick and easy coupling with different target depths. The sac contacts the dura through a thin film of saline solution and conforms to the shape of the brain. There is thus little chance of air bubbles being trapped in the path of the sound. The hydrostatic pressure in the cone and the sac is normally atmospheric but, if necessary, can be raised to prevent pulsation of the exposed brain. The rubber is thin and does not reduce the sound intensity measurably unless it lies in the focal plane - a situation that cannot arise in placement of deep lesions.

Presence of bone in the ultrasonic path not only distorts the field by reflection, but may also destroy the underlying tissues in contact with it by absorbing ultrasonic energy and dissipating it as heat. The skull also poses some difficult problems in ultrasonic diagnosis of intracranial pathology, as we shall see later. A craniotomy, adequate in extent to permit unimpeded passage of the cone of sound, is imperative. The extent of the craniotomy depends on the solid angle of radiation and the depth of the target from the cranial surface. The larger the angle and the deeper the target the larger the size of the craniotomy needed. An optical projector, interchangeable with the irradiation head, projects a cone of high intensity cold light of the same dimensions as the cone of ultrasound and is of considerable assistance in determining the extent of needed craniotomy on curved or irregular surfaces [26]. A further check on the adequacy of craniotomy and of the coupling is provided by the pulse-echo lesion detection system discussed later.

For stereotaxic procedures the irradiation head is mounted on a X-Y-Z positioner carrying the stereotaxic head holder so that each of the coordinates reads zero when the ultrasonic focus is exactly at the Horsley-Clark zero. To assist in irradiation of structures not amenable to stereotaxis, such as the spinal cord where the coordinates of the target have to be determined with reference to the posterior median sulcus, a rotating retractable pointer is set to indicate the focal length.

Radiation pressure, particle velocity and intensity of ultrasound are interrelated and Heuter [27] has shown that any of the three may be used to describe adequately the ultrasonic conditions of the focal point of a single focusing transducer. Radiation pressure measurements can be performed easily and rapidly by radiation pressure gauges calibrated to read the output in watts directly [28]. The measurements can be converted into units of particle velocity or intensity for comparison with results of other workers. In practice, by actually setting the ultrasonic output to the level desired, it is possible to reproduce the irradiation conditions accurately. Lesion size is found to be related to radiation pressure in a reproducible and consistent manner.

Functional Tests

To test rapidly the proper functioning of the equipment, and to establish the repetitive stability of the acoustical output, the results of irradiation of a stable and relatively homogeneous medium such as methacrylate (plexiglas, lucite,Perspex) are analyzed [29: 412]. Transparent, clear and stainfree bars of methacrylate, 18mm square and about 45 cm long, buffed and polished on all surfaces, are used. Irradiation of such a bar with an adequate dosage of ultrasound and with the focus placed beneath the surface results in the development of stable, trackless, discrete, egg-shaped areas of stress within the bar, readily visible in polarized light. For the same irradiation parameters, the dimensions of these 'lesions' in the three planes are remarkably constant. There is a consistent and reproducible relationship between the ultrasonic dosage and the size of these lesions comparable to that obtained in the brain of the experimental animals. The use of methacrylate as a preliminary test medium permits rapid evaluation of varying ultrasonic dosage parameters as well as the repeatability and long term stability of the ultrasonic output. Tests of ultrasonic generators (with different harmonic content in their output, for instance) show that the plastic can indeed be used as 'phantom brain' in screening of ultrasonic equipment.

Results in Animals

Over the last 15 years the brains or spinal cords of more than 2,600 animals have been subjected to over 20,000 irradiations by the author and his colleagues, the total number of lesions placed being well over 10,000. The time required for each operative procedure, involving stereotaxis, in the cat or monkey, from the administration of intravenous anaesthetic to wound closure is about 20 minutes of which ultrasonic procedures account for approximately 3 to 5 minutes; the irradiation itself lasting 0.5 to 2.0 seconds. The set-up time for ultrasonic procedures, namely filling and closure of the applicator cone, temperature stabilization and radiation pressure calibration take about 10 minutes. Many of the irradiation procedures have been carried out by a nurse-technician, without assistance except during the administration of anaesthesia. For non-stereotaxic procedures - experimental or therapeutic - the amount of time required increases sharply - most of it being consumed by preoperative positioning of the target organ and its immobilization for the alignment of the focus with the target and irradiation. The postoperative mortality and morbidity are low, which is attributed to the fact that the irradiation is transdural and thus the chances of infection are reduced. The intracranial fluid dynamics are also relatively undisturbed since there is no less of the cerebro-spinal fluid.

In an extensive study of the effects of irradiation on the CNS of experimental animals [11:513], with suprathreshold irradiation dosages, a trackless, pan-necrotic, focal lesion restricted to the irradiation target was evident in every single instance. The outstanding feature of the results obtained was the reproducibility of the effects of irradiation at each of the dosage levels, with single or multiple pulses, the coefficient of variation being about 10 percent. The lesion dimensions were directly related to dosage levels in a statistically significant manner. Smallest lesions are almost spherical, larger lesions are progressively more elongated corresponding to the shape of progressively lower isointensity-lines. At a frequency of 2.7 MHz the smallest lesion that could be made with certainty was approximately 1.2mm in length and 0.4mm in diameter. The length of the largest lesions that could be consistently made was 11mm with single pulses and 17mm with multiple pulses. Like those in the plastic bar the lesions were egg-shaped, the sphericity (diam.length) of those from multiple pulses being greater than that of single pulse lesions. For maximum sphericity the optimum interpulse interval was about 1.0sec. Sphericity was also related to the solid angle of radiation; the smaller the angle, the less spherical the lesion. All lesions were found to be sharply demarcated, discrete, friable areas of coagulative change. No evidence of any hemorrhage was even encountered in lesions of up to the maximum size referred to above. But all attempts to increase the size of the lesion - particularly by the use of a single pulse of high intensity - invariably produced hemorrhages distending the lesion, spreading intracerebrally and often rupturing into the lateral ventricles. The volume of the tissue necrotised can be increased safely by juxtaposition (with slight overlapping) of two or more individual lesions. No damage was detected in the brain in the path of the sound nor in the periosteum or bone underneath the deepest lesion.

Factors Affecting Lesion Size

The attenuation sound in the 'normal' dura mater of the cat is about 5 percent. Presence of dural adhesions, as well as of scars, was found to be one of the sources of variation in the results of irradiation. In patients with intracranial pathology this is likely to be a bothersome-though a minor-problem unless the dural attenuation can be determined without opening it.

The target depth can influence the dosage-lesion relationship by attenuation and refraction. However, both of these can easily be estimated for and corrected. It must be remembered that attenuation is frequency dependent whereas refraction is velocity and temperature related. In the cat the presence or absence of the superior sagittal sinus and the lateral ventricles in the path of ultrasound was not found to influence the dosage-lesion relationship. However, the presence of air or other radio-opaque material within the ventricles might have profound influence on the ultrasonic beam geometry.

With the same ultrasonic dosage, the lesions in grey matter were found to be significantly smaller than those in the white matter. Lesions in grey matter, surrounded by white matter, showed constriction in the region of grey matter. This exemplifies the common statement that the threshold of the white matter is lower than that of the grey matter. On the other hand, with planar lesions in the brain the grey matter of the cortex was found to remain intact and the underlying white matter was extensively destroyed. When the spinal cord is irradiated similarly it is found that it is the grey matter that is completely destroyed leaving the surrounding white matter almost intact. Thus the thresholds appear to be reversed. Work in progress indicates that the direction of the fibres in the white matter, relative to the direction of the main radiation axis, is an important factor governing the apparent sensitivity of the white matter to ultrasonic destruction. In the CNS the intrathecal parts of the spinal nerve roots appear to be the elements most resistant to ultrasonic destruction.

The state of local circulation also affects the lesion formation. Arrest of cranial circulation during irradiation with near threshold dosages resulted in larger lesions both in grey and white matter. At suprathreshold dosage levels, the effect was smaller and restricted to grey matter alone; the blood perfusion through which is much higher than that in the white matter. The lesions then tended to approximate to those in white matter in size [11:513]. The constriction in the grey matter, referred to above, disappears in lesions made with cranial circulation arrested. Examination of control specimens by routine histological methods did not reveal any damage attributable to anoxia.

The temperature of the tissue at the time of irradiation has a profound effect on the dosage-lesion relationship. The lesion size is directly related to the temperature (Table 1). In addition to the evident implications of this observation to the mechanism of lesion formation it is of considerable practical importance in surgery since so many of the surgical procedures are now carried out under hypothermia.

Fry and Myers [30:315] have reported irradiation through intact scalp after previous craniectomy, the advantages of which are obvious. However, data on the ultrasonic properties of human scalp

Table 1. Relationship of ultrasonic lesion size in brain to temperature of the tissue: cat and dog. Average focal intensity=42x10W/cm^2; pulse duration=1.3sec., single pulse.

Core temperature [°C]	Number of observations	Lesion Volume [mm^3]	Length [mm]	Diameter [mm]	Sphericity d/l
37	5	4.94	6.74	1.18	0.175
31	5	1.25	4.36	0.74	0.170
24	5	0.14	1.23	0.41	0.333
22	10	No lesions produced			

are not readily available. The scalp of furry animals - lacking sweat glands - apparently is not comparable to the human scalp. Scalp of the adult cat was found to attenuate as much as 66 percent of the ultrasonic energy. Lesions placed deep within the cat brain by irradiation through the scalp were found to be variable in size and, in each instance, the scalp showed signs of damage. The scalp and the skull of the foetus and the newborn, on the other hand, have acoustic properties similar to other soft tissues and lesions of controlled size could consistently be placed within the brain of foetal rabbits by irradiation through intact scalp and skull [12:502].

It is evident that any factors which alter the acoustic properties or the thermal characteristics of the tissue in the path of ultrasound would influence the dosage-lesion relationship. Currently the ultrasonic dosage is determined empirically whether it be in the plastic bar or the brain. Temperature gradients in the focal region in plexiglas have been studied by incorporation of thermochromic materials [31] and by embedded thermocouples [32]. The resulting heat calculations indicate the lesions in plexiglas are probably thermal in origin. A study was therefore undertaken to determine if thermal mechanisms alone can explain the development of the lesions and all of their measurable characteristics in plastic as well as in brain. A purely thermal model was assumed and analytical prediction of lesion development and lesion size and shape for varying values of ultrasonic and thermal constants and controllable variables (frequency, focusing, dosage, target depth, etc.) was attempted [32]. An empirical equation to describe the axial and radial ultrasonic energy distribution at the focus in water was derived. Appropriate heat transfer equations were developed for temperature distribution resulting from ultrasonic irradiation. The computed temperature profiles were plotted against non-dimensionalized parameters. Temperatures at the lesion center and the lesion boundary were determined experimentally and were found to rise up to 400°C and 150°C in plexiglas and 100°C and 65°C in brain respectively. Expected axial and radial lesion dimensions were read off the curves of temperature distribution at the measured lesion boundary temperature. Comparison of these predicted lesion dimensions with experimental data indicates that lesion development in the brain (as in the plastic) at the frequency and power levels used in these studies can be explained by purely thermal considerations. The range of field variables (frequencies and power levels) as well as state variables (base temperature of the tissue, circulation, type of tissue and consideration of anisotropy, etc.) over which the above observations of a relatively constant lesion boundary temperature is valid is currently being studied. It is expected *a priori* that the duration over which the temperature rise is sustained will influence to some degree the extent of damage to the tissue. These ultrasonic experiments are being repeated using purely thermal sources for elevating the tissue temperatures. Preliminary results indicate that the extent of damage can be best correlated with the magnitude of temperature rise and its duration whether produced ultrasonically or thermally.

The irradiation system described here permits accurate lesion placement in any preselected target of white or grey matter regardless of its location within the CNS. In a large series of blind experiments, using six different target sites, it was found that the probability of lesion placement is restricted only by the variability in the stereotaxic location of the structure itself [11:513].

The histopathological features of the ultrasonic lesion in adults and foetuses have been studied starting immediately after irradiation and afterwards at intervals up to 2 years [9:105, 10:484, 12:502]. The lesions are pan-necrotic, all tissue elements including the blood vessels being completely destroyed within the lesion. They are sharply demarcated from the surrounding healthy tissue. All but the smallest lesions have a characteristic moat-island form consisting of a coagulated central core resistant to lysis and a margin of liquefaction. Healing occurs by phagocytosis of necrotic tissue by microglia and macrophages and the concurrent production of astrocytic gliosis and, in about 4 months, the lesion is replaced by a glial scar. There is no delayed local or remote spread of the lesion except by Wallerian degeneration. The ultrasonic lesion and the scar are non-irritative and are electro-encephalographically 'silent'. The earliest change is difficult to see except as a reduction in the staining ability of the tissue [13]. But with E.M. it is restricted to the mitochondria which appear swollen and lack electron density [14]. This only indicates that the tissue has undergone some change which may or may not be reversible and can also be seen in muscle fibres after exercise. In later specimens at the same dosage level, however, large pores are seen in the plasma membranes indicating the presence of irreversible damage leading to the death of the cell. In the early specimen the structural organization of the tissue shows no signs of disruption such as could be attributed to any mechanical effects of ultrasonic irradiation. Ultrasonic changes, comparable in every respect to these are also produced by thermal irradiation.

Surgical Applications

With all these virtues, it is not surprising that in basic neural sciences (and for non-stereotaxic superficial lesion placement in clinical practice) single beam focused ultrasonic radiation is the tool of choice for the researcher (or clinician) who needs precise lesion placement but no tracks to confuse the neuronantomic or neurophysiological observations-- and who is prepared to make a trephine hole instead of a burr hole in the skull. And it is this factor, the size of the craniotomy, that has prevented the routine use of the technique in surgery of deep structures. In man many of the targets of neurosurgical importance lie 6-10cm from the cranial surface and craniotomies of up to 10cm were required. But now, with improvements in the focusing, the solid angle of radiation can be reduced to 35° and still produce lesions of acceptable shape. This reduces the size of craniotomy needed to a more practicable diameter of up to 6.5cm. In man craniotomy, of even this reduced size, leads to a shifting of the brain from its normal position which adds to the difficulties in accurate anatomical localization of the target areas. Even without craniotomy, stereotaxic location of brain structures, that is their location relative to external cranial landmarks, is not as precise in man as it is in the cat. Thus, injection of some radio-opaque material into the ventricular system is necessary to determine the exact location of internal landmarks with respect to the skull. Since the acoustic impedance of radio-opaque materials (air to iodine compounds) is very different from that of the brain, ultrasonic irradiation has to be delayed until they are completely excreted--with the possibility that the brain structures may shift again. Fry and Fry [33:171] have reported on the use of a water soluble iodine compound (ConrayMallinckrodt Chemical Works) which is apparently excreted within 10 mins. Use of this compound in suitable dilutions may improve this situation considerably.

In radio-frequency or cryogenic lesion techniques the electrode is inserted towards the target under radiological control. The same or another adjacent electrode is then used to record the electrical activity (resting or evoked potentials) from the tissues at or near the advancing tip, until the correct response is obtained. If a satisfactory response is not obtained in one track the electrode is reinserted a little distance away and the tip advanced gradually. Each track is small but necessary, and after a number of insertions, the total damage may not be inconsiderable, though it is not clinically manifest, at least with the current state of knowledge of the function of the CNS and the techniques of examination. The ultrasonic beam cannot be inserted under radiographic control. Unless the trackless beam itself can be used for anatomical and or functional localization of the target, the raison d'etre of the technique is defeated and it cannot--in spite of all its virtues--compete with the current techniques for placement of deep focal lesions stereotaxically. Hense the importance of the so-called 'reversible lesions'; these are ultrasonically-induced temporary functional changes, for which no structural changes can be demonstrated, at least by the time when the function is fully restored. Only one single instance of this phenomenon in man is on record, supported by studies on peripheral nerves [34:47] in the visual system of the cat [35:281,21:858] and the sympathetic ganglia [36]. The need for extensive and critical statistical studies aimed at confirming the occurrence of reversible functional changes in each of the clinically important target systems and establishment of the dosimetry cannot be overemphasized. The problem is complicated by the requirement of certainty of the absence of any lesion. But, presumably, a lesion which escapes careful examination of serially cut sections by a competent neuropathologist, would not be clinically disastrous--certainly no more so than the track of an electrode. In a current study the approach has been to develop a technique of detecting the lesion as it is formed by incorporation of a pulse-echo system connected to the same transducer that is used for irradiation [37:451,38]. The detection of the system can be set to signal incipient lesion formation. This permits rejection of the results in instances where there might be even a remote chance of lesion existence and to concentrate on only those results and specimens where the irradiation is known to have been at subthreshold dosages. The system is also useful in confirming the adequacy of the craniotomy and of the coupling and provides, in addition, a rather simple and accurate method for rapid determination of the focal length of the system.

The answer to the question of reversible functional effects is crucial to the future of focused ultrasonic radiation in surgery and unless it is answered satisfactorily the efforts to marry the rather heavy irradiation head to a stereotaxic apparatus would appear to be premature and possibly futile. In experimental sciences focused ultrasonic irradiation is the method of choice for placement of focal lesions and is being successfully used in studies of the connections of otherwise inaccessible brain structures such as the anterior commissure [39] and the insula [40]. A confirmation of the exisence of the more exotic effects, such as size--or component-- differential lesions, would only enhance its usefulness. But so far as the surgery of deep structures is concerned, its status is still sub judice. The reason for this, of course, is the presence of the skull.

Now, going back to the diagnostic applications, you may recall that of all the body regions, I did not mention two--viz the head and the chest. In the head, the existing ultrasonic systems for detection and localization of intracranial pathology, e.g. hematomas, aneurisms, tumors, etc. are severely restricted in accuracy and reliability because of the unpredictable effects of the skull on the acoustic transmission. These effects are due to the variable thickness and irregular shape of the skull, the differences of sound speed in skull ($4080 m.sec^{-1}$) and brain ($1540 m.sec^{-1}$), the relatively large and variable absorption ($a/f = 20 dBcm^{-1} MHz^{-1}$) in the skull compared to that in the brain ($a/f = 0.85 dBcm^{-1} MHz^{-1}$), and the inhomogeneous structure of the bone. Attempts are being made to compensate for the skull effects on the echoencephalograms by simultaneous measurement of local skull characteristics [41]. In addition, phase-coherent Doppler technique is integrated into this system to determine the location and size of pulsatile structures. This combination of pulse-echo and Doppler techniques also appears promising in detection of myocardial infarctions [42]. The frequency contents of echoes from infarcted and normal myocardium are being studied to determine if the former has a characteristic 'signature' which could be correlated with the state of pathology.

In yet another study, focused ultrasound is being used to create regions of focal necrosis within the myocardium to serve as models of myocardial infarction [42] more reliably and reproducibly than is possible by techniques based on coronary occlusion.

Lung, composed of tiny air sacs presents a different problem. Tissue-Air interface presents a large mismatch of acoustic impedances. In addition, lung has very high acoustic absorption ($a/f = 41 dB cm^{-1} MHz^{-1}$). Thus, pulse-echo techniques are not very useful except in detection of pleural thickening; but resonant absorption and sonic scattering techniques under development [43] may prove very useful in early detection of lung disease without the hazards of radiography.

In contrast to such sophisticated systems, it is feasible to use the simple pulse-echo system to detect non-invasively the formation of ice intracellularly or extracellularly in tissues or during their freezing for preservation[44] or to use the focused beam to place occluding lesions in the epididyms for non-surgical sterilization of the male [45]. Thus indeed, ultrasound promises many an exciting application in biomedicine. The question of toxicity of ultrasound in its diagnostic applications is still sub judice and will not be completely answered until the mechanism(s) of action of ultrasound-- at the frequency, power levels and the modes actually used-- are known with less ambiguity [46]. But meanwhile pracmatically it should be considered safe as in histopathological [47,48:65] and behavioural studies [49:296] no untoward effects have been observed.

Acknowledgments

This work is supported by U.S.P.H.S. Grant No. NS 08571 and No. 5-S05-RR07047-05. Thanks are due to the Editors of the Journal of Physiology, Experimental Neurology, Environmental Biology [50:207] and Academic Press for permission to use previously published data.

References

1. WELLS, P.N.T., Physical Principles of Ultrasonic Diagnosis, London and New York: Academic Press(1969).

2. BROWN,B. and D. GORDON, eds., Ultrasonic Techniques in Biology and Medicine, Springfield:Charles C. Thomas Publisher(1967)

3. GORDON, D., Ultrasound:As a Diagnostic and Surgical Tool, Edinburgh: E. and S. Livingstone Ltd. (1964).

4. GROSSMAN,C.C.,J.H. HOLMES,C.JOYNER,and E.W.PURNELL,Diagnostic Ultrasound,New York:Plenum Press(1966).

5. EL'PINER,I.E.,Ultrasound-Physical,Chemical and Biological Effects, New York:Consultants Bureau(1964).

6. GITTER,K.A.,A.H.KEENEY,L.K. SARIN,and D. MEYER,eds.,Ophthalmic Ultrasound,St. Louis:C.V. Mosby Co.(1969).

7. GOLDBERG,R.E. and L.K.SARIN,Ultrasound in Ophthalmology,Philadelphia and London:W.B. Saunders Co.(1967).

8. KELLY,E.,ed.,Ultrasonic Energy,Urbana:University of Illinois Press (1965).

9. LELE,P.P.,Production of Deep Focal Lesions by Focused Ultrasound-Current Status,Ultrasonics,5:105(1967).

10. ASTROM,K.E.,E. BELL,H.T.BALLANTINE,JR.,and E.HEIDENSLEBEN,J. of Neuropathology and Experimental Neurology,20:484(1961).

11. BASAURI,L. and P.P.LELE,J. of Physiology,160:513(1962).

12. YOUNG,G.F. and P.P.LELE,Experimental Neurology,9:502(1964).

13. LELE,P.P.,*Time Course of Development of Ultrasonic Lesion*,In preparation.
14. SOMEDA,K. and P.P.LELE,*An Electron Microscopic Study of Ultrasonic Lesions in the Brain*,In preparation.
15. MANLAPAZ,J.S.,K.E.ASTROM,H.T.BALLANTINE,JR.,and P.P.LELE,*Experimental Neurology*,10:345(1964).
16. LELE,P.P.,To be published.
17. LELE,P.P. and J. HAYES,Conference on Biologic Effects of Laser Radiation,Boston Laser Conference,The Institute of Electrical and Electronics Engineers and Northeastern University (1964).
18. FRY,W.J. and F.DUNN,*Physical Techniques in Biological Research*, Vol.4,New York:Academic Press,p.261(1962).
19. LELE,P.P.,*J.of Physiology*,160:494(1962).
20. LELE,P.P. and H.T.BALLANTINE,Jr.,To be published.
21. BALLANTINE,H.T.JR,E.BELL,and J.MANLAPAZ,*J.Neurosurgery*,17:858 (1960).
22. STEVENS,S.S.,ed,*Handbook of Experimental Psychology*,New York and London:John Wiley and Sons,Inc.(1951).
23. KAHN,H.A.,*A Theoretical Analysis of Ultrasonic Focusing Systems*, Master's Thesis, Massachusetts Institute of Technology(1970).
24. LELE,P.P.,*A Highly Stable, Sensitive Thermocouple Probe for the Determination of Energy Distribution in Ultrasonic Fields*,In press.
25. HSU,W.L.,*An Improved System for the Coupling of an Ultrasonic Transducer to the Brain of a Cat or Monkey*,Master's Thesis,Massachusetts Institute of Technology (1970).
26. BELLOWS,A.H.,*Design and Construction of an Optical Adjunct to Ultrasound Neurosurgical Equipment*,Bachelor's Thesis,Massachusetts Institute of Technology (1962).
27. HEUTER,T.F.,Report on Progress on Contract B-816(C) to the National Institutes of Health,U.S.Public Health Service(1956).
28. LELE,P.P. and D.S.ALLES,*A Radiation Pressure Balance for the Measurement of Ultrasonic Power*,In press.
29. LELE,P.P.,*J. Acoustical Society of America*,34:412(1962).
30. FRY,W.J. and R. MEYERS,*Confinia Neurologica*,22:315(1962).
31. HSU,W.L.,J.D.MCGILL,G.A.WANEK,and G.D.WIGHT,Projects Report No. 2.671,Department of Mechanical Engineering,Massachusetts Institute of Technology (1966).
32. ROBINSON,T.C. and P.P.LELE,An Analysis of Lesion Development in the Brain and in Plastics by High Intensity Focused Ultrasound at Low Megahertz Frequencies,*J.Acoust.Soc.Amer.*,51:1333-1351.

33. FRY,W.J. and F.J.FRY,Anatomical Record,147:171(1963).
34. LELE,P.P.,Experimental Neurology,8:47(1963).
35. FRY,W.J.,Advances in Biological and Medical Physics,Vol.4,New York:Academic Press,p.281(1958).
36. LELE,P.P.,S.SHIBATA,and J.RUNNING,The Physiologist,13,No.3(1970).
37. LELE,P.P.,Medical and Biological Engineering,4:451(1966).
38. MATISON,G.,Sonar Detection of Ultrasonic Lesions During and After Production,Doctoral Thesis,Massachusetts Institute of Technology (1971).
39. PANDYA,D.,E.E.KAROL,and P.P.LELE,The Distribution of the Anterior Commissure in the Squirrel Monkey,Brain Research,49:177-180 (1973).
40. LELE,P.P. and D.PANDYA,Unpublished data.
41. LELE,P.P.,A.D.PIERCE,E.A.WOODIN,and H.E.EGERTON,Unpublished data.
42. LELE,P.P. and J.C.CAULFIELD,Unpublished data.
43. MURPHY,R.L.H.,E.A.WOODIN,and P.P.LELE,Unpublished data.
44. LELE,P.P. and E.CRAVALHO,Unpublished data.
45. LELE,P.P.,Unpublished data.
46. LELE,P.P. and A.D.PIERCE,Unpublished data.
47. LELE,P.P.,Unpublished data.
48. POND,J.B. and R.WARWICK,Megahertz Irradiation of Pregnant Mice, IEEE Ultrasonic Symposium Abstracts,IEEE Trans. on Sonics and Ultrasonics,SU-17:65(1970).
49. SMYTH,M.G.,JR.,Animal Toxicity Studies with Ultrasound at Diagnostic Power Levels,In Diagnostic Ultrasound,C.C.Grossman,C. Joyner,J.H.Holmes, and E.W.Purnell,eds.New York:Plenum Press, p.296.
50. LELE,P.P.,Environmental Biology,P.L.Altman and D.S.Dittmar,eds. Federation of American Societies for Experimental Biology,Bethesda,Maryland,p.297(1966).

SAFETY OF ULTRASOUND IN DIAGNOSIS

C.R. Hill

Physics Division, Institute of Cancer Research

Royal Marsden Hospital, Sutton, Surrey, U.K.

ABSTRACT

This paper extends and updates previous reviews of this subject, including two by the present author. Any hazard that may arise from diagnostic ultrasound is presumably related to physical levels of exposure. Information on this point is sparse but some order-of-magnitude figures are given. Evidence for hazard at a given exposure level comes from three complimentary types of investigation : biophysical studies, screening investigations and epidemiology. In each case some of the data that is available is open to criticism and certain gaps and inconsistencies remain. Nevertheless the concensus of evidence from systematic studies provides strong indication that current practice is without hazard. Reports to the contrary must always be taken seriously but none so far has been substantiated by independent work.

INTRODUCTION

Over the past few years the use of ultrasound for medical diagnosis has undergone a transformation from being a pastime for enthusiasts to becoming a widespread, routine diagnostic tool. It has happened that some of its first major applications have been in obstetrics and this, in the light of the radiation-like qualities of ultrasound, has led to some questioning of its safety. Radiologists are particularly conscious of the problems of balancing risk and benefit in the use of radiation in medicine, and of the long delay that occurred before the recognition that even diagnostic levels of X-ray exposure may involve some statistically significant hazard when very large populations are at risk.

In 1968 the present author reviewed the question of ultrasound safety (1) and drew the conclusion that there was no clear evidence that current practice in the use of ultrasound was inherently hazardous. Subsequent developments have tended to confirm both this conclusion and the detailed arguments put forward in that and a subsequent review (2). In the present paper therefore only a brief recapitulation of the previous discussion is given and attention is concentrated on some of the more recent data that bear on the question.

An essential consideration in assessing the safety of ultrasonic diagnostic techniques is that of the physical conditions of exposure that are involved. This is a matter on which systematic information is still notably scarce but the outlines of the situation can be reasonably well defined. If physical conditions of exposure can be assumed to be known, assessment of safety can in principle be made from any or all of three rather different approaches : (a) quantitative investigations of the fundamental biophysical interactions of ultrasound and tissue, (b) empirical screening procedures specifically designed to investigate aspects of safety under defined conditions, and (c) epidemiological studies on human populations exposed to ultrasound.

DIAGNOSTIC EXPOSURES

In the field of ionizing radiation, knowledge of radiobiology has reached the point where it is possible to demonstrate reasonably good correlation between levels of absorbed "dose" (energy per unit mass) and resulting biological change. With ultrasound no such effective dosimetric concept has been established and reporting of dosimetry therefore entails the provision of a complete description of all the independent beam parameters.

The carrying out of such measurements is not in principle difficult and a set of survey measurements was in fact made by the present author in 1969 on the diagnostic instruments available commercially in the UK at that time (3). Rather surprisingly little else in the way of systematic measurements has been reported subsequently and the following data are therefore based on this source.

At present (and this is likely to remain true for some time to come) the great majority of patient exposures to diagnostic ultrasound arise from one or other of two basic types of machine: pulse-echo and continuous wave Doppler. Although some of the exposure parameters (acoustic frequency, average acoustic power, beam width) are comparable for the two cases, others, such as peak amplitude, may be very different. In Table 1 is listed the range of values found for some parameters of possible biological significance. For comparison purposes some typical data for ultrasonic therapy instruments is also included (4).

BIOPHYSICS OF ULTRASOUND INTERACTIONS

The main biophysical evidence on ultrasound safety has been outlined in a companion paper (5). The two biophysical mechanisms whose effectiveness has been established are cavitation and heat generation.

The extent to which cavitation can be made to occur in the human body *in vivo* even under extreme ultrasonic

TABLE I Measured Ultrasonic Exposure Parameters (Ref.3,4)

Parameters	Range of Values Measured		
	Pulse-Echo	C.W. Doppler	Therapy
Nominal Acoustic Frequency (MHz)	1 - (15)	2 - 5	1 - 3
Average Acoustic Power (mW)	0.3 - 21	19 - 24	0-25,000
Peak (Space-Time) Intensity (W.cm^{-2})	1.4 - 95	0.003-0.023	0 - (25)
Peak Pressure Amplitude (bar)	2 - 17	0.1 - 0.3	0 - (8.5)
Pulse Duration (μs)	~ 1	∞	∞

exposure conditions is still very uncertain. Diagnostic conditions of exposure (low intensities and/or very short pulse durations) are generally insufficient to induce cavitation even in nonviscous liquids and it is thus particularly unlikely to occur in vivo. The biological effects of cavitation in liquid cell suspension systems is predominantly that of cell death by disintegration. Associated mechanical damage to surviving cells has been demonstrated but no clear evidence has been found to indicate that this is in its nature either mutational or otherwise indicative of significant hazard.

Heat generation can be an effective mechanism for ultrasonically induced biological change in intact tissue systems, which are characterized by relatively high ultrasonic attenuation coefficients (of the order 0.1 $(MHz)^{-1}$ for soft tissues) and low thermal mobility. Here again however the low average intensities involved in diagnostic exposures are insufficient under practical conditions to lead to temperature rises amounting to more than a small fraction of a degree.

Biophysical evidence for other possible mechanisms of action of ultrasound is insufficient for any useful assessment of hazard and further evidence in this direction must at present be sought from the results of empirical, screening type investigations.

SCREENING INVESTIGATIONS

A problem that runs through the whole question of the possible "hazard" associated with the use of ultrasound is that there is no a priori indication of the nature of the hazard that might be found to occur. The most serious type of consequence would seem to be that of genetic or teratogenic change and most of the screening work has been in this direction. Some such investigations were referred to in the earlier reviews. (1,2). More recently systematic investigations have been carried out on mice irradiated under conditions very greatly in excess of those used in diagnosis, with no resulting evidence for effects either on specific genetic damage (6) or on such factors as gestation time, fetal weight, litter size and incidence of resorptions and abnormalities in pregnant mice and their litters (7). Some investigators have reported contrary findings but in general these are difficult to interpret. One recent investigation for example yielded results which led the authors to suggest that ultrasonic exposure (at 40 mW cm^{-2} for 5 hours) of pregnant mice at the 9th day of gestation increased the frequency of maldevelopment or intra-uterine death of mouse embryos (8).

EPIDEMIOLOGY

In any discussion of the safety of patients, regardless of the nature of the particular agency that may be under suspicion, no completely satisfactory conclusion can be drawn that does not rely on evidence from humans: "the proper study of mankind is man". Epidemiology is a demanding science and to produce fully satisfactory evidence on the safety of medical ultrasonic exposures would call for a study involving large numbers of subjects, extending over a period of a number of years and preferably designed on a prospective basis. No such study has yet been carried out although one, at least, is now being planned. Meanwhile the only evidence of this nature comes from a retrospective study on 1114 apparently normal pregnant women examined by ultrasound in three different centres and at various stages of pregnancy (9). A 2.7% incidence of fetal abnormalities was found in this group as compared with a figure of 4.8% reported in a separate and unmatched survey of women who had not had ultrasonic diagnosis. Neither the time in gestation at which the first ultrasonic examination was made, nor the number of examinations, seemed to increase the risk of fetal abnormality.

CONCLUSION

The problem of assessing the possible dangers involved in exposing the human body to any new or unfamiliar form of stress is generally difficult and in this respect ultrasound is no exception. The medical uses of ultrasound suggest certain analogies to those of X-rays and this factor itself tends to induce an extra degree of caution. The concept of absolute proof of safety however is a chimera: the best one can expect is a strong indication in terms of reliable and repeatable evidence, and it is only realistic to expect that, where complex scientific questions are being investigated on a large scale and under conditions subject to human frailty, occasional inconsistencies will arise.

Seen in this light, and drawing on evidence from fundamental studies, screening investigations and epidemiology, diagnostic ultrasound has a strong claim to be considered safe. There is no evidence from systematic and independently repeated studies to suggest otherwise: a small number of contrary findings have been reported but none has yet received independent confirmation, and where confirmatory investigations have been attempted they have proved negative.

The potential importance of the subject, in terms of the likely future scale of exposure of patients, is such that further work will be fully justified. The establishment of reliable epidemiological data is one important aspect of this and it is also to be hoped that a more self-consistent pattern of results from independant investigators can be built up in relation to some of the more important biological endpoints, and particularly those indicative of genetic effects in mammalian systems. At the same time it will be important to establish adequate means for recording the physical levels of exposure of patients in various diagnostic procedures so that any possible future epidemiological evidence for hazard can be placed in its proper quantitative relationship to patient exposure.

REFERENCES

1 HILL, C.R.: The possibility of hazard in Medical and industrial applications of ultrasound. Brit.J. Radiol. 41 (1968) 561.

2 HILL, C.R.: Biological effects of ultrasound, Ultrasonics in Clinical Diagnosis (Wells, P.N.T. Ed.) Ch.9. Churchill Livingston, London (1972).

3   HILL, C.R.: Acoustic intensity measurements on ultrasonic diagnostic devices. Part 2, <u>Ultrasonographia Medica</u> (Bock, J. Ed.). Vienna Academy of Medicine. Vienna (1969).

4   STEWART, H.F. <u>et al</u>: Survey of use and performance of ultrasonic therapy equipment in Pinellas County, Florida. US Govt. Publication DHEW(FDA)73-8039 (1973).

5   HILL, C.R.: Action of ultrasound on isolated cells and cell cultures (proceedings of this meeting, 1974).

6   LYON, M.F. and Simpson, G.M.: An investigation into the possible genetic hazards of ultrasound. Brit. J. Radiol. (in the press, 1974).

7   WARWICK, R., POND, J.B., WOODWARD, B. and CONNOLLY, C.C.: Hazards of diagnostic ultrasonography - a study with mice. IEEE Trans. Sonics Ultrason SU-17 (1970) 158.

8   SHIMIZU, T. and SHOJI, R. An experimental safety study of mice exposed to low intensity ultrasound. Excerpta Medica, International Congress Series No. 277 (1973) p.28.

9   HELLMAN, L.M., DUFFUS, G.M., DONALD, I. and SUNDEN, B. Safety of diagnostic ultrasound in obstetrics. Lancet $\underline{1}$ (1970) 1133.

-DISCUSSION-

<u>OSEPCHUK</u> - What was the duration of exposure during the ultrasonic diagnostic procedure? And, also the exposure durations in those two screening tests?

<u>HILL</u> - A typical duration of exposure in a diagnostic examination may vary from two minutes, to seldom more than 20 minutes. The durations of exposure in these tests would be on the order of ten to thirty minutes, as I recall.

<u>OSEPCHUK</u> - You reported a wide range of power densities. I presume 10 to 30 minutes will not be applicable in that full power density range.

<u>HILL</u> - Yes. This is an important parameter.

SAFETY OF ULTRASOUND IN DIAGNOSIS

SUESS - When you speak about safety, we have to consider the patient, the operator and the physicians. Did you refer only to the patient?

HILL - Yes.

SUESS - Is there any hazard to the operator and the physician as you would have more than one person attending?

HILL - One of the great features of ultrasound is that air is a very good screening material. That takes care of a large part of the problem. The other aspect of this is, that the requirements for designing transducers simply for good diagnosis require that you must strongly attenuate the backward propagating portion of the beam. Any doctor who holds a probe in his hand and gets irradiated must be using a very badly designed probe. So, I think there is no danger in this. Nobody has measured it that I know of, but I would be very surprised if you found ultrasound floating out of the back side of a transducer.

SUESS - Another point is the responsibility of the manufacturer and the designer of the equipment; and the need and requirement for the manufacturer to indicate on his equipment the restrictions on use and the safety precautions. By analogy with the X-ray machine, while you can operate one in a real safe manner, even in the United States there still exists the problem of a physician who is not trained to operate an X-ray machine and therefore exposes himself.

HILL - There are two parts to this problem. First of all, one should certainly require manufacturers to produce safe equipment. But, before you can really do that, you in fairness have to be able to have some agreement as to how measurements of safe performance should be made. And we are still at the stage of trying to work this out.

One international organization concerned with this is the International Electrotechnical Commission. This has a Working Group of which I am the Secretary, concerned with just this problem. And at the moment we are trying to agree on a way in which the main parameters of performance of diagnostic equipment, including acoustic output, can be measured and expressed. The next stage after that would be to say what performance is right. Once we know how to measure it, let's get together and think of what we ought to specify as maximum permissible outputs and so forth, if this is really necessary. But remember that this is not done with X-rays. No one says that a diagnostic X-ray set must not emit energy toward the operator. There, the regulations are concerned with shielding.

FLOOR - Other ultrasonic equipment is potentially more dangerous than the two types that you discussed. We have cause to believe that the power output meters are inaccurate. Would you care to comment on this?

HILL - You are probably referring to the study which was done in Pinellas County. That was a very worthwhile study. The Bureau of Radiological Health here looked at the outputs of all the ultrasound therapy machines that could be found in Pinellas County, Florida. Almost all the machines were giving outputs that were more than plus or minus 20% outside the reading on the dial. Some of the meters were down at zero and some were up to 200% of the output -- is that right?

One thing that really came clear out of the talk that Dr. Lehmann gave today is that if ultrasound therapy is to be done properly and intelligently, just for positive therapeutic purposes, the dosimetry should be done properly. This requires enough attention to take care of the safety aspect.

CZERSKI - Was any explanation offered for the production of sterility in mice by ultrasound? What is the mechanism and what exactly does sterility mean?

STEWART - I read an article recently in Medical World News where they were using ultrasound, microwaves, and other types of heating to treat male mice and then the mice were mated and the females followed, to look at the capability of producing pregnancy. Ultrasound was found to produce temporary sterility. It was suggested as an approach to male contraception. Another similar article was in the May 11, 1974 issue of Science News.

GRAMIAK - I stood next to the air conditioner for many years and it didn't work.

STEWART - According to the article, cooling can enhance sperm production.

GRAMIAK - Mr. Mitchell is going to tell us about the electromagnetic incompatibility of medical prosthetic devices. My interest rises from the fact that as a local part-time user of ultrasound in the area I get many phone calls from people who want to know - may I have ultrasound, I have a pacemaker. Should I have ultrasound diagnoses? And, I don't really know the answer but I am very cautious.

EMC DESIGN EFFECTIVENESS IN ELECTRONIC MEDICAL PROSTHETIC DEVICES

John C. Mitchell, William D. Hurt, Terry O. Steiner

USAF School of Aerospace Medicine

Brooks Air Force Base, Texas

ABSTRACT

The increasing use of electronic prostheses in a society where the numbers and intensities of radiofrequency (RF) radiation sources are ever increasing requires special attention by the manufacturers of the medical devices and the practicing physicians who prescribe these devices. One such device, the artificial cardiac pacemaker, was tested extensively to assess the extent of radiofrequency electromagnetic radiation interference (EMI) possible from a variety of RF sources. Pacemaker responses were measured on twenty-one different types (manufacturers and models) of devices, exposed in "free-field" and "simulated-implant" configurations. Relative interference thresholds were vastly different with the most sensitive pacemaker being adversely affected at electric (E) field levels as low as 10 volts per meter and the least sensitive pacemaker being relatively free of interference at levels as high as several hundred volts per meter. In many cases the real time E-field level around radiofrequency radiation (RFR) emitters manifests itself as a pulsed or pseudo-pulsed (changing E-field level) signal which can adversely affect cardiac pacemakers and is potentially hazardous for other types of medical prosthetic devices. These empirical findings demonstrate the need for continuing awareness of potential RF interference situations and provide reasonable evidence that through such awareness many of the potential EMI problems can be effectively circumvented.

Introduction

Technics for designing equipment for electromagnetic radiation compatibility (EMC) are applied rather extensively throughout a large segment of the electronics industry. Recent studies indicate such technics are now being effectively incorporated in the design of medical prosthetic devices such as the artificial cardiac pacemaker. The test results reported in this paper demonstrate the success of many manufacturers in eliminating or circumventing the unwanted electromagnetic interference sensitivity common in many of the earlier pacemaker designs.

The fact that external interference can disrupt the normal operation of some cardiac pacemakers has been recognized almost since the first unit was placed in service, but at the same time, practicing clinicians have generally maintained that such interference is not clinically significant (15). Notwithstanding these facts, the manufacturers have apparently recognized the sources of potential interference are ever increasing and they have included EMI as one of many design considerations in their newer devices.

In regard to the overall interference aspect, EMI tests generally establish the "relative sensitivities" of the pacemakers on the market at some particular time, under certain test conditions. Examples of such studies are referenced (4, 9, 10, 13, and 16). A current study of the relative EMI characteristics of the newer pacemakers as compared to previous designs for several different types of radiation sources follows.

Materials and Methods

Seventy pacemakers including 10 manufacturers and 21 different designs as listed in Table 1 were tested. Radiation sources included laboratory generators operating at pulsed frequencies of 450 MHz and 3100 MHz; such electric devices as sabre saws, variable speed drills, food mixers, hair dryers, pocket calculators, garage door openers, and razors; and the RF emission from an automobile ignition. Additional devices such as lawn mower and motorcycle ignition systems, portable radio transmitters, outboard motors, diathermy machines, and certain communication and radar devices will also be used in this test series in the next few months.

The pacemakers were tested in both free-field and simulated-implant configurations for each of the radiation sources. For the free-field configuration the pacemakers and leads were mounted on a lucite stand. For the simulated implant configuration the lucite stand was placed in the phantom (20 x 30 x 30 cms) filled with 0.03 molar saline solution, being careful to locate the pacemaker to

TABLE I. Cardiac pacemakers included in these tests

Manufacturer	Model	
American Optical	281003	
American Optical	281013	
American Optical	281143	Predicta Series
Biotronik	IDP44	
Cordis	133C6	Atricor
Cordis	133C7	Atricor Jr.
Cordis	143E7	Stanicor
Cordis	162C	Omni-Stanicor
Cordis	164A	Omni-Atricor
General Electric	A2072D	
General Electric	A2075A	Sentry Series
Medcor	3-70A	
Medtronic	5842	
Medtronic	5942	
Medtronic	5943	
Medtronic	5944	
Medtronic	9000	Nuclear
Pacesetter	BD101	Rechargeable
Starr-Edwards	8114	
Starr-Edwards	8116	Ventrac
Stimtech	3821	
Vitatron	MIP40RT	

position one cm of solution between the pacemaker and the wall of the phantom. The phantom wall thickness was 1.5 mm and its measured attenuation of the RF field was negligible. Pacemaker response was recorded via a fiber optics monitoring system consisting of a light emitting diode (LED) mounted in a subminiature audio plug, loaded with a network to maintain 600 ohms, ten feet of sheathed light pipe coupled to a photoresistive voltage dividing network, a Mennen Greatbatch amplifier, and a dual channel strip chart recorder.

The 450 MHz and 3100 MHz tests were conducted in an anechoic chamber at the Georgia Institute of Technology Engineering Experiment Station (GIT), Atlanta, Georgia. The 450 MHz tests were conducted with pulse widths of one microsecond to one millisecond and pulse repetition rates of 2, 10, 20, 40, and 50 pulses per second (pps) to circularly polarized E-field intensities up to 292 volts per meter (V/m). The 3100 MHz tests were conducted at pulse widths of 10-120 microseconds and pulse repetition rates of 7, 10, 20, 40, 100, 200, and 400 pps to vertically polarized E-field intensities up to 320 V/m (rms). The E-field levels to which the pacemakers were exposed were measured by both GIT personnel and by personnel from the Air Force Communication Service, 1839 Electronic Installation Group, Keesler AFB, Mississippi.

The appliance tests were conducted in the USAFSAM Radiation Science Laboratory, Brooks AFB, Texas. Measurements of the radiofrequency radiation emission of the appliances have not been completed so the pacemaker effects were recorded as a function of distance from the respective appliances.

Test Results

Table II is a summary of the adverse effect thresholds of each pacemaker model tested in the simulated-implant configuration at 450 MHz. An adverse effect is defined as a pacemaker rate which falls below 50 beats per minute (bpm) or exceeds 120 bpm as a direct result of RF radiation interference. In most instances the value at which the most sensitive of so-called identical pacemakers cut off completely was selected as the adverse effect threshold. In cases where the threshold is based on an increased rate, it was generally observed that the pacemaker rate continued to increase with increasing E-field level. Where no adverse effect was observed at the maximum E-field level available, it is noted by >292 V/m. Blank spaces indicate the other data points are adequate to describe the effect.

The test data summarized in Table II serve to illustrate the wide range (8 V/m to >292 V/m) of EMI susceptibility thresholds among the 21 pacemaker models tested. Comparing the relatively new A.O. pacemaker (item No. 3) with the older A.O. models (Nos. 1 and 2) shows a dramatic improvement in EMI characteristics. The same is true for the new Starr-Edwards model 8116 compared to their model 8114. It is also noteworthy that the Pacesetter pacemaker marketed this past year was not affected by the maximum E-field available indicating that EMI characteristics were considered during the design stages. Again as in tests conducted two years ago, the Biotronik pacemakers (obtained just prior to these tests) maintained good EMI characteristics. Although the improvements in EMI characteristics were much greater for some models, it appears that all of the manufacturers are including EMI as a design consideration and in essentially every case the newer models show improvement in this respect.

The data in Table II also illustrate that some of the pacemakers revert to their interference rejection mode (fixed rate) upon sensing interference at pulsed rates as low as 10 pps while some others revert at a much higher pulse rate. Very few effects were noted at 3100 MHz since the implanted adverse effect thresholds were all greater than 200 V/m with only 4 of the 21 pacemaker types being significantly affected at 320 V/m.

TABLE II. Summary of adverse effects thresholds recorded during simulated-implant tests

	450 MHz, 1 msec PW			
	Pulse Repetition Rate (pps)			
	2	10	20	40
	V/m(bpm)	V/m(bpm)	V/m(bpm)	V/m(bpm)
1. A.O. 281003	13(0)		15(0)	243(0)
2. A.O. 281013	23(0)		26(0)	243(0)
3. A.O. 281143	>292	>292		>292
4. Biotronik IDP44	141(0)	>292		>292
5. Cordis Atricor	>292	>292		141(172)
6. Cordis Omni-Atricor	>292	>292		>292
7. Cordis Stanicor	15(0)	15(0)	243(24)	>292
8. Cordis Omni-Stanicor	8(0)	9(0)	9(0)	>292
9. G.E. A2072D	29(0)	207(122)		
10. G.E. A2075A	23(0)	141(125)		
11. Medcor 3-70A	29(0)	141(0)	141(0)	141(0)
12. Medtronic 5842	15(0)		15(0)	13(0)
13. Medtronic 5942	12(0)			12(0)
14. Medtronic 5943	23(0)		19(0)	>292
15. Medtronic 5944	26(0)	36(0)	207(400)	207(400)
16. Medtronic 9000	10(0)	10(0)	12(0)	>292
17. Pacesetter BD101	>292	>292	>292	>292
18. Starr-Edwards 8114	23(0)	26(0)	>292	
19. Starr-Edwards 8116	>292	>292	>292	>292
20. Stimtech 3821	107(0)	114(0)	>292	>292
21. Vitatron	93(0)	107(0)	243(0)	243(0)

Although the "free-field" results are not presented, the "free-field" to implant attenuation factors using the phantom with one cm of solution were ~3 at 450 MHz based on the pacemaker response data and ~5 at 3100 MHz based on antenna measurements by GIT personnel.

The effect of "pulse-width" was also studied indicating that one might expect a higher E-field response threshold for shorter pulse widths (less than one millisecond) for some pacemakers. In this regard it should be noted that the 3100 MHz data were taken with a 120 microsecond pulse width.

A cursory evaluation of the effect of leads was also made. For instance, the data presented for the Medtronic and A.O. pacemakers were taken using the Medtronic model 6914 epicardial leads.

Switching to the model 5818 endocardial leads appears to raise the E-field thresholds somewhat for the Medtronic pacemakers, but did not change the effect on the A.O. devices. However, the lower threshold values are reported based on the fact that many of the model 6915 epicardial leads are probably still in use and because we believe it is highly probable that many of these model pacemakers would likely have response thresholds as low as the three devices making up our basic test sample.

Table III is a summary of the response of the pacemakers when subjected to the indicated sources of RF radiation emission. These tests were conducted with the pacemakers in the saline solution phantom and the sources located at 2, 10, 25, 50, and 100 cms from the phantom. These data represent the worst case effect, and in all instances it was within 25 cms of the source. The Roman numerals grade the pacemaker response:

    I - No apparent change in pacemaker rate;
   II - Intermittent change in rate, e.g., missing 1 or 2 beats periodically;
  III - Steady rate between 50 bpm and 120 bpm;
  IV - Rate is less than 50 bpm or greater than 120 bpm; and
   V - Cut off, misses more than five consecutive beats.

The fact that many different types of RF radiation emitters can disrupt the normal operation of some pacemakers is well known and is covered in most of the manufacturers' literature provided to patients. These tests confirm the fact that, in general, sources of interference such as elecric razors, drills, and food mixers must be very close to the pacemaker to result in any significant interference.

CONCLUSIONS

Electromagnetic RF radiation having a field intensity above a certain threshold value (dependent on the specific pacemaker) can mimic the ventricular activity (R-wave signal) of the heart, thus resetting the demand pacemaker timing circuit, so that a pacemaker impulse is not provided until a certain escape interval has elapsed. The extent or significance of such interference (EMI) is primarily dependent on the envelope of the E-field gradient as a function of time. If the E-field intensity is changing in such a manner to mimic a pulse repetition rate of $\sim$1 to $\sim$10 pps with the peak of each pulse above the pacemaker's interference threshold, the pacemaker will inhibit (cut off). If the effective pulse repetition rate is greater than some inherent value (specific to each device), the pacemaker may revert to its interference rejection mode (fixed rate). Reversion to fixed rate is judged nonhazardous. Inhibition

TABLE III. Summary of pacemaker response to indicated sources in the simulated-implant test configuration

	Sabre Saw	Variable Speed Drill	Food Mixer	Hair Dryer	Pocket Calculator	Garage Door Opener Alliance	Electric Razor	Volkswagen Automobile Ignition
A.O. 281003	II	I	II	II	I	II	V	V
A.O. 281013	II	II	II	II	V	II	III	V
A.O. 281143	I	I			I	I		I
Biotr. IDP-44	II	II	I	I	I	I	I	III
Cordis 133C7	III	III	III	III	I	III	III	IV
Cordis 143E7	II	II	II	II	I	II	V	V
Cordis 162C	III	V	II	II	I	II	II	V
Cordis 164A	I	I	III	III	I	III	III	III
G.E. A2072D	II	IV	I	I	I	I	III	IV
G.E. A2075A	IV	V	II	II			V	V
Medcor 3-70A	II	I	V	I	I	I	III	IV
Medtr. 5842	V	II	II	II	I	II	V	V
Medtr. 5942	II	II	II	II	I	I	I	V
Medtr. 5943	III	I	II	I	I	I	I	V
Medtr. 5944	I	I	I	I		I	I	III
Medtr. 9000	I	I	II	I	I	I	I	II
Paces. BD-101	I	I			I	I		
Starr-Ed. 8114	II	V	V	II	I	I	II	IV
Starr-Ed. 8116	II	V	II	II		II	II	III
Stimtech 3821	I	I			I	I	I	V
Vita. MIP-40-RT	I	I	I	I	I	I		II

is judged hazardous. Some of those pacemakers tested reverted to their fixed rate at any pulse repetition rate above 5 pps, while others would not revert at pulse rates as high as 40 pps.

Although these tests reflect remarkable abilities of some manufacturers to essentially solve the potential interference problem, the data also show some currently marketed devices still have adverse effect thresholds at E-field levels likely to be found around RF sources in areas accessible to the general populace. These current tests validate previous test results; i.e., the most effective interference frequency appears to be between 200 MHz and 600 MHz. The adverse effect thresholds at 450 MHz ranged from ~8 V/m to >292 V/m.

A cursory examination of the effect of the pulse width of the incident radiation indicates at 3100 MHz that decreasing the pulse width from 120 microseconds to 10 microseconds raises the E-field threshold on some pacemakers by a factor of ~3; and at 450 MHz decreasing the pulse width from 1 millisecond to 1 microsecond increases the E-field threshold for some pacemakers by a factor of ~25-35.

Measured shielding factors in the simulated-implant configuration (one cm of 0.03 molar saline solution) as compared to free-field are ~3 at 450 MHz and ~5 at 3100 MHz. Furthermore, the type of pacemaker leads used and the lead geometry can alter the E-field threshold by a factor of ~2.

At the 3100 MHz frequency, using 120 microsecond pulse width, none of the pacemakers tested under simulated-implant conditions were seriously affected at an E-field level of 200 V/m.

At 450 MHz, the American Optical (A.O.) model 281143, Starr-Edwards model 8116, and Pacesetter model BD-101 were not affected at 200 V/m in the simulated-implant tests. The General Electric (G.E.) model A2072D, Biotronik model IDP-44, and Cordis Atricor pacemakers demonstrated no serious effects at 200 V/m for pulse repetition rates greater than 10 pps. All other pacemakers tested were seriously affected at E-field values below 200 V/m and pulse repetition rates greater than 10 pps.

In general, the pacemakers being marketed today as compared to those of two years ago offer considerably more resistance to electromagnetic interference. Also, it appears the total number of the more sensitive pacemakers in service two years ago has been reduced about 80%. Continuing effort by the manufacturers will ultimately resolve most of the potential pacemaker EMI problems, and it is hoped that the manufacturers of other medical instrumentation and electronic prostheses will incorporate good EMI rejection technics in all new devices.

REFERENCES

1. CRYSTAL, R.G., KASTOR, J.A., and DESANCTIS, R.W.: Inhibition of discharge of an external demand pacemaker by an electric razor. Am. J. Cardiol. 27 (1971) 695-697.

2. D'CUNHA, G.F., NICOUND, T., PEMBERTON, A.H., ROSENBOUM, F.F., and BOTTICELLI, J.T.: Syncopal attacks arising from erratic demand pacemaker function in the vicinity of a television transmitter. Am. J. Cardiol. 31 (1973) 789-791.

3. ESCHER, J.W., PARKER, B., and FURMAN, S.: Influence of alternating magnetic fields on triggered pacemakers. Supplement II to Circulation XLIV (1971) 162.

4. FURMAN, S. et al: The influence of electromagnetic environment on the performance of artificial cardiac pacemakers. Am. Thorac. Surg. 6 (1968) 90.

5. KING, G.R., HAMBURGER, A.C., FOROUGH, P., HELLER, S.J., and CARLETON, R.A.: Effect of microwave oven on implanted cardiac pacemaker. JAMA 212 (1970) 1213.

6. KOHLER, F.P. and MACKINNEY, C.C.: Cardiac pacemakers in electrosurgery. JAMA 193 (1965) 855.

7. MEIBON, J. and ANDERSON, J.: Inhibition of demand pacemaker by leakage current from electrocardiographic recorder. Brit. Heart J. 33 (1971) 326.

8. MICHAELSON, S.M. and MOSS, A.J.: Environmental influences on implanted cardiac pacemakers. JAMA 216 (1971) 2006.

9. MITCHELL, J.C., RUSTAN, P.L., FRAZER, JW., and HURT, W.D.: Electromagnetic compatibility of cardiac pacemakers. Presented at the 1972 IEEE International Electromagnetic Compatibility Symposium, Arlington Heights IL, and published in the Symposium Record (July 1972).

10. MITCHELL, J.C., HURT, W.D., WALTER, W.H., III, and MILLER, J.K.: Empirical studies of cardiac pacemaker interference. Aerosp. Med. (February 1974) 189-195.

11. PARKER, B., FURMAN, S., ESCHER, D.: Input signals to pacemakers in a hospital environment. Ann. NY Acad. Sci. 167 (1969) 823.

12. PICKERS, B.A. and GOLDBERG, M.J.: Inhibition of a demand pacemaker and interference with monitoring equipment by radiofrequency transmissions. Brit. Med. J. 2 (1969) 504.

13 RUSTAN, P.L., HURT, W.D., and MITCHELL, J.C.: Microwave oven interference with cardiac pacemakers. *Medical Instrumentation* 7 3 (1973).

14 SANCHEZ, S.A.: When you transmit, you can turn off a pacemaker. *QST* (March 1973) 53.

15 SMYTH, N.P.D., PARSONNET, V., ESCHER, D.J.W., and FURMAN, S.: The pacemaker patient and the electromagnetic environment. *JAMA* 227 12 (1974) 1412.

16 SOWTON, E., GRAY, K. and PRESTON, T.: Electrical interference in noncompetitive pacemakers. *Brit Heart J.* 32 (1970) 626.

17 WALTER, W.H., III, MITCHELL, J.C., RUSTAN, P.L., FRAZER, J.W., and HURT, W.D.: Cardiac pulse generators and electromagnetic interference. *JAMA* 224 (1973) 1628-1631.

-DISCUSSION-

GRAMIAK - Did you test any therapeutic ultrasound devices?

MITCHELL - Very early in the program we checked several pacemakers near an ultrasound source at one of the local clinics. On the basis of these qualitative evaluations the ultrasound devices do not appear to represent any significant problem.

GLASER, Z. - Do you foresee any problems with those automobile, anti-collision radar devices that are being talked about?

MITCHELL - From what I know about these radars they operate at sufficiently high frequencies and should not represent any threat to pacemaker patients.

VOGELMAN - I just thought you might be interested in knowing that one of our patients has a pacemaker. It is the older model Medtronics, and at one foot from his power saw or at one foot from his calculator, the pacemaker is adversely affected. He has had some weakness problems as a result until we found out what was doing it.

MITCHELL - Two years ago we predicted and published the threshold level of interference on a Medtronics model 5842 at 450 MHz (pulsed) was a half volt per meter. A recent article in Circulation magazine discussed a case of an individual passing out in a parking lot due to interference from a TV tower, and when they measured the level it was a half volt per meter.

VOGELMAN - Well, when this patient is more than an arms length from the source he is all right.

MITCHELL - I have numerous similar reports. It is real to the people who use pacemakers, and most particularly to those persons using the older, more sensitive models.

SUESS - I have two questions. In the slides you have shown, you have demonstrated or at least presented test data on deleterious effects from the pacemakers due to several types of electrical equipment. Would these effects result from a continuous or from a momentary failure?

MITCHELL - I should point out that as soon as any interference is removed or turned off the pacemaker resumes normal operation. It can interpret external signals as heart activity and properly inhibit when they are present.

SUESS - My second question relates to a short activity which took place last October in our office. One of the problems discussed was whether a pacemaker may hurt the man?

MITCHELL - Are you talking about possible electrical pick-up and subsequent harm to the user?

SUESS - Well, we may now have an increasing problem in urban areas created by electromagnetic fields from various sources. A pacemaker, even when well protected, still has the wiring and so on. Now, if a magnetic field is being produced there, and considering the probably growing number of people who may use pacemakers, we may face, to some extent, a public health problem. How would you look at this problem?

MITCHELL - The leads contribute to the interference but do not concentrate sufficient electrical fields to be harmful to the user.

SUESS - But what about the field created by the pacemaker and its wiring?

MITCHELL - Like a metal implant situation? We have considered that and do not consider it to be any problem under normal environmental exposures.

DUNN - A stupid question, but how does a calculator interfere?

MITCHELL - It has an oscillator that puts out a high frequency (pulsed) signal, but, it must be very close to the pacemaker to cause the effect.

VOGELMAN - Are there other electronic medical devices which might be EMI susceptible?

MITCHELL - Yes, devices like the brain pacemaker, the pain stimulation device, perhaps artificial organs, maybe even to some extent elaborate electrical control of arms and legs could possibly be affected.

FLOOR - But the blocking of one of those devices should be less serious than the blocking of the cardiac pacemaker. Don't you think?

MITCHELL - Yes.

FLOOR - Is there a reason for your choice of 450 and 3100 MHz or is it pure convenience?

MITCHELL - These frequencies were originally selected to meet an operational concern, but since then we have run tests for a broad frequency range down to about 30 kiloHertz, and up to 8 or 10 GHz. We find that the most sensitive frequency is somewhere between a hundred and 600 MHz. So the 450 MHz turns out to be a useful frequency. From a practical standpoint most pacemakers that show low EMI at this frequency do fairly good at most of the other frequencies.

ELY - Do you get a feeling for measurement or estimation of field strengths of any of these other interfering devices that would allow planning of the construction of a device? Suppose you are going to build a new drill or food mixer? Would you design it to emit less than so much?

MITCHELL - Probably not.

ELY - Or do you think the whole way to go is immunity of the patient --

MITCHELL - Well, it appears to me that the manufacturers are really solving the problem.

OSEPCHUK - I would like to comment that in the last few years the immediate reaction of some people to such interference is that the source is bad. Of course, as time goes on and almost every source becomes incriminated, the point finally dawns on some people that there may be something wrong on the other side, described by a word called susceptibility. You know, the military would never procure pacers the way they were designed with the transparent potting and absolutely no attention to RFI. Susceptibility doesn't need to be there. It

OSEPCHUK - turns out it is fairly easy to remove.

The reaction of some medical doctors to engineers' new pacer designs for reduced susceptibility is a fear of overkill. They imagine that the pacer will be so immune to RFI that they can't do things like using a transistor radio to pick up the pacer signal. I presume manufacturers all know that they still can do that, even with a shielded model - isn't that correct?

FLOOR - There is no fear for the MD's that you are going to have overkill or the pacer manufacturer is going to overkill this thing and prevent the MD's from doing what they want with it. There have been MD's who feel that the engineers are going to force this on them.

MITCHELL - I think there are some valid points on both sides of that argument. You never please everybody, but we hope somewhere along the way this thing will get resolved.

Occupational Aspects

ANALYSIS OF OCCUPATIONAL EXPOSURE TO MICROWAVE RADIATION

P. Czerski and M. Siekierzyński

Natl. Res. Inst. of Mother and Child, Warsaw
and Inst. for Postgraduate Study, Military Medical
Academy, Warsaw, Poland

ABSTRACT

Principles of analysis of environmental conditions and of comparison of various microwave worker groups are discussed. Early and actual findings concerning the health status of microwave workers, published by various Polish authors, are compared and discussed in the light of personal experiences. Advantageous effects from enforcement of safety rules may be documented by the comparison of present results with those published ten years ago.

Introduction

The present paper is an attempt to summarize briefly personal experience and that of our colleagues, who worked in close contact with us during the last sixteen years. All the comments, however, express personal opinions of the present authors only. The list of references was restricted to the essential minimum, as an almost complete list of pertinent papers can easily be had by referring to the bibliographies in references 4, 13, 14, 24, and 29.
Analysis of occupational exposure to microwave radiation is fraught with many difficulties, the main being the assessment of the relationship between the microwave exposure levels and the health status of the examined groups of workers. The possible role of other environmental factors and of socio-economic conditions must be taken into account. As it often happens in clinical work, it is difficult to demonstrate a causal relationship between a disease and the influence of environmental factors, at least in individual cases. Large groups must be observed to obtain

statistically significant epidemiological data. The problem of adequate control groups is controversial and hinges mostly on what one considers "adequate."

Analysis of Environmental Conditions

Precise quantitation of human exposure is possible only in the case of therapeutic or diagnostic applications of microwaves. In view of the lack of adequate instrumentation, especially of individual dosimeters, the quantitation of occupational exposure is extremely doubtful. This is particularly true where personnel move around in the course of their duties and are exposed to non-stationary feilds (i.e., moving beam or antenna), as well as to near- and far-fields alternatively. It is impossible to quantitate the exposure over a period of several years within reasonable limits. Attempts to present detailed data as to the source of microwave radiation, effective area of irradiation, position of the body in respect to the field, etc. for an individual worker for a period of several years would be misleading to an extreme degree. In the present authors' opinion, it is far better to present approximate evaluations, than to create an impression of accuracy, where none can be had.

Gordon (14) divided the microwave exposed workers examined by her into 3 groups, according to exposure levels:

1. periodic exposure to "high energy density" levels, i.e. 0.1-10 mW/cm^2,

2. periodic exposure to "low energy density" levels, i.e. 0.01-0.1 mW/cm^2,

3. systematic exposure to low energy density levels.

The first group consisted of technical maintenance personnel and workers of repair shops and certain factories (montage). This group consisted of production, montage, technical maintenance and repair of microwave equipment personnel. It should be mentioned that a large part of this personnel was periodically exposed to near-zone fields. The second group consisted of technical maintenance personnel as well as certain categories of personnel engaged in the use of microwave apparatus, research workers and others. The third group consisted of personnel engaged in the use of various microwave equipment, mainly radar.

In our investigations we adopted a similar rule of division of the personnel examined into high, mean or low exposure groups, or later, into two groups - high and low exposure. In regard to environmental conditions, analysis of such factors as air temperature, movement and humidity, noise, lighting and exposure to ionizing radiation generated incidentally by electronic equipment were taken into account. It should be stressed that exposure to these factors is carefully controlled, according to Polish laws.

Safe microwave exposure limits and regulations enforcing safety measures and precautions were introduced in Poland in 1961.

The exposure limits adopted at this time were identical to those introduced two years earlier in the USSR (0.01 mW/cm^2 unlimited exposure, 0.1 mW/cm^2 permissible for 2 hr per day and 1 mW/cm^2 for 20 min per day). It was possible to enforce this law only gradually. In view of this, all examinations of personnel carried out before 1962 concerned persons subjected to uncontrolled exposure. The probable exposure levels could be evaluated only ex post, and only approximately. In the period 1962-1968 data on exposure levels based on power density measurements and analysis of working conditions became available; most examined individuals had, however, a shorter or longer uncontrolled exposure history.

In view of this, the Polish publications concerning the health status of personnel professionally exposed to microwaves may be divided into papers concerning:

1. persons having a history of longer or shorter periods of work under uncontrolled conditions, exposure levels undetermined or calculated retrospectively;
2. persons with a history as above and a period of work in controlled environment;
3. persons who were examined medically before work was undertaken, found fit for work under microwave exposure conditions, and working under controlled conditions.

Numerous Polish papers concern the first and second groups (1, 6, 9, 11, 15, 17, 19, 23) or the second group and third groups jointly (3, 12, 16, 18, 20, 30) and only very few exclusively the third group (7, 8, 26, 27).

Early Investigations

Selected groups of microwave workers were first examined at intervals of 3 months, later 6 months, and, finally, annually. All the persons examined were usually divided into 3 groups:

1. low-level exposure of the order of tens of μW/cm^2 usually in far-field conditions in open space or in complex fields in closed rooms where only very low power equipment was installed;
2. "mean" exposure levels in far- and near-field zones; the measured levels being of the order of hundreds of μW/cm^2 up to about 1 mW/cm^2;
3. high level exposures of the order of 1 mW/cm^2 up to 10 mW/cm^2; in certain instances even more.

No attempt was made to differentiate between exposures at various microwave frequencies, as most individuals were exposed at one time or another over the entire microwave range. No adequate control group could be found because of difficulties in finding individuals working under sufficiently similar conditions (temperature, noise, humidity, time of day, etc.) of work and of sufficiently similar economic and social position, having similar everyday living habits, as well as belonging to the same age group. It was decided, therefore, to analyse the material within the group according to the total period of work and according to the exposure level by making comparisons between groups. It was possible to collect such groups which were reasonably similar in all respects (socio-economical and psychological) except the level of exposure.

All cases, where pathology related to any known etiology could be found, were excluded from the analysed material. Medical documentation on earlier examinations and physical check-ups was obtained in all cases and compared with actual findings. This documentation was nevertheless in most instances unsatisfactory in respect to eye examination and laboratory data.

The results obtained confirmed the findings of USSR authors (14, 24, 29). Complaints were analogous and demonstrated a periodicity of occurrence in relation to the duration of occupational exposure, stressed in the Soviet literature (24).

The presence of complaints was characteristic persons subjected to periods of uncontrolled exposure, before safety rules were introduced. Headaches and fatigue disproportionate to effort occurred respectively in 47% and 45% in group 3, 30% and 34% in group 2, and 30% and 30% in group 1 during the first year of work. These complaints disappear for two years, recur during the period 3-5 years of work and may reappear in certain individuals after 5 or 10 years. Abnormally excessive sweating (during the night) had a similar time dependence and was found in 68.8% in group 3, in 33.4% in group 2 and in 22.5% in group 1 during the first year of work; in the period of 3-5 years of work, the respective values being 14.5%, 15%, and 7%. Later on this symptom was not observed. Changes in blood pressure occurred only in group 3; the percent of hypotonia being 18 during the first year, 14 in the period 1-3 years, 6 in the period 3-5 years, 8 in the period 5-10 years and 11 after 10 years of work. In the remaining groups this percentage was less than 1. It should be added that no correlation with changes in heart rate could be demonstrated.

The peripheral blood picture did not demonstrate any abnormalities. In group 1 diverse WBC responses were seen during the first year of work. After ten years of work in the same group 10.5% of workers show absolute lymphocytosis usually accompanied by monocytosis; the total WBC being over 10,000 per mm^3.

Neurological examinations are difficult to evaluate. Many of the physicians, who carried out these examinations, differed in the evaluation of reflexes, dermographism, signs of irritability, etc. In view of this, the only means of obtaining objective results are electroencephalographic studies. These do not demonstrate any abnormalities in group 1. In groups 2 and 3 depending on duration of work and degree of exposure, a definite decrease in the number and amplitude of alpha waves occurs. Theta and delta waves and spike discharges may occur. The response to photostimulation is decreased. The most impressive finding is the poorly expressed bioelectric activity after more than 5 and, especially after 10 years of work.

It should be stressed that a rather specific phenomenon occurs in microwave workers (3, 12). Intravenous administration of cardiazole (metrazole) may be used for provocation of discharges (preconvulsive discharges) in the EEG, convulsions or shock. According to the literature this phenomenon is dose dependent and a cardiazole threshold exists; the dose of 7 mg/kg body weight being without any effect. Intravenous administration of 500 mg cardiazole in 10 ml (1 ml/30 s at 30 s intervals) does not provoke any effects in the normal adult male. In microwave workers with greater than 3 years of exposure, theta waves, theta discharges, spike discharges and even convulsions occurred. Twelve persons were examined, in 8 the test could not be completed. The study was discontinued, as it was considered dangerous for the patient. It should be pointed out that this phenomenon was studied extensively in rabbits (2, 3), and a decrease of cardiazole tolerance in irradiated animals may be considered as established.

Cases of what was considered a "microwave sickness" were reported in the Polish literature (5, 6). It may be doubted if a specific nosologic entity may convincingly be demonstrated in the present state of understanding of microwave bioeffects. The abnormal findings consisted mainly of a syndrome of autonomic nervous system disturbances with bradycardia, hypotonia and a disabling neurotic syndrome with typical EEG changes as described above.

In the course of work connected with health surveillance and risk analysis, the present authors encountered "clusters" of certain deviations from normal in factories or other working places, where exposure levels were exceptionally high, i.e. about 10 mW/cm^2 or more during about 1 hr/day. In such places also uncontrolled exposure by leakage could be expected. The abnormalities were present in 0.5-2% of otherwise healthy persons with deep bradycardia (less than 50/min) and signs of impairment of myocardial conductivity as represented by ECG, variable percentage of workers with gastric ulcers or peripheral blood picture changes (slight anaemia, lymphocytosis, granulocytopenia, or persisting unexplained granulocytosis). Such groups were usually too small to draw any valid conclusions, so only an impression remains that working conditions had "something to do" with these phenomena. This impression is strengthened by the fact that after introduction of rigorous health and microwave exposure surveillance, as well as partial exchange of personnel, no further cases were noted.

It should be added that no cases of "microwave cataracts" were described in the Polish literature or found by us. A higher incidence of lenticular opacities was reported in groups with histories of uncontrolled exposure periods and may possibly be related to poorly controlled exposure conditions (15, 17, 30).

These findings were used to develop the principles of medical examinations for selection of candidates for work under microwave exposure (8), and as an empirical basis for setting up the new Polish microwave safe exposure limits (7), which were introduced in 1972 (25).

Actual Findings

A special study was undertaken to determine if work under conditions conforming to actual Polish safe exposure limits (25) may be considered as truly safe. Detailed results are to be published shortly in English (8, 26, 27); therefore, the results will be presented very briefly.

An analysis of the incidence of disorders considered as contraindications for occupational microwave exposure among 841 males aged 20 to 45 years and exposed occupationally to microwaves for various periods was made. The analysed population was subdivided into two groups differing only in respect to microwave exposure - low i.e. below 0.2 mW/cm^2 and high i.e. between 0.2 mW/cm^2 and 6 mW/cm^2. No dependence of the incidence of disorders, considered contraindications for occupational microwave exposure, on the exposure level or duration of occupational exposure could be demonstrated (8). The incidence of lenticular opacities was compared between both these groups, as well as analysed within each group, subdivided according to age or duration of occupational exposure. No dependence of the incidence of lenticular opacities on the exposure level, nor on duration of occupational exposure was found. Significant correlation with age was demonstrated (27).

The incidence of functional disturbances (neurotic syndrome, gastro-intestinal tract disturbances, cardiovascular disturbances with abnormal ECG) was also analysed and no dependence on the exposure level or duration of occupational exposure (years) could be demonstrated (26).

Conclusions

Uncontrolled occupational exposure leads to the appearance of autonomic and central nervous system disturbances, asthenic syndromes and other chronic (prolonged exposure) effects, which are well documented by early Soviet, Polish and Czech reports. Similar observations were made by Miro (22) and Deroche (10) in France and in the United Kingdom and USA, according to a personal communication made by Mumford to Seth and Michaelson (28).

Controlled occupational exposure of healthy adults seems to have no untoward effects if the (new) Polish safe exposure limits (25) or, even more so, the conservative Soviet standard (24) are observed.

The last point, which is most important and cannot be sufficiently emphasized, is that all available data concern healthy human adult exposure, mostly men. The effects of intermittent or continuous exposure of children living near radar installations or TV transmitters is completely unexplored. Children may be expected, because of body size and geometry, to absorb microwave energy differently from adults. Exposures to 4-8 min/day microwave irradiation at low mean (tens or hundreds microwatts per cm^2), and very high peak power densities are sufficiently real for children, as to cause concern. This is one of the many reasons to adopt lower safe exposure limits for the general public. It seems reasonable to differentiate between occupational exposure (adult, healthy men under medical surveillance) and general public exposure (the aged, the sick, pregnant women and children). It seems evident that in the former case higher values are acceptable, in the latter - a safety margin must be introduced.

REFERENCES

(Russian names, book-titles and journals were transliterated according to International Organization for Standardization recommendation IO/RG-1954 E)

1. BARANSKI, S.: Badania biologicznych efektów swoistego oddzialywania mikrofal, Inspektorat Lotnictwa. Warsaw (1967).

2. BARANSKI, S. and EDELWEJN, Z.: Acta Physiol. Polonica 19 (1968) 37.

3. BARANSKI, S. and EDELWEJN, Z.: Pharmacologic analysis of microwave effects on the central nervous system in experimental animals, In "Biologic Effects and Health Hazards of Microwave Radiation" (P. Czerski et al. eds). Polish Medical Publishers. Warsaw (1974) in press.

4. BARANSKI, S., and CZERSKI, P.: Biological Effects of Microwaves. Polish Medical Publishers. Warsaw (1975) in press.

5. CIESLIK, Z., KORZENIOWSKI, K. and KAFLIK, I.: Problemy Lekarskie 12 (1973) 483.

6. CZERSKI, P., HORNOWSKI, J. and SZEWCZYKOWSKI, J.: Medycyna Pracy 15 (1964) 251.

7. CZERSKI, P. and PIOTROWSKI, M.: Medycyna Lotnicza 39 (1972) 104.

8. CZERSKI, P., SIEKIERZYNSKI, M. and GIDYNSKI, A.: Aerospace Med. (in press).

9. DENISIEWICZ, R., DZIUK, E. and SIEKIERZYNSKI, M.: Polski Archiwum Med. Wewn. 15 (1970) 19.

10. DEROCHE, M.: Arch. des Malashes Professionnelles 62 (1971) 679.

11. DZIUK, E., DENISIEWICZ, R., SIEKIERZYNSKI, M. and SYMONOWICZ, N.: Lekarz Wojskowy 42 (1970) 20.

12. EDELWEJN, Z. and BARANSKI, S.: Lekarz Wojskowy 46 (1970) 781.

13. GLASER, Z.R.: Bibliography of Reported Phenomena (Effects) and Clinical Manifestations Attributed to Microwave and Radio-frequency Radiation. BUMED Report, National Technical Information Service, Springfield Va., A.D. 750271, 1972 and Bureau of Medicine and Surgery, Dept. of the Navy, Washington (1973).

14. GORDON, Z.V.: Voprosy Gigieny Truda i Biologiceskogo deistvija elektromagnitnyh polei sverhvysokih castot. Medicina, Moskva (1966).

15. HORNOWSKI, J., MARKS, E., CHMURKO, E. and PANNERT, L.: Medycyna Pracy, 17 (1966) 213.

16. KOLAKOWSKI, Z.: Lekarz Wojskowy 47 (1971) 309.

17. MAJEWSKA, K.: Klinika Oczna 38 (1968) 323.

18. MAZURKIEWICZ, J., MILCZAREK, H., ZALEJSKI, S. and SIEKIERZYNSKI, M.: Lekarz Wojskowy 42 (1966) 9.

19. MILCZAREK, H.: Lekarz Wojskowy 47 (1971) 442.

20. MILCZAREK, H., ZALEJSKI, S., MAZURKIEWICZ, J. and SIEKIERZYNSKI, M.: Polski Tygodnik Lekarski 50 (1967) 1924.

21. MINECKI, L.: Medycyna Pracy 12 (1961) 329.

22. MIRO, L.: Rev. Med. Aeronaut 1 (1962) 16.

23. ORNOWSKI, M.: Medycyna Lotnicza 20 (1967) 47.

24. PETROV, I.R. (Ed.): Influence of Microwave Radiation on the Organism of Man and Animals. NASA TT-F-708. Natl. Technical Information Service. Springfield (1972).

25. Rozporzadzenie Rady Ministrów z 25.05, 1972. Dziennik Ustaw PRL 21, poz. 153.(Order on Polish safe exposure limits in Polish statute journal).

26. SIEKIERZYNSKI, M., CZERSKI, P., MILCZAREK, H., GIDYNSKI, A., CZARNECKI, C., DZIUK, E. and JEDRZEJCZAK, W.: Aerospace Med. (in press).

27. SIEKIERZYNSKI, M., CZERSKI, P. ZYDECKI, S., CZARNECKI, C., DZIUK, E. and JEDRZEJCZAK, W.: Aerospace Med. (in press).

28. SETH, H.S. and MICHAELSON, S.M.: Aerospace Med. 35 (1966) 734.

29. TJAGIN, N.V.: Kliniceskie Aspekty Oblucenija SVC-diapazona. Medicina. Leningrad (1971).

30. ZYDECKI, S.: Assessment of Lens Translucency in Juveniles, Microwave Workers and Age - Matched Groups. In "Biological Effects and Health Hazards of Microwave Radiation" (P. Czerski et al. Eds). Polish Medical Publishers. Warsaw (1974) in press.

-DISCUSSION-

McAFEE - Whenever electronic devices operate at high voltages, there is the possibility of significant deleterious by-products, emanating in the vicinity of the tubes. These by-products include X-radiation, ultraviolet radiation, ozone and oxides of nitrogen. Is it possible that chronic exposure to these toxic substances may explain the results of the provocative cardiazol tests as well as the altered blood and neurologic picture reported among microwave workers?

CZERSKI - Certainly not. Why? Because as I tried to say in the introduction, these things were controlled in human exposure. As concerns Cardiazol tests, the reaction of this drug was explored under experimental conditions. Rabbits were exposed in an anechoic chamber in far field conditions of course in the absence of X-rays. These animals demonstrated the reaction to Cardiazol. So I feel that I can easily say that none of the factors mentioned by Dr. McAfee could influence the response.

KAUFMAN - In talking about subjective complaints, as you are aware, it is often very hard to determine the cause of such complaints. I wonder whether complaints that occur in the first year and then go away might be due to the psychological stress associated with starting a new job, particularly if these people were told that their work might be hazardous.

CZERSKI - Such a possibility certainly exists, but we try to explain that really there is no work which is not hazardous,

CZERSKI - because as it is said "whatever you do - work is dangerous to your health." Seriously speaking, I think one question has to be considered. We had some effect at very low levels as I tried to show in the example of the influence of microwaves on the circadian rhythm of adrenalin-noradrenalin levels. Then you have an appearance and disappearance of complaints in the same individual. They complained the first year and then in the second and third year they had no complaints, and they were perfectly happy. Afterward, they started the complaining once more about the fifth year.

I am just wondering if we shouldn't try to evaluate the whole thing according to the classical concept of stress, adaptation, and fatigue, and not look for any more complicated explanations when we may find an answer in well-known physiologic mechanisms.

POSTOW - Is there a specific reason why tuberculosis is considered a contraindication for occupational exposure to microwaves?

CZERSKI - There is really a non-specific reason on general grounds. This is a socially dangerous disease from a transmission point-of-view. This is a precaution. Especially, perhaps under Polish conditions, where just after the war for many years tuberculosis was a serious problem. I am happy to say it is no longer; we are apt, however, still to remain oversensitive just in this respect.

TOMPKINS - In your chamber exposure situation at the end, would you repeat the exposure conditions? I am sorry; I missed them.

CZERSKI - You mean those I showed on the slides. This is really no anechoic chamber exposure, but simply a real open-space exposure. The situation was such that there were some people and installations at a certain distance in the open space.

Measurements were made at these points where the examined individuals worked and the exposure level was 0.2 mW/cm^2 and at certain times up to 6 mW/cm^2 during two minutes per 4 hours. What I said is based on actual timing. The men worked four hours and rested four hours, respectively during 24 hours.

We found another situation where another group of men had done exactly the same work, but there were no microwaves detectable by measuring equipment. We looked especially for any microwave irradiation at the second location, and none could be detected. And, as I said, the men performed exactly the same work in the same cycle, and the adrenalin-noradrenalin

CZERSKI - secretion was compared.

THOMSON - Were there any differences in the incidence of absence from work in the people exposed occupationally to microwaves?

CZERSKI - No, but this is another question because when we introduced this very strict medical surveillance, in certain instances we had such a situation, where we had no complaints, no absence from work, and so on. After medical surveillance and strictly controlled exposure conditions were introduced, the people were, I would say, more healthy, in comparison to matched groups. Of course, this was a result of selection and good medical care.

CONTROL OF OCCUPATIONAL EXPOSURE TO NON-IONIZING RADIATION

Thomas S. Ely

Assistant Director, Health and Safety Laboratory

Eastman Kodak Company, Rochester, New York

ABSTRACT

Potential non-ionizing radiation hazards exist in many occupational settings. Welding arcs, radio frequency heaters, incandescent filaments, gas discharge lamps, lasers, and large hot surfaces can cause injury, and may need to be evaluated. Some of the situations are covered by state and federal regulations. Evaluation and control often involve practical "rules of thumb" whose ancestry can be traced to scientific data and practical experience.

Report

Much of what you have heard and will hear during the course of this Conference is of a much more academic and scientific nature than what I am about to say. The development of scientific information from basic research is essential in the control of occupational hazards, but it is only the first step in a sequence of activities that is necessary in this control. The information needed for control purposes must evolve through several stages including basic research, applied research, the development of a consensus position leading to a practical standard. This may then evolve into a practical regulation. It is a fact of life that a regulation is often necessary to accomplish control that a consensus standard could not.

It would be as wrong to use the techniques of basic research to achieve control as it would be to use the techniques of control to conduct basic research. In basic research one strives for extensive and accurate data. For control purposes, such would almost always be impractical and unnecessary. I have frequently responded to proposed federal regulations requiring 5% accuracy with the comment that 20 or 25% should be plenty good enough. I have been too timid to say what I really think -- that factor-of-two accuracy is usually sufficient for control purposes. However, the realities of regulatory control are that a rigid set of requirements must lead to a rigid conclusion about compliance. Regulatory people are very uncomfortable with factor-of-two accuracy.

Unfortunately, sometimes there are missteps in the transitions between the steps I have mentioned. Good basic research can be followed by bad applied research. Good applied research can lead into a bad practical standard. A good practical standard can result in a bad regulation. In this last case we then have two kinds of hazard, a health hazard and a regulatory hazard, and one needs to be careful to specify which he means. This is a most lamentable situation, where getting into compliance with the law actually increases the health hazard. Fortunately, I am aware of only a few such examples.

More common is the case where the standard and the regulation are different but not mutually exclusive. For example, there are biological carcinogens and regulatory carcinogens, and the lists are not the same. Another example is the frequent case in which the regulation is of a "specification" rather than a "performance" type. Here it is frequently possible to have a safe situation but be out of compliance with the regulation. The point I am trying to make is that health hazard and regulatory hazard should not be confused in one's thinking.

Another general thought I would like to introduce with respect to the development of occupational hazards standards is that it is immensely useful to have a "natural" source of whatever agent is under consideration for comparison purposes. For example, our sun is a familiar source of a wide range of non-ionizing radiation to which the human race has become more or less accustomed. The ability to relate quantitatively the various "artificial" sources in the occupational environment is a great help in the development of practical standards. For instance, it becomes a little silly to agonize over an ultraviolet radiation exposure standard that represents a small fraction of the dose the person gets on his way to or from the parking lot.

I would now like to discuss briefly some practical issues involving potential non-ionizing radiation hazards in the occupational setting.

Arc Welding

Among several other non-radiation welding hazards such as the release of toxic chemicals, ozone production, and eardrum perforation, the significant non-ionizing radiation hazards are those from infrared, visible, and ultraviolet. The near infrared and visible portions of the spectrum could constitute a retinal burn hazard if intensity were adequate, but this is probably not a real problem for ordinary welding. The retinal dose seems to be less than that required to produce injury, and the warning properties are pretty good in that the welding arc is uncomfortable to look at.

The classic non-ionizing radiation hazard from arc welding is that of "flashed eyes" or ultraviolet keratoconjunctivitis. This painful but temporary condition leaves a lasting impression on its victim, and it rarely happens to him the second time. It is interesting to note that the condition often occurs in a nearby associated person instead of the welder. In my Navy experience it was almost always the "fire watch" -- the man posted nearby to observe any fires the welder might start. At Kodak Park we can recall five cases of flashed eyes in the last 20 years or so. All five were welder's helpers; none was the welder himself. It is interesting to note that in the same period, we have had no case of injury by infrared, visible light, or ionizing radiation.

Radio Frequency Heaters

Radio frequency power has been found very useful for its heating potential and many industrial applications such as glue heaters, plastic sealers and welders, and dryers. The adverse effects of overexposure include the direct biological hazard, and indirect hazards such as the influence on electronic cardiac pacemakers. All of these are treated extensively in other parts of this Conference.

The applicable standard is American National Standards Institute (ANSI) C-95.1-1966, and the applicable regulation is Occupational Safety and Health Administration (OSHA) 1910.97 or its State equivalent. The standard covers a large frequency range, and was designed around the most hazardous part of that range. Thus, for many RF heaters in the lower frequencies, the standard is overly restrictive.

In making a practical evaluation of a potential RF hazard, I have found that an extremely useful first step is to determine the frequency. If this is not in one of the Industrial, Scientific, and Medical (ISM) bands of approximately 13, 27, 40, 915, 2450, 5800, or 22,000 MHz, the problem disappears. This is because the amount of leakage radiation permitted by the Federal Communications

Commission because of communication interference is well below the level that would be of concern from an occupational health standpoint. If the frequency is in one of the ISM bands, the situation will need to be evaluated. The lower frequencies are rarely likely to be a problem as a direct hazard because of their poor absorption and diffuse absorption. For the higher frequencies, we have a good standard.

One RF heater that has appeared on the scene lately in great numbers is the microwave oven which is used in industry mainly to enable the employee to heat his lunch efficiently. In the occupational setting, the applicable standard is the same OSHA 1910.97, which is an exposure standard. Covering all microwave ovens in addition is a Bureau of Radiological Health (BRH) manufacturing regulation that is an emission requirement. In essence, it requires a maximum of 10 W/m^2 (1 mW/cm^2) at the time of manufacture and 50 W/m^2 (5 mW/cm^2) subsequently.

In Kodak in Rochester we now have some 70 microwave ovens. At the time the first oven was purchased, we decided to monitor radiation leakage from the ovens on a six-month interval basis. A few years ago, early in this experience, we found an occasional oven leaking more than it should, but certainly not a hazardous level. The leakage could always be fixed by hinge, latch, or seal maintenance. Our more recent experience has been that significant leakages are rarely found, and we are thinking of decreasing the frequency of measurement. There is no regulation which specifically requires oven monitoring.

I feel that even an annual frequency of monitoring constitutes hyperscrutiny insofar as the actual potential for hazard alone is concerned. However, there are other considerations here such as confidence engendered in regulatory agencies and reassurance of users that justify the program.

Incandescent Filaments

Electrically heated wires in transparent envelopes can represent non-ionizing radiation hazard in infrared, visible, and ultraviolet wavelengths. In addition to the obvious case of whole body heating from large incandescent lamp arrays, the question of retinal hazard occasionally has arisen. In general, the bulb with frosted glass or some other diffusing arrangement is not a retinal hazard. The bare tungsten filament may be. Low wattage bulbs with thin filaments are probably not a retinal hazard, but we know that high wattage bulbs, particularly those with concentrated filaments such as projection lamps, can be. We know this not only from theoretical calculation, but there have been instances of retinal burns from such bulbs. Sometimes the question asked has been whether an unfrosted bulb operating at less than nominal voltage represents a hazard. On separate occasions, I have had to run out the calculation at several different temperatures, and would suggest that at temperatures above 2000 or 2100 kelvins a retinal hazard should be considered. At this temperature, the radiant emittance is of the order of a recommended maximum (not an injury threshold). Although not injurious in this range, such temperatures would certainly be uncomfortable. In the situation where these filaments must be watched for inspection or other purpose, a filter could be placed over the bulb, or the employee could wear standard goggles that would remove essentially all the infrared, and enough of the visible to make the operation comfortable. Since the infrared contributes nothing to the task, it might as well be removed.

High temperature filament operation combined with an envelope transparent to ultraviolet such as the relatively recently developed quartz-halogen lamps emit potentially injurious amounts of UV, and have caused "sunburn" and scratchy eyes. The control is easy if the problem is recognized. Suitable glass or plastic windows with low UV transmittance is all that is necessary.

Gas Discharge Lamps

Low pressure lamps are rarely a concern from the retinal burn consideration, low pressure mercury vapor lamps with the appropriate envelope can be an ultraviolet hazard however.

"Black lights", those low pressure mercury arc lamps with an envelope transmitting only near ultraviolet, are not a significant hazard. The most comforting and memorable reference material on this issue was an article on Go-Go dancers appearing in the American Industrial Hygiene Association Journal in 1969. This is one of the few "black and white" situations we enjoy in the occupational health profession. It allows us to give commendably brief and authoritative answers to the occasional question we get on the subject.

Lasers

Plenty has been said about laser hazards already, and little more needs to be offered here. There have been standards for laser hazard protection almost as long as there have been lasers. Naturally, these have increased in complexity over the years as more and more information has become available and as more and more thought has been given to the development of recommendations. The preeminent standard in the field now is ANSI Standard Z-136.1. It is a good standard. Verbatim it would be a bad regulation because of its complexity and detail. More than a smattering of optical physics is required to understand it. The only federal occupational laser standard exists in the Part 1926 (OSHA construction) regulations. It is good as far as it goes, but is not inclusive enough, and of course covers only the construction occupations. I anticipate a Part 1910 (**OSHA general industry**) laser standard sometime based on (but I hope not identical with) ANSI Z-136.1. In the meantime, there are operable state regulations in some of the states. Some of the state codes are good and some are terrible. I fear that in some cases the regulation represents an overreaction to a new and mysterious hazard and is long on busywork and short on practicality.

Radiation from Large Hot Surfaces

This sort of radiation (primarily infrared) can be a problem in whole body heating, particularly in some industries, such as steel and aluminum. Although there are no specific regulations currently, there has been a considerable effort to develop a heat stress standard at the federal level, and there has actually been a NIOSH criteria document published on the subject. It proposes to measure the wet bulb-globe temperature (WBGT), which quantity is subject to radiant heat load in addition to dry bulb temperature, humidity, and air velocity. Considerable "heat" has been generated over the heat stress document, and there is much sentiment that such a sweeping regulation is not really necessary.

Electronic Cardiac Pacemakers

My philosophy about these devices in the industrial setting is approximately the following:

If an employee is wearing an electronic cardiac pacemaker we in the larger industries with medical programs will know it. This is because we will find it out when he is hired if he acquired it before that time. If he acquires it after that time, we will know about his reason for the absence when he had it installed.

Therefore, we will be able to ascertain whether the employee was counseled by his cardiologist or cardiac surgeon with respect to electromagnetic interference, and we will be able to advise him about potential sources in his workplace environment. If he has not been counseled by his private physician, we can see that he is or do it ourselves.

In any case, there are probably some areas, for example around some RF heaters, that this man should not go.

The issue of access by a non-employee is somewhat more difficult. Such person may be a visitor, a salesman, an outside maintenance man, or perhaps a family member attending an open house. In these cases, it may be necessary to post signs, ask the question, or make certain areas off-limits.

There have been occasional expressions of concern over microwave ovens as potential sources of electronic cardiac pacemaker interference. This sometimes takes the form of suggestions that warning signs be placed on the ovens. There actually was such a military regulation promulgated, and last year there was a petition that such signs be required by federal regulation.

On this issue, I fully subscribe to the position taken by BRH in 1971 that the generalized use of warning signs at microwave oven installations would be impractical and unnecessary. This position is based on:

1. Such signs would focus attention on a single source and fail to warn about other sources of interference such as electric tools, household and industrial appliances, ignition and lighting systems, radio, television and radar systems, that could not be effectively delimited by signs.

2. The signs would label all microwave ovens as hazardous regardless of the quality of the oven. Microwave ovens have leaked less and less as time goes on, probably because of a combination of the BRH regulation and improved manufacturing techniques that would have occurred anyway.

3. Electronic pacemakers are becoming less and less sensitive to interference.

4. In the real environment, microwave ovens have not been shown to be a serious cause of electronic pacemaker interference.

-DISCUSSION-

McAFEE - I would like to remark on the criticism of bureaucratic standards, and to say that this is easy to do and sometimes telling, but in the case of the ultraviolet standard you mentioned comparing ultraviolet from sunlight with ultraviolet produced indoors, we find important differences. For example, ultraviolet sterilization devices produce ozone, which stays indoors. It is perhaps not fair to criticize the bureaucrat on the basis that this occurs in nature since sunlight-produced ozone stays outdoors.

In the same vein, although we are exposed to electromagnetic

McAFEE - radiation from the sun, very little of that is microwave radiation.

ELY - Yes, I agree with that. On your first point though, the place for an ozone standard is in an ozone standard, not in an ultraviolet standard. If you need an ozone standard, for heaven's sake let's have one. But don't bury it in an ultraviolet standard.

ZIGMAN - There is some recent evidence that near ultraviolet light can be damaging to animals, and cultured cells, although the work hasn't been done in humans. One of the ways that near UV works is by photo-oxidizing aromatic substances either of biological origin or possibly chemicals that cells are in contact with, so as to convert them into photo-sensitizers. These photo-products do have inhibitory effects on enzymes and do change protein structure to some degree.

It has been found by our group and by others that near ultraviolet light, as is emitted by black light lamps, can over a long period of time with exposure for many hours a day, lead to cataracts and retinal degeneration in mice. So, I don't think that we should say a priori that it's safe radiation and that we shouldn't take any precautions to avoid it.

ELY - If there is not evidence that ultraviolet in industry is causing a significant amount of detrimental effects to people, then why have a regulation. And, there is great doubt that there is that evidence. Every once in awhile a welder's helper gets his eyes flashed, but having a regulation isn't going to change that -- that's an accident.

MILLER, M. - In mutagenesis work, where 254 nanometers is the wave length used to induce mutations, there is a rather direct correlation between the absorbance of 254 nanometers by DNA, and mutagenesis. The effect of this UV light in terms of what it does to the DNA is the induction of pyrimidine dimers which bungs up the functioning of the DNA, rendering the organism incapable of duplication or of duplicating mistakes and so forth. The process of UV-induced mutations is reversible in organisms from bacteria to marsupials - but not in "higher" organisms such as placental mammals - by longer UV wave length.

McAFEE - Sterilizers that produce ozone used to be made for hospitals. They operate on the principle that ultraviolet radiation at specific wave lengths, interacting with O_2, produces ozone. The point is that ultraviolet produced indoors stays indoors and the individual indoors in a working situation is exposed to the by-produces of ultraviolet radiation, including ozone.

ELY - So, there is an ozone standard and there is an ozone regulation, and there isn't an ultraviolet one. Another example of that is that maybe we ought to have an X-ray addendum to our microwave standard, because microwave generators sometimes make X-rays.

MILITARY ROLE IN SAFE USE OF MICROWAVES

Lawrence T. Odland, Colonel, USAF, MC

USAF Radiological Health Laboratory (AFLC)

Wright-Patterson Air Force Base, Ohio 45433

ABSTRACT

The military services of the United States have been, and continue to be, the major users of microwave energy in this country. By the very nature of the necessity that these systems be operational, safe use is mandatory if the missions of the services are to be completed. To this end the military medical departments have assumed the leadership role in studying the biological effects of microwaves, and in setting and evaluating the adequacy of occupational exposure standards. To date, no member of the military service has been shown unequivocally to have been injured because of occupational exposures. Cataracts are the most dramatic effect of microwave injury, but current standards and safety precautions, in the military services, are entirely adequate for protection against this injury. All services are supporting continuing programs of research and clinical studies to insure that the past record of safety continues, and that many of the unknown aspects of the program are properly and thoroughly investigated. A plea is made for the development of a personnel dosimetry and permanent program for recording exposure doses on a permanent basis.

(Opinions expressed in this paper are those of the author, and in no way reflect US Air Force or Department of Defense official policy.)

Introduction

Military organizations of this country, and probably most others in the world, are the major users of microwave energies. But, as these applications are perfected, many are readily adopted by the civilian components of the societies for the improvement of the quality of life, through advancement of technology. Since this pattern has evolved, it has been, and, certainly, still is, incumbent on the military services to insure that such hardware developed can be used without injury to operators or to the general public. Of equal, or perhaps more importance to the military services, is the absolute requirement that their uses of microwaves be conducted in a safe manner. Certainly, military missions can only be successfully carried out by personnel who have complete trust in their equipment, and have reliable information on hazards and safety precautions. A system, however effective, cannot be used if, in so doing, operating personnel or the general public are being injured because of overexposures.

The growth in all uses of microwave energy since WWII has paralleled, in many respects, the increase in world population. In 1946 it is estimated that 1.2 million radar and microwave units were in operation. Today, that number, in the military alone, has grown to a figure of about 20 million. Exact numbers are unobtainable because of security restraints, and the immensity of the of the inventory problem. This phenomenal growth has been made possible largely by the constant surveillance, research and evaluation by the military medical departments, of the associated hazards and medical problems. It has been, and continues to be, mandatory that the safety aspects of each new system be thoroughly evaluated before its large-scale use. Thus, we find a "closed loop" mechanism by which growth in use of microwaves is tempered by medical and safety considerations. Based on the growth numbers mentioned above, it becomes obvious that if these systems were not safely used, their numbers would be much, much smaller.

History of Microwave Developments

Hertz's studies[1] in 1886 showed that radio waves could be reflected by metallic and dielectric bodies. Thus, the basis for the great majority of use of microwave energy was laid nearly 100 years ago. Hulsmeyer,[2] working in England, foresaw the vast potential for application of radio waves in the maritime industries, and in 1904 obtained a patent for an obstacle detector and ship navigation device. Even though the patent was granted by the British government, a demonstration was held for the German Navy, which, at the time, concluded that the device was little better than a visual observer. Marconi, in 1922,[3] clearly predicted the vast potential of radio waves for detection of objects, and stimulated by his suggestion, two members[4] of the Naval Research Laboratory detected a wooden ship using a cW wave interference radar in the same year. Their proposals to conduct further work on this question were disapproved. The first detection of aircraft using the wave-interference effect was made in June 1930 by Hyland[1] of the Naval Research Laboratory.

By 1936 the US Army Signal Corps sensed the importance of this discovery, and encouraged largely by the Navy's work, the first operational radar used for anti-aircraft fire control was developed two years later. Col W. R. Blair,[5] then director of the Signal Corps laboratories, is credited as the legal originator of pulse radar by reason of a patent granted to him in 1937. Even though the early developments of pulse radar were primarily concerned with military applications, radio altimeters were used on commercial aircraft as early as 1936,[4] and this was probably the first nonmilitary application of the principle.

Stimulated by the survival instinct, the British made marked improvements in radar detection systems in the late thirties and early forties, and developed the magnetron[4] which was the most important contribution to the realization of microwave radar.

With the advent of space exploration and travel, another spurt in development of radar tracking and detection systems occurred, characterized by huge increases in power and antennae size.

The above brief historical review is presented to illustrate the point that military need was the primary motivating force for the development of microwave for many of the contemporary applications. With this in mind, it should be obvious to the reader that, had not the safety aspects of these "tools" been investigated in a manner parallel to their development, we could not have advanced the state-of-the art to what it is today.

Evaluation of Medical Surveillance

Studies on the biological effects of that part of the electromagnetic spectrum known as radio-frequencies, were reported during the latter portion of 1890's.[6] From then until 1945 the biomedical literature is sprinkled with accounts of related work. However, dosimetry techniques and equipment during this period were rather crude, and the experimental findings are considered by most as gross indicators of effects rather than quantitatively reproducible phenomena.

Just as the US Department of the Navy pioneered the early developmental work on radar, it was that service's medical department that published one of the first clinical studies[7] of results of exposure of laboratory personnel to radar and high-frequency radio waves. Two years later (1945) the Air Surgeon's Bulletin[8] of the US Army carried an account of the effect of radar energies on the hematopoietic system.

The US Navy provided leadership in studies on biological effects of microwaves during the 1945-1954 time period. In 1954 the responsibility for tri-service coordination of matters relating to biomedical aspects of microwave exposures was transferred to the US Air Force, and in the same year the Air Materiel Command (now AF Logistics Command) published[9] pertinent information to all major divisions of the US Air Force on the biological hazards associated with microwave exposures.

On 15-16 Jul 57 the US Air Force sponsored the first tri-service conference on biological aspects of microwave radiation[10] largely as a result of efforts of Col Knauf, USAF, MC. Representatives from the Department of Defense, industry, and selected universities, participated in the discussions. Additional meetings of this group were convened in the succeeding three years, all sponsored by the US Air Force. During the next decade the group functioned in an informal manner. However, in 1969 the Bureau of Radiological Health, US Public Health Service, recognized the need for a common platform for exchange of data on biological effects of microwaves, which would reflect and interface problems common to the military and the civilian sectors of society. Accordingly, under the guiding hand of Capt A. Wheeler, USPHS, the Joint Services Ad Hoc Committee on Microwave Ocular Effects was formed. It includes members of the uniformed services, and internationally-recognized civilian experts in ophthalmology as permanent consultants. This group coordinates investigative efforts of all member organizations, exchanges information, and, when appropriate, addresses specific problems.

Research Efforts and Standards

In order to give the reader an insight for the role played by the military medical departments of the United States, to insure safe use of microwave energy, a brief summary of research activities is necessary.

Prior to 1954 each service devoted resources to problems as they pertained to their immediate operational needs, and, as mentioned earlier, Daily,[7] of the US Navy, published the first clinical evaluation in 1943. With the formation of the Tri-Service Committee, and, later, the Joint Services Ad Hoc Committee, research efforts were coordinated in such a way as to permit those services most experienced in a given area of the overall problem, to concentrate on these aspects. Since the early 1950's, all services devoted substantial funding to both in-house and contractual research with selected civilian institutions. Our host for this meeting, the University of Rochester, provided the US Air Force and Department of Defense with valuable research data during the late 1950's and early 1960's.[11] While the principal goal of these studies was to insure personnel safety in the use of microwave systems, significant effort was encouraged in the basic science aspects of the question of biological responses to microwave energies.

In the mid 60's our fourth uniformed service, USPHS (commissioned corps) initiated several research programs, both in-house and under contractural arrangements.

At present (Jun 74) the US Navy is funding studies on dosimetry methods and devices, as well as the operation of special laboratories to study biological effects of selected energies in the nonionizing portion of the electromagnetic spectrum. US Army medical officials are conducting research on ocular effects of microwave exposures, and, as will be discussed below, have developed an excellent technique for ophthalmological screening of workers whose occupations require them to work with microwaves. As an excellent example of how the services have cooperated on these matters, the US Navy funded the purchase of special cameras for use by all services in conducting evaluation of lens changes. The US Air Force is devoting significant resources at its School of Aerospace Medicine, Brooks AFB TX, on problems associated with the microwave environment and its relation to certain bioelectronic prosthetic devices, as well as basic studies on effects of selected portions of the nonionizing radiation spectrum.

While much of the research effort has been, and is being, devoted to refining standards for occupational exposures, it was the military services that recognized the need for an exposure standard, nearly 20 years ago, to insure viability and safety of microwave systems.

Based on information, opinions and judgments of members of the Tri-Service Committee and their consultants, the Rome Air Development Center (RADC) of the US Air Force, published a regulation[12] in 1957, setting the hazardous microwave radiation level at 10 milliWatts per cm^2 over the entire microwave spectrum. Two weeks later a second[13] technical order was published, establishing this standard, Air Force-wide.

The limit of 10 mW/cm^2 served as a practical exposure guide for several years. However, it became apparent that duration of exposure should be considered, and that higher levels could be tolerated for shorter periods without injury. Applying the premise basic to all toxicology problems (i.e., the product of time of exposure to a toxic agent multiplied by concentration of that agent during exposure, represents the hazard), new guidelines were developed and adopted by the US Army and Air Force in 1965.[14] Exposures to personnel within limited occupancy areas were permitted only for the length of time given by the following equation:

$$T_p = \frac{6000}{W^2}$$

where T_p = permissible time of exposure in minutes during any 1-hour period

W = power density in area to be occupied in mW/cm^2.

The equation is useful only for power densities up to 100 mW/cm^2 and because exposures of less than two minutes are operationally impractical, its use for power densities above 55 mW/cm^2 is not recommended.

Operational Safety Guides

Since the first instructions issued for safe use of microwaves, the military services have continually monitored the applicability of standards and safety guides. The Army and Air Force policies are contained in a joint publication, TB MED 270 and AF Manual 161-7, "Control of Hazards to Health from Microwave Radiation," Dec 1965. The US Navy guidelines are contained in Bureau of Medicine Instruction 6470.13, "Control of Microwave Health Hazard," 10 Nov 72. Both documents set the exposure limit at 10 mW/cm^2 and furnish instructions on preplacement and periodic physical surveillance of personnel whose duties require working with radiations of frequencies from 100 mega to 100 gigahertz. Each presents general outlines of hazards involved, and precautions to be followed by operating personnel. In the US Air Force, additional guidance on these matters is contained in Technical Manual T.O. 31Z-10-4, "Electromagnetic Radiation Hazards," dated 1 Aug 66. Similar guidance for US Navy Personnel is found in NAVSHIPS 0900005-8000, Technical Manual for Radio-frequency Radiation Hazards, and NAVORD OP3565/NAVAIR 161-1-529, Technical Manual, Radio-frequency Hazards to Ordnance, Personnel and Fuel. In addition to the guides and instructions relating to use of microwaves, the services have organized special units[15] to serve as consultants to the field on matters pertaining to personnel hazards and protective measures.

Evaluation of Standards and Control Measures

While publication of standards for microwave exposures and operational safety guides are a necessary part of the overall safety program, these factors are perhaps no more than 20% of the total effort necessary to realize the goal of an acceptable program. Studies must be conducted to insure compliance with directives, and to evaluate the effectiveness of the standards. Each service has supported, on a continuing basis, special surveys and clinical studies to evaluate effectiveness of safety guides. In 1954-56 the School of Aerospace Medicine cooperated with officials at Griffiss AFB NY in studying individuals with work histories in electronics and radar environments. Both the US Air Force and the US Navy supported work by civilian experts[16] in clinical examination of personnel and research designed to evaluate operational standards. Additionally, several independent observers[17,18] conducted valuable clinical studies on civilian workers and veterans of the Department of Defense.

One of the most significant aids in evaluating possible eye changes associated with microwave exposures was that developed by Col B. Appleton, USA, MC, Senior Consultant in Ophthalmology to the Surgeon General, Department of the Army. Dr. Appleton designed an examination protocol[19] which permits evaluation of all physical changes which are known to occur in the human eye lens, and at a stage long before they become clinically significant. Using a biomicroscope and a mydriatic agent, the protocol requires recording the presence of lens vacuoles or opacities or posterior subcapsular iridescence, or all three. Each eye is studied separately. The examinations are conducted on both control and study groups, without the examining physician knowing to which group any particular subject belongs. Using this examination procedure, the Army and Air Force have conducted several thousand examinations on controls and study groups.

Cataractogenesis

Many tissues and organ systems of the mammalian body respond to exposure to microwave energies, and the response is roughly proportional to the exposure dose. However, biological response must be clearly distinguished from biological injury. Thus, the cardiovascular system will respond to heat loads imposed by microwave energy by increased heart rate, but this is certainly not injury any more than exercise which increases heart rate can be considered injurious.

Given sufficiently high power for a long enough period, microwave exposures can kill an animal. Reduction in the dose brings responses of lesser magnitude. Thus, the germinal cells of the testicle are damaged by low doses of microwave in much the same manner as when their temperature is increased from any cause, and recovery is complete unless the entire organ is heated to the "fry" stage—an unlikely occurrence in a conscious subject.

While controversy exists on the subject of thermal effects from microwave exposures versus nonthermal effects, the current consensus is that thermal effects predominate, and until more information or better data becomes available, practical discussions centered around thermal injury and effects are the most productive from an operation safety viewpoint. Since the lens of the eye does not have an adequate vascular system for the exchange of heat, even slight temperature elevations, from whatever cause including microwaves, can cause protein coagulation. If unrepaired because a certain threshold for damage has been exceeded, an opacity may form, microscopic in size initially, and later progressing to a grossly visible cataract.

Technically, any opacity in the lens can be termed a cataract, however in clinical practice it is usually when the opacity interferes with visual acuity that the term cataract is considered appropriate. Thus, as used herein, a cataract is defined as an opacity of the lens or its capsule which obstructs the passage of light to a sufficient degree to interfere with normal vision.

Because of the seriousness of any eye injury, and the fact that experimentally, cataracts in animals can be produced by microwave exposures, our programs for assessing safe use of this energy form have placed emphasis on the eyes. In addition, the lens is readily available for detailed examination with a minimum of patient discomfort. But, a sharp distinction must be made between experimentally-produced cataracts and those which have been alleged to result from occupational exposures to microwaves. In the experimental environment, the animals are anesthetized or so restrained that the eye is practically coupled to the wave guide. In fact, one large government group[20] is experimenting with a lens system to focus microwave energy on the eye for cataract-induction research. The justification for this procedure is "Without such devices, whole-body exposures would result in substantial animal mortality before cataracts could be induced." In all experiments wherein the animal is free to move within limits of the cage or box, cataracts have not occurred, regardless of the exposure dose or duration.[21] The occupational environment much more closely resembles the "free-moving" type experiment than that in which the animal is restrained. Therefore, the relevance of cataract-induction experiments to the occupational or safety aspects of microwave eye injury is minimal.

Some observers[22] feel that repeated exposures of the lens to low doses of microwaves (i.e., doses which, taken individually, cause no detectable changes) result in an accumulation of injury, and after a certain total "dose" is absorbed, opacities and even cataracts will appear. Other workers[23] explain that, based on the data presented to support the cumulative injury concept, the interval between sub-threshold doses was too short to permit full repair; therefore, repeated injuries add up to detectable changes, whereas, only one such insult would completely repair if given sufficient time. It appears that the "residual injury" concept described[24] as a characteristic of ionizing radiation injury has been erroneously translocated to the radio-frequency portion of the electromagnetic spectrum.

Experimental evidence indicates that all cataractogenic levels for one-hour exposure are above 100 mW/cm^2 in frequency range of 19 MHz to 24.50 GHz, and in the rabbit, single and multiple exposures below 120 mW/cm^2 do not result in cataract production.[25]

With the above discussion as background, we will now review the major studies conducted or sponsored by the military departments of the United States relating to eye injuries or changes in personnel occupationally exposed to microwaves.

Following the examination protocol developed by Appleton,[19] both the Army and Air Force have conducted thousands of examinations during the past four years, and found no difference between the lens of the eyes of control and study groups.[19,26] In 1954 a large study[10] conducted by the School of Aerospace Medicine, found no significant ocular changes when the study group was compared with 100 controls. Zaret,[16] under contract to the Air Force, has published several reports on retrospective studies on occupationally exposed individuals, and found slight, but statistically significant, increase in lenticular effects. A report by Cleary, et al[17] is particularly significant in that records of WWII and Korean War veterans were evaluated with the cooperation of the Veteran's Administration. Medical histories of 1945 veterans being treated for cataract were compared with those of 2164 without cataracts. No increased risk of cataract was found in the group whose occupational histories indicated work with microwave radiation.

A review of data on cataract incidence in the US Air Force was conducted in 1970 by the staff of The Surgeon General's Office,[27] Hq USAF, in conjunction with its continuing program of evaluating radiation exposure limits and adequacy of control measures. The incidence of reported cataracts from all causes in the US Air Force personnel during the period 1959-1968, inclusive, is shown in Fig 1. It will be noted that the incidence has remained relatively stable over the 10-year period, and the minor variations are well within the statistical limits of random variation. While these data are preliminary and gross in nature, they are significant in that no peaks are found in the early and middle 60's. At the onset of the war there was a great increase in the size of the Army, and many individuals remained on for the 20-year retirement benefit. Of these, a number worked in radar and communication specialties for a great majority of their careers, and these careers spanned the period when virtually no standards for occupational exposures existed. In addition, many under pressure of combat necessity ignored what few safety precautions may have been in force. Thus, if occupational exposures of a nearly unrestricted nature do contribute to incidence of cataracts, the data in Fig 1 should show a sharp increase in the 1960-68 period when the retirements were being made. The current safety standard of 10 mW/cm^2 and greatly increased awareness of all personnel on safety precautions gives us an occupational environment which is safer, by several orders of magnitude, than that during the period of 1940-60.

Based on available evidence, it appears to this observer that the present standard of 10 mW/cm^2 is entirely innocuous insofar as injury to the lens of the eye is concerned.

Current Studies and Future Programs

The US Navy, besides its very significant in-house research

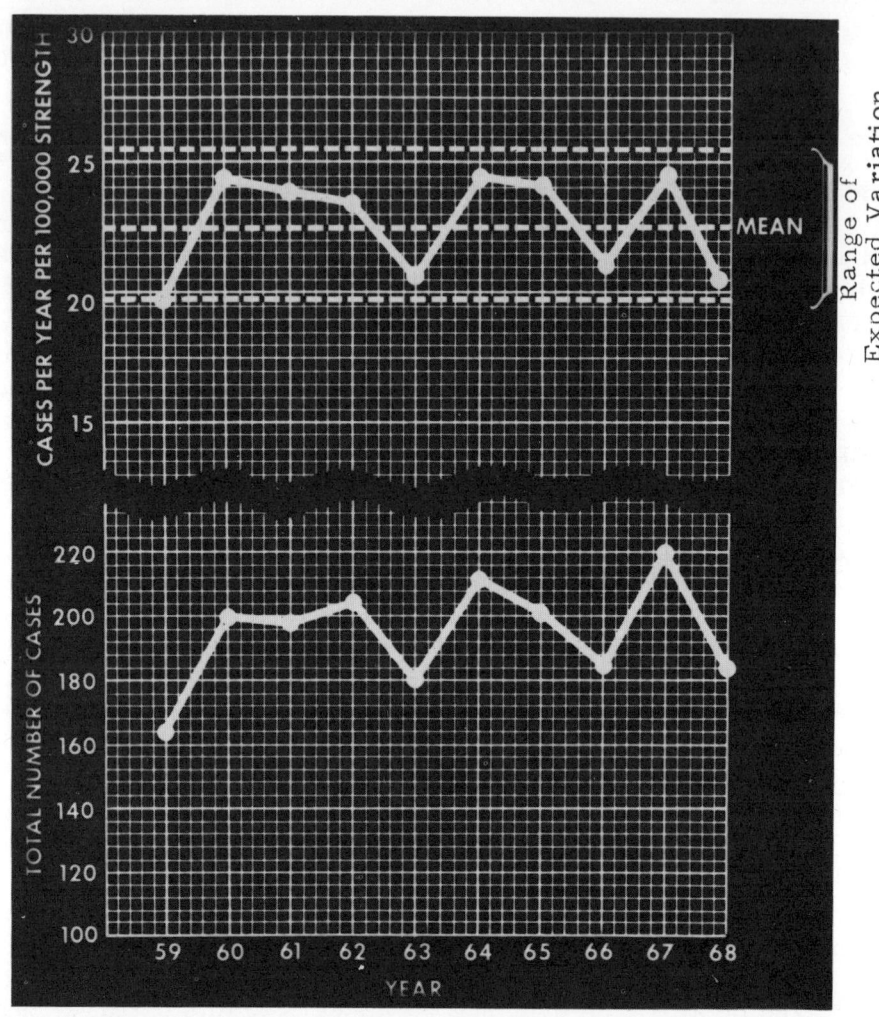

FIGURE 1
CATARACT INCIDENCE RATE, AND TOTAL, FOR
USAF PERSONNEL, 1959-68

program, is cooperating with the US Public Health Service on a long-term epidemiological study on personnel entering the radar and microwave career fields. The Army continues its ophthalmologic studies on groups occupationally exposed to microwaves, along with its research program. The Air Force is rewriting its Manual, AFM 161-7, on hazards of microwaves, instituting an intensive training program for base level personnel having responsibilities for radiation control programs, and purchasing several of the very latest survey meters for accurately evaluating the work environment. All services have consultant groups readily available for investigating overexposure incidents/accidents and documenting the physical and clinical aspects for permanent retention and long-term followup, as may be indicated. Emphasis is being placed, not only on operational microwave equipment, but on the industrial and domestic uses, such as traffic control and home cooking.

The one important area largely ignored is that of personnel dosimetry for nonionizing radiation. Using current technology,[28] a small device has been built which summates exposure doses to microwaves over time. The long-term effects of microwave energy may not involve the eye, nor become apparent unless data is collected on the magnitude of exposure. By recording exposure doses over a career, as we do now with ionizing radiation, in time a meaningful relationship between cause and effect would emerge, if such exists. Our current efforts and safety programs fall short since we rarely have accurate exposure data on the individual(s) involved. It seems rather basic, but many have yet to realize that in any toxicological problem there are two factors—the size, magnitude or dose of the agent, and the effect. In microwave work we have no reliable information on dose; therefore, in the practical sense we can make few, if any, creditable predictions on effect. This fundamental concept was reiterated by the emminent Czechoslovakian scientist, Pazderova,[29] who stated, "In order to judge the significance of the occupational hazard of electromagnetic radiation more accurately, it will be necessary to correlate medical findings obtained from long-term observations of workers exposed to electromagnetic radiation with the extent of the exposure." One international working group[30] has formally recommended increased emphasis on personnel dosimetry for microwaves.

Summary

The military services of the United States have been, and continue to be, the major users of microwave energy in this country. By the very nature of the necessity that these systems be operational, safe use is mandatory if the missions of the services are to be completed. To this end the military medical departments have assumed the leadership role in studying the biological effects of microwaves, and in setting and evaluating the adequacy of occupational exposure standards. To date, no member of the military service has been shown unequivocally to have been injured because of occupational exposures. Cataracts are the most dramatic effect of microwave injury, but current standards and safety precautions, in the military services, are entirely adequate for protection against this injury. All services are supporting continuing programs of research and clinical studies to insure that the past record of safety continues, and that many of the unknown aspects of the program are properly and thoroughly investigated. A plea is made for the development of a personnel dosimetry and permanent program for recording exposure doses on a permanent basis.

References

1   GUERLAC, H. E.: "OSRD Long History," Vol V., Division 14, "Radar." OTS, US Dept of Commerce.

2   British Patent 13,170, issued to Hülsmeyer, Sep 22, 1904, entitled "Hertzian-wave Projecting and Deceiving Apparatus Adapted to Indicate or Give Warning of the Presence of a Metallic Body, Such as a Ship or a Train, in the Line of Projection of Such Waves."

3   MARCONI, S. G.: Radio Telegraphy. Proc. IRE, Vol 10-1922.

4   SKOLNIK, M.: Introduction to Radar Systems. McGraw-Hill. New York (1962).

5   US Patent 1,981,884: "Object Locating System," issued to W. A. Blair. Aug 20, 1937.

6   THOMPSON, W. D., and BOURGEOIS, A. E.: Effects of Microwave Exposure or Behavior and Related Phenomena. ARL-TR-65-20. USAF Systems Command, Holloman AFB NMex.

7   DAILY, L.: A Clinical Study of Results of Exposure of Laboratory Personnel to Radar and High-Frequency Radio. US Naval Med Bull 41:1052 (1943).

8   LIDMAN, B.L. and COHN, C.: Effect of Radar Emanations on the Hematopoietic System. Air Surg Bulletin 11:448 (1945).

9   Microwave Radiation Hazards. Urgent Action Tech Order 31-1-18. Rome AF Depot, Griffiss AFB NY (1 Nov 54).

10 PATTISHALL, E.G.: Ed, Proceedings of Tri-Service Conf on Biol Hazards of Microwave Radiation, 15-16 Jul 57. Air Res & Development Command, Hq Griffiss AFB NY. ASTIA Document No. AD 115-603.

11. MICHAELSON, S.M., et al: Biological Effects of Microwave Exposure. Griffiss AFB NY. Rome Air Develop. Ctr, ASTIA Document No. AD 824-242 (1967).

12 Industrial Hazards, RADC Reg No. 160-1, 31 May 57. Hq Rome Air Development Ctr, Griffiss AFB NY.

13 Microwave Radiation Hazards. Urgent Action Tech Order 31-1-511, Rome AF Depot, Griffiss AFB NY, 17 Jun 57.

14 TB MED 270/AFM 161-7: Control of Hazards to Health from Microwave Radiation, Depts of Army and Air Force (Dec 1965).

15 US Army Environmental Hygiene Agency, Edgewood Arsenal MD 20104. Chief, Bu of Med & Surgery (Code 74), Wash DC, EMC/Measurements Div, 1839 Electronics Instal Gp, Keesler AFB MS 39534.

16 ZARET, M., et al: A Study of Lenticular Imperfections in the Eyes of a Sample of Microwave Workers and a Control Population. RADC TDR 63-125. New York Univ NY (1963) (ASTIA 413-294).

17 CLEARY, S.F., and PASTERNACK, B.S.: Lenticular Changes in Microwave Workers—A Statistical Study. Arch Environ Health 12 (1966).

18 BARRON, C.I., and BARAFF, A.A.: Medical Considerations of Exposure to Microwaves (Radar). JAMA 168 (1958).

19 APPLETON, B., and McCROSSAN, G.C.: Microwave Lens Effects in Humans. Ophthalmology Svc, Walter Reed General Hospital, Wash DC.

20 DHEW Pub (FDA) 74-8022: Progress in Radiation—1973 Protection. USDHEW PHS/FDA, BRH, Rockville MD 20852.

21 MICHAELSON, S.M.: Personal communication.

22 CARPENTER, R.L., and VON UMMERSEN, C.A.: The Action of Microwave Radiation on the Eye. J. Microwave Power, Vol 3 (1968).

23 MICHAELSON, S.M.: Human Exposure to Nonionizing Radiant Energy—Potential Hazards and Safety Standards. Proc of IEEE, Vol 60 (Apr 72).

24 BLAIR, A.A.: The Constancy of Repair Rate and of Irreparability During Protracted Exposure to Ionizing Radiation. Ann NY Acad Sci, Vol 114 (1964).

25 MICHAELSON, S.M.: Relevancy of Experimental Studies of Microwave-Induced Cataracts in Man. Univ of Rochester Rpt 3490-103, Rochester NY.

26 ODLAND, L.T.: Radio-frequency Energy—A Hazard to Workers? Inter Jr Indus Med & Surg, Vol 42 (1973).

27 Biometrics Division, Dir of Plans and Hosp, Office of The Surgeon General, Hq USAF, Wash DC (Feb 70)

28 BRODIN, M.E.: Northwestern Univ, Dept of EE, personal communication (May 72).

29 PAZDEROVA, J.: Effects of Electromagnetic Radiation of the Order of Centimeter and Meter Waves on Human Health. Pracovni Lekarstui 20:10, 1968. Translated by Alice Marosi, M.S., Lovelace Foundation, by Med Ed & Res, Albuquerque NMex.

30 Report of Working Group EURO 4701. Health Effects of
    Ionizing Radiation. The Hague 15-17, Nov 71. Reg Ofc for
    Europe WHO, Copenhagen (1972).

-DISCUSSION-

LIN - You mentioned something about the formation of cataract
due to, among other things, the precipitation of lens proteins.
I am looking for information. Do you have any reference for
that?

ZIGMAN - I don't think that it has yet been proven that the
opacities always represent the precipitation of proteins.
Aggregated proteins have only been found after lens homogenization,
and the artifactual nature of the aggregated proteins needs to be
resolved into what is in the lens and what is formed during
chemical work up.

LIN - What is it?

ZIGMAN - There are many other things that it could be, especially
lakes of water or spaces between lens fibers leading to the
scattering of light. It seems from the position of the opacity
that there is a physical change rather than a metabolic one.
Damage at the anterior aspect of the lens would be related to
abnormal metabolism of epithelial cells or young fiber cells.
These cells are active metabolically. But you don't see the
problems in the anterior portion of the lens. The heat seems
to be built up at the back of the lens, and therefore, the damage
is expected to occur here. What the damage represents still needs
to be determined biochemically. I don't believe that this
damage can be taken to be protein precipitation. Proteins
denature at fairly high temperatures, which I don't think have
been found. You need at least 60 to 80°C. Thus, the damage may
represent open structural changes leading to light scattering
at the posterior pole of the lens. There are few metabolically
active cells in this region of the lens.

ODLAND - Wouldn't protein be involved in whatever process
happened?

ZIGMAN - Not necessarily. It could be a secondary thing.

ODLAND - Calcium deposits?

ZIGMAN - Calcium is known as one factor.

LIN - I heard a lot about protein denaturization as far as lens

LIN - opacity is concerned. However, we did look quite extensively and we weren't able to locate any data that would permit one to come out and say this. In fact, biochemists are usually very, very reluctant in pointing out the exact temperature at which protein denatures. This depends a great deal on the protein involved. Apparently, at this time, not enough is known about the temperature dependence of protein denaturation.

ZIGMAN - Does anybody know what temperature is generated at the back of the lens when microwaves are radiated?

TYLER - I think Dr. Lin has some data, which is probably relevant only for rabbits subjected to a specific insult under controlled conditions.

LIN - In our experiments we used 8 month old albino rabbits. They were irradiated with a diathermy C-director at a distance of 5 centimeters from the radome to the corneal surface. The temperature immediately behind the lens is around 42° for the cataractogenic threshold. The temperature in the lens near the posterior pole is estimated to be 43-45°C.

ZIGMAN - That in itself, of course, is not capable of denaturing proteins -- but it could have affected the rate of the reaction taking place.

CZERSKI - From the point-of-view of epidemiologic and statistical methodology I believe that the validity of data presented by Dr. Odland should be discussed in a more detailed manner.

I am just wondering, concerning Dr. Appleton's study, and I think he has now studied additionally 1,000 or 1,500 men more, would it be possible to re-analyse the results according to age, duration of occupational exposure and exposure levels. This was a cross-section study (Arch. of Opth., 1973). To have any real evidence at all from this type study, the slides you were kind enough to show, are insufficient. No data are available on the age distribution in the control and exposed groups. What you need is a comparison of matched age groups examined in a "longitudinal type" of study - the same individual examined successively several times in the course of a few years.

This was done in our earlier studies, and it was shown that you get a difference in incidence of opacity in older age groups which were, of course, exposed for longer periods. A difference between exposed and unexposed people of the same age group has a meaning. An attempt at correlating the incidence with the age, and a second time at correlating the incidence with the duration of occupational exposure and exposure levels should be made. This

CZERSKI - is necessary to obtain any data which can be discussed. Perhaps you have such data, but they were not on the slides you have shown.

ODLAND - We have lots of data but ---

CZERSKI - But I think simply we can't have any discussion on incidence or non-incidence in exposed or non-exposed groups just on the basis of these two slides. First of all, a longitudinal study, as I said, of the same individual examined several times successively, is needed. Moreover, in our experience, and this is a fact which may have some importance, when we had people with progressive changes in the lens we had them transferred to another type of work, and the opacities regressed. This shows the importance of follow-up examinations and what I call longitudinal studies. This is one of the reasons why we suspect that there may be a related causal relationship, between microwave exposure and lens opacities. After a change of work, and a no-exposure period, regression of changes may be seen. I am afraid I am speaking about a piece of work which is not as yet finished. We have a five-year follow-up study of persons who were transferred to another type of work because of progressive opacities on eye examination. I will be very happy to present the results sometime next year because then we will have the complete results. But I am afraid that on the basis of results presented by Dr. Odland, you can't have any conclusions without an additional analysis, correlation with age and duration of occupational exposure. We have done detailed analyses of microwave workers' eyes and we feel reasonably comfortable about our safe exposure limits. I would suggest that you do the same type of analysis on your subjects.

TYLER - I would like to end this session with one plea to the scientists. We need to insure that our research is relevant and can be related to our environment, be it occupational exposures or population exposures.

I am not asking for a decrease in laboratory research, but for increased efforts in providing laboratory simulation which can be more closely equated to exposure of people in the environment. It is very nice to present a paper and talk about joules per gram in some tissue, but it is very difficult to relate this directly to people working in an occupational environment.

We should all keep in mind that the sole reason for our research is to better understand what happens to man from his environmental exposure to electromagnetic radiation and to be better able to protect him.

Future Applications and Controls

PROSPECTS FOR EXPANSION OF INDUSTRIAL AND CONSUMER USES OF MICROWAVES

John M. Osepchuk

Raytheon Research Division

Waltham, Massachusetts 02154

ABSTRACT

The present trends in industrial and consumer uses of microwaves are reviewed with the special purpose of assessing potential hazards of microwave radiation. The particular interest in the microwave portion of the electromagnetic spectrum is related to its property of optimum coupling of energy to macroscopic objects of human interest. A survey of sources shows that electron tubes, though mature in technology, will continue to be the main sources of microwave power. There will be a large growth in low power solid-state sources, however, for a growing number of low-power applications. The biggest growth market in non-communications type of microwave applications is that of consumer microwave ovens. A long list of industrial applications is reviewed and a few are selected for favorable expectations in market growth. The potential for microwave radiation hazards in this entire field is considered small and under strict control by government regulations. As an aid in perspective the amount of typically radiated energy is compared to the radiated energy of deliberate microwave radiators in society. The problem of incidental interference, as exemplified by that of cardiac pacemakers is distinguished from direct biological hazards and is shown to require standards on device susceptibility. With appropriate relation of personnel exposure to emission of microwave sources a high degree of safety is expected for microwave products and systems directed toward industrial and consumer use.

MICROWAVES AND THEIR UNIQUE PROPERTIES

The purpose of this paper is to survey the qualitative and quantitative features of microwave products directed towards industrial and consumer use and assess their potential as microwave radiation hazards.

The definition of the microwave portion of the electromagnetic spectrum is somewhat arbitrary but can be based on the property of maximum coupling of electromagnetic energy to macroscopic objects of common use and interest to man. On this basis one can defend the choice of 100 MHz (3 meter wavelength) to 300 GHz (1 millimeter wavelength) as the microwave range. It is expected that biological effects due to non-ionizing radiation depend on the "penetrating power" of the radiation and the latter is expected to peak in the microwave region.

One can determine the wavelength dependence of penetration capability from the complex propagation constant. The decay constant is given as

$$\alpha = \frac{2\pi}{\lambda} \left[\frac{\epsilon'}{2} \left(\sqrt{1 + \tan^2 \delta} - 1 \right) \right]^{1/2} , \qquad (1)$$

where λ is the free space wavelength, and $\tan \delta = \frac{\epsilon''}{\epsilon'}$ is the ratio of imaginary and real parts of the dielectric constant - i.e., the loss tangent. The inverse of α can be interpreted as penetration depth. Because $\tan \delta$ is of the order of unity for biological tissue, when ϵ' is large the penetration depth can be considerably smaller than a wavelength. The penetration depth becomes insignificant at or above 10 GHz when the wavelength is about 3 cm because of the high absorption rate. At very low frequencies $1/\alpha$ is not small and the penetration depth is greater than the dimensions of the human body. However, another effect, dielectric shunting, reduces the internal field. As shown by Schwan, (25) at very low frequencies, the value of ϵ is very high, in the millions; in the principal microwave range, $\epsilon \sim 60$ to 80, and then drops off at the upper end of the microwave range. If the low-frequency dielectric constant is as high as 10^6, then the electric field in the body is very small compared to the "incident" electric fields. The internal power density is also small. Thus body shielding exists at low frequency. This shielding has been analyzed by Bridges (2) with respect to interference with implanted pacemakers.

The net result is that the penetration capability of non-ionizing radiation is peaked in the microwave range as shown in Fig. 1. The "window-like" nature of this penetration dependence on frequency is sketched in that figure on a hypothetical scale. Clearly

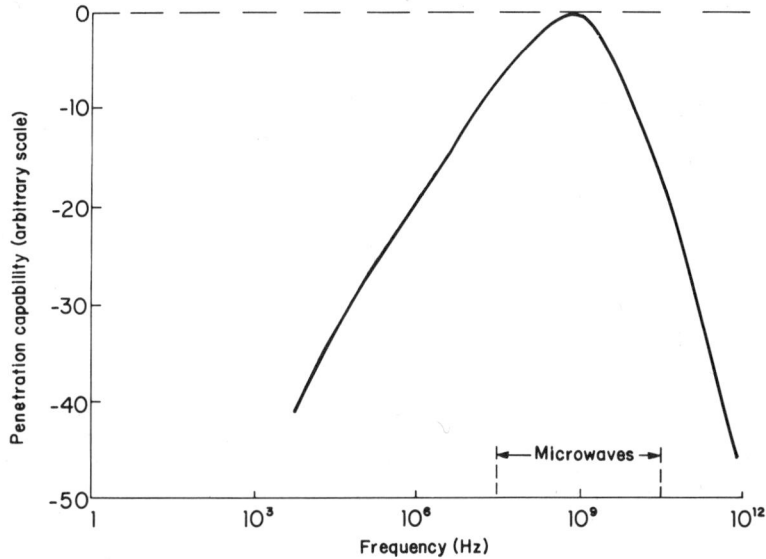

Figure 1. Sketch of Penetration Capability as a Function of Frequency

this marked dependence on frequency should influence one's approach and priorities in analyzing research on non-ionizing radiation bioeffects as well as searching for useful applications of non-ionizing radiation. Clearly whole-body heating of objects smaller than man by radiation is most effectively accomplished in the microwave range. This points to not only medical diathermy and microwave cooking but also a host of potential industrial applications to different materials and products.

SURVEY OF MICROWAVE POWER SOURCES

The scope and potential of microwave applications depends on the technological and economic features of available microwave power generators. These include electron tubes and the newer solid-state sources.

In Fig. 2 there are shown curves which bound the achieved power capability of various types of microwave sources as a function of frequency. Both gridded tubes and microwave tubes (magnetrons, klystrons, traveling-wave tubes, crossed-field amplifiers etc.) exhibit a fundamental technical limit (18) on power which has a fifth power inverse dependence on frequency. (This results mostly from basic limits on electron space-charge density and to a lesser extent on increasing circuit loss at higher

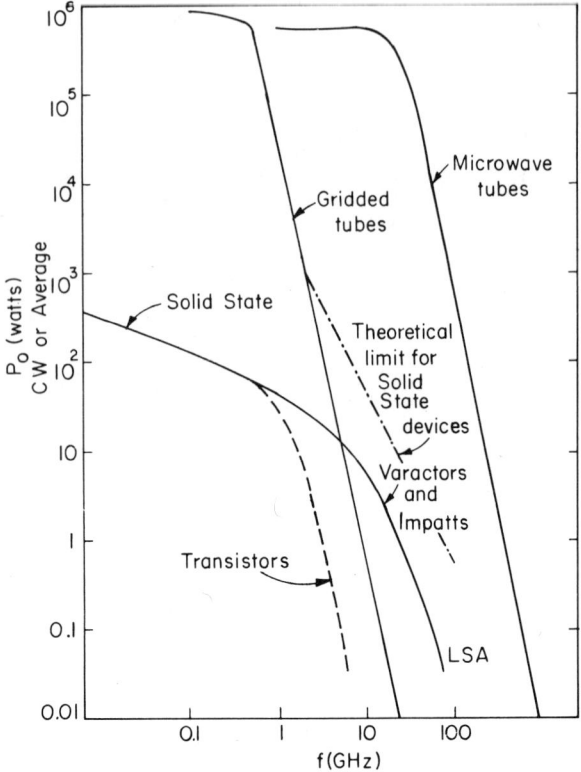

Figure 2. Maximum Limits of Average Output Power from Various Microwave Power Sources (1970)

frequency.) At frequencies below the $P \propto \frac{1}{f^5}$ curves the limits are economic in nature.

The microwave tube technology is considered relatively mature (having developed in the last thirty-five years) and little extension in power capability is foreseen except for the speculative possibility of a breakthrough in "fast-wave" tubes (18) which could extend the frequency boundary upwards by a factor of ten at a given power level.

The microwave solid-state device technology has evolved only in the last fifteen years but is now also considered to be relatively mature, at least as far as power capability is concerned. Thus the curves in Fig. 2 apply today for power available from a single device.

It is natural to expect a cost premium for devices at the technological frontier so that applications in which cost is a large

EXPANSION OF INDUSTRIAL AND CONSUMER USES

factor can be expected to utilize devices well below maximum capabilities. Thus in industrial and consumer applications electron tubes of power capability up to 50 kilowatts are used whereas applications in which solid-state sources are envisaged are generally limited to several watts.

The efficiencies of most devices are high, > 50 percent, below 1 GHz but efficiency decreases to about 10 percent for solid-state devices at 10 GHz. Magnetrons and crossed-field amplifiers are generally the most efficient with efficiencies of over 80 percent demonstrated at 1 GHz to over 30 percent at 10 GHz. The efficiency of high-power klystrons has been developed to the level of crossed-field devices but this is not maintained in low-power (e.g., < 1 kW) klystrons.

The cost of microwave power sources is of prime importance in industrial and consumer applications. The history of power tube costs in general is not very indicative of what is possible, partly because most of the microwave tube market heretofore has been military and has involved stringent performance objectives in priority higher than cost and has been limited to relatively small production volumes of less than several thousand units per year typically for a given tube. One, nevertheless, can expect on sound physical grounds that the cost in $/watt to decrease with power level and with production volume. The only market where production volumes have approached the order of 100,000 tubes/year for an individual manufacturer is that of magnetrons for consumer microwave ovens. Here the selling price to oven manufacturers is below $50 per unit.

For industrial applications at 915 MHz tube prices range from $200 per tube at the 1 kW level, $1000 at the 5 kW level to $3000 at the 30 kW level. This corresponds respectively to $.20/watt, $.20/watt and $.10/watt and for small quantities. At 2450 MHz, magnetron prices range from less than $50 at the 1 kW level to $1000 at the 5 kW level or less than $.05/watt to $.20/watt.

Solid-state devices of interest include primarily transistors for < 1 GHz, and avalanche (IMPATT and TRAPATT) diodes above 1 GHz. These devices can yield a few tens of watts at 1 GHz to a few watts at 10 GHz but on a $/watt basis are still prohibitively expensive if operation over 100 watts is considered. Small quantity prices are over $100/unit. A long-range projection (by Stanford Research Institute in 1970) suggests that in volume production the cost of solid-state microwave power will be down to $1/watt at 2.45 GHz by 1980. This clearly suggests that for the foreseeable future applications of solid-state sources in industrial and consumer applications will be limited to those requiring only a few watts of power.

(In principle, it is possible to combine many individual solid-state sources to produce substantial total power. This approach is being intensively pursued in phased-array studies for military radar. A considerable amount of technical problems and cost factors to date prevent this approach from being competitive with microwave tubes.)

The annual dollar volume of the microwave tube market has been rather steady at about $150 million for the last fifteen years. Though the military market is slightly declining the consumer microwave oven market has made up the difference. In 1964 Herold (10) predicted a slow growth in the microwave tube market based on growth of industrial "power conversion" applications. Although this has not materialized as anticipated the trend is in the anticipated direction.

The nature of the microwave power source is important in determining potential hazards. Microwave power tubes all operate at potentially lethal high voltage (> 1 kV) and are so designated. Although many microwave power tubes present potential x-ray hazards these are serious only at the high peak powers used in military and radar applications where megawatts of peak power and peak voltages of 50 - 100 kV are common. Consumer microwave ovens employ magnetrons operating at less than 5 kV and present no x-ray hazard. In only a few industrial applications of over 50 kW are tube potentials over 20 kV. In all cases microwave tube manufacturers control such radiation and so inform the user of any required protection.

Solid-state sources generally operate at well below 100 volts d.c. bias. Only in speculative high-power arrays of solid-state devices are potentially unique hazards found - in this case that of high magnetic fields from supply current flow of thousands of amperes required in the generation of high power.

POWER CONVERSION APPLICATIONS - ISM

The non-communication applications which Herold(10) described are that of interest in this paper - industrial as well as consumer applications.

In 1964 Herold, in an excellent and perceptive article (10), predicted that the microwave power field would comprise 40 percent of the microwave tube market by 1984 and would offer the greatest long-range promise for microwave tubes. This optimism was given unexpected support from the great Russian physicist P. L. Kapitsa in technical and popular publications with predictions (14) of growth in microwave power applications or "high-power electronics".

EXPANSION OF INDUSTRIAL AND CONSUMER USES

At least four major companies were involved in microwave power equipment development in the sixties. From a purely technical viewpoint there appeared sound reasons for expecting growth in industrial applications, even more so than that in consumer applications. To quote from Herold (loc cit): "There is no a priori reason why the low operating cost of thermal ovens for food preparation cannot be equalled by some form of microwave device. However, as already learned by those in the field, one must sell the service, not the tube; we are still a long way from either the innovation in the electron tube or the innovation in marketing that will make this form of cooking universal. Perhaps the same thing is true of microwave heating in industrial processes. However, in the latter instance, there are applications in which no other process will do, and even present-day equipment will serve."

What has happened since is a surprise - at least to the microwave experts. The consumer microwave oven has arrived with a U. S. market(15) estimated as around 500,000 for 1974 but the industrial applications have shown only slow growth and some notable failures. In Fig. 3 the growth of the U. S. microwave oven market is depicted.

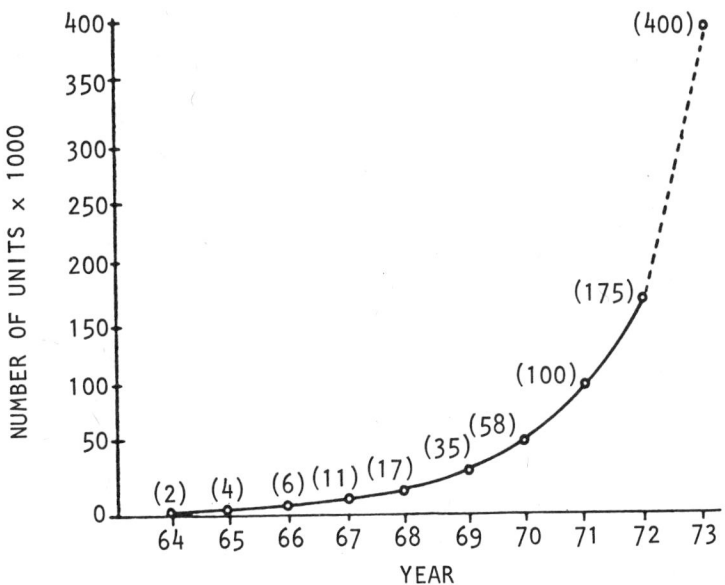

Figure 3. Domestic Microwave Oven Sales (U.S.)
Courtesy Journal of Microwave Power

In the late sixties the microwave oven market bloomed in the U. S. and Japan and to some extent in Sweden. There may be unique explanations for the case in Japan related to nature of diet but in the U. S. it appears to be the injection of expertise of appliance manufacturers, notably the Amana Refrigeration Corp. after its acquisition by Raytheon. The considerations that brought this market to a reality after twenty years involve much more than microwave engineering clearly.

There have been successful microwave industrial applications in Europe and Japan as well as in the U. S. But there have been tragic failures such as the "Great Potato Chip Incident of the 1960's" (see articles in the Journal of Microwave Power, Vol. 8, July 1973). The microwave finish drying of potato chips was recognized (8) as economically and practically feasible in 1966 and was heralded as a success when about twenty microwave equipments, typically at 50 kW, were installed in the potato chip industry. Today "most of this equipment has been sold, scrapped, put into storage or lies idle". (17)

The reasons for this and other failures are mostly non-technical and elusive (17) and in part related to economics. There is more to these problems than this, however, and in part can be described as resulting from the difficulties of communication between two different worlds or two different types of people. George Freedman (6) in some penetrating reviews of the industrial microwave field describes the problem of marriage of food industries with aerospace industries by the following question, "Can a homespun (but rich) girl from a farm in the Mid-West make it with the mustachioed college-trained city slicker from a sea coast metropolis?" There is little doubt that Freedman senses the basic part of the problem.

In addition, there is some suggestion (9) that part of the problem may be related to inadequate scientific knowledge on microwave interaction with foods.

In the past ten years there have been some successes that remain such as with meat tempering and processing of rubber tires. Activity in this field continues within IMPI and also IEEE but with increased emphasis on microwave bioeffects and RFI suppression - reflecting increased governmental regulatory activity.

Despite the difficulties in market success, technical activity in industrial applications of microwaves is considerable and is reported primarily in the Journal of Microwave Power and in increasing extent in other publications such as those of the IEEE. (The Journal of Microwave Power is published by the International Microwave Power Institute (IMPI) which was founded in 1966 as a

Canadian organization at the University of Alberta "for the purpose of furthering the use and understanding of microwave energy for non-information, non-communication application". As such IMPI attempts to bridge the gap between users and developers.)

Over the years many applications have been studied, developed and marketed - all so far limited to small scale use. In recent years some special surveys (4, 26, 28) have been published including a special issue of the Proc. IEEE on ISM applications.

Some of these applications are listed in Table I.

TABLE I. A List of Some Industrial Applications of Microwaves

Food Industry

Potato chip drying, poultry processing, freeze drying, meat tempering, thawing of frozen food, donut proofing, reheating (reconstitution) of foods, cooking, sterilization, oyster shucking.

Forest Products Industry

Hardwood drying, paper drying, newsprint drying, destruction of fungus, woodworm in timber.

Mining Industry

Curing of concrete, breaking of concrete, heating of oil shale.

Chemical Industry

Plasma chemistry processes, curing of resins, sealing of plastics, curing of rubber products.

Agriculture

Treatment of seeds, destruction of insects, protection of plants against frost, drying of grain.

Other

Drying of match heads, film, leather. Manufacture of drugs. Melting of explosives. Repair of asphalt pavements.

The applications listed in Table I are those that typically might be recognized as "industrial". Of course, the rest of the "microwave power" field includes the significant areas of medical applications (diathermy, blood-warming, thawing of frozen tissues, diagnostic microwave techniques), scientific applications (particle accelerators and plasma heating), the special field of microwave power transmission, and lastly the microwave oven - both in commercial and industrial applications.

In addition to the general class of "microwave power" applications there are more ambiguously classified microwave products such as small boat radar, microwave burglar alarms, and non-destructive microwave testing systems - all of which are of relevance to exposure of the "general population".

Furthermore, there are a wide array of conceivable low-power microwave products utilizing solid-state sources that are envisaged (7) for the future ranging from auto-collision avoidance radar to microwave toys.

Most of the applications mentioned here have not led to substantial market growth. The one exception is the microwave oven. The reasons are in part technical but in large part the problems are those of interdisciplinary communication. Nevertheless, in several areas at least there is considerable optimism at present. One of these areas is that of meat tempering. In Fig. 4 a 50 kW conveyor system at 915 MHz for meat tempering (heating frozen meat to a temperature close to but still below the freezing point) is shown. The quality, in terms of uniform heating, and economics of the system appear attractive. Other areas presently considered promising include curing of rubber products, particularly large tires.

One of the broad technical problems encountered in many applications is that of uniformity of heating or the lack thereof. This is particularly severe in the process of thawing by microwave heating. It is well known (1) that the dielectric loss of water and most food products is very low in the frozen state but rises sharply to a high value in passing through the melting point. This means that any non-uniformity of heating is exaggerated by a runaway process at points which melt before the bulk of the material.

Another aspect of this problem is the adequacy of "mode stirring" techniques. It is clear that use of a finite bandwidth signal could be beneficial. In addition, use of more than one ISM frequency could be beneficial (27). Some flexibility is provided by the existence of several authorized ISM frequencies in the microwave range - viz, 918, 2450, 5800 and 22,500 MHz even though their derivation is somewhat arbitrary. The authorized bandwidths correspond to something less than 4%, 4%, 2.6% and 1.1% respectively at these frequencies. Not much has been done heretofore to utilize the major part of the bandwidths by

Figure 4

electronic tuning or other means to attain the randomizing effect of variable frequency.

Some work is underway to exploit advantages of multi-frequency heating. Figure 5 shows an experimental heating chamber developed for the U. S. Army Natick Research Laboratory. It permits simultaneous application of several kilowatts of energy both at 915 and 2450 MHz as well as infrared, steam and hot air.

At present the market in the U.S. for the large strictly industrial systems has been static for several years at a few million dollars per year. Optimistic projections foresee $18 M per year by 1980. Despite the uncertainties and the intangible aspects of the market problems the reasons for optimism include:

  a. the validity of the physical basis for optimum heating of macroscopic objects in the microwave range;

  b. an increase in related scientific research such as measurement of microwave properties of materials with account of temperature dependence;

  c. recent and potential changes in environmental and/or energy conservation policy which may benefit the microwave heating option;

  d. the encouraging precedent of the microwave oven industry, which was static for over twenty years before a marked growth began.

Those with ability and motivation to cooperate and communicate across disciplinary boundaries, called "linkers" by Jolly (13), will play an important role in stimulating a substantial growth.

A REVIEW OF POTENTIAL MICROWAVE HAZARDS IN INDUSTRIAL AND CONSUMER MICROWAVE PRODUCTS

To assess potential radiation hazards of microwave ISM equipment in perspective one should recognize the totality of sources producing such non-ionizing radiation. As various reports (22), (12) point out the number of deliberate radiators like broadcast stations is increasing. The trend (22) has been about 2000 new radio broadcast stations per decade and 350 new TV stations per decade. In addition, there are thousands of fixed civilian and military radar installations along with a multitude of smaller radiating sources like communication systems, small-boat radar etc.

Figure 5

A reasonable estimate of the total effective radiated power of TV broadcast stations in the U.S. is 1000 megawatts, or 20,000,000 kilowatt-hours per day. In contrast, leakage power of modern microwave ovens is well below a watt so that total radiated power and energy of microwave ovens is negligible by comparison.

A reasonable correlation exists between the level of total radiated power and the minimum accessible distance to the radiator by the general public. Thus the general public can not approach closely the high power radar antennas or the high power TV broadcast antennas which radiate kilowatts to megawatts. In the other extreme consumers have ready access to microwave ovens which typically radiate less than one watt and are expected (10) in the future to be exposed to many low-power microwave systems as in auto collision-avoidance radar and toys. Even here though, the expected level of exposure is very low. The concern expressed in the form of a technological assessment (10) reflects today's awareness of microwave hazards.

In most cases, including that of small-boat radar potential microwave exposure is well below 1 mW/cm^2. In the case of broadcast TV (22) it is reported that power density levels in some high buildings near TV transmitters have exceeded 1 mW/cm^2. Both because of power density level and exposure time (transmitter on time), this situation represents the greatest microwave exposure permitted the general public under today's regulations.

In order to properly assess potential hazards it is useful to understand the simple relationships in a radiated microwave beam between total radiated power, beam cross section and average power density. This is shown in Fig. 6. We see that if the total radiated power is less than one watt as in a microwave oven then an <u>emission</u> power density of several mW/cm^2 may exist at 2.0" from the oven surface but at several feet where the diffuse leakage radiation spreads to at least 10 square feet that the power density is less than 100 μW/cm^2.

If whole-body exposure corresponds to 10 square feet, then whole body exposure at greater than 10 mW/cm^2 requires at least 100 watts radiated power. Another useful benchmark is the fact that 1 megawatt of radiated power can yield a power density of 1 mW/cm^2 over an area of more than 0.1 km^2 - e.g., an area of 1 km $\times$ 0.1 km which may roughly describe the annular area of a radiated teelvision beam a couple of hundreds of yards from an antenna.

The typical decrease of leakage density with distance (inverse square law) is also depicted in Fig. 7. This holds only for small localized leakage sources which is a valid description of a leakage

EXPANSION OF INDUSTRIAL AND CONSUMER USES

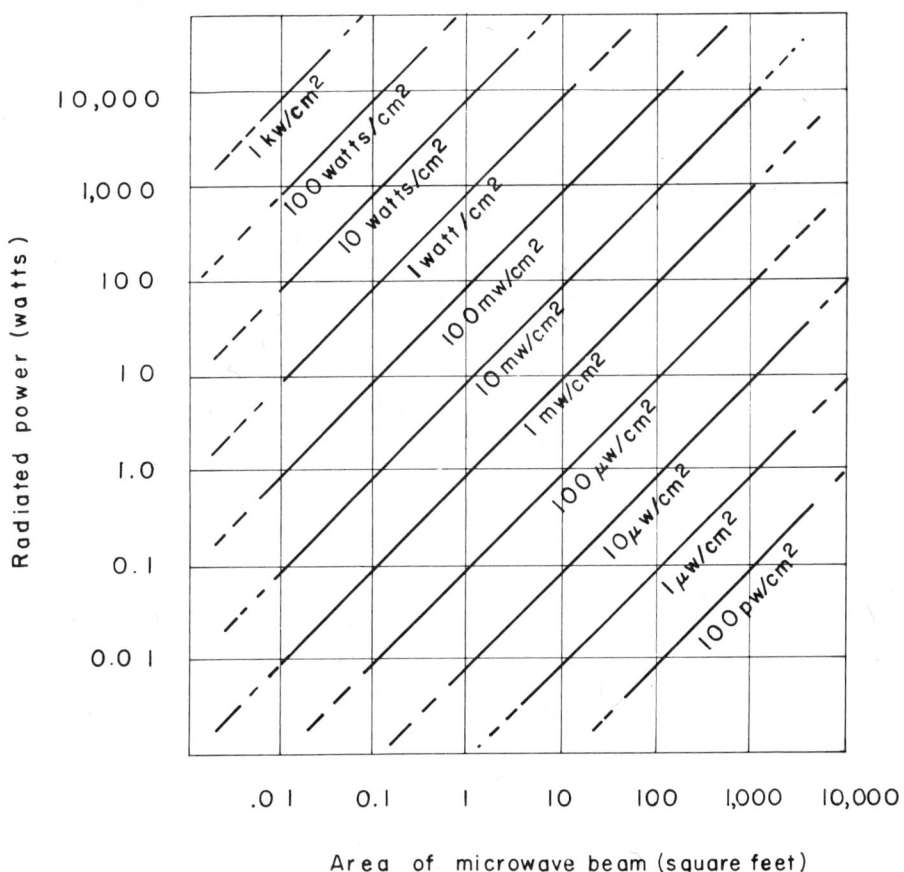

Figure 6. Relationship of Radiated Power Density to Total Radiated Power and Area of Radiated Beam

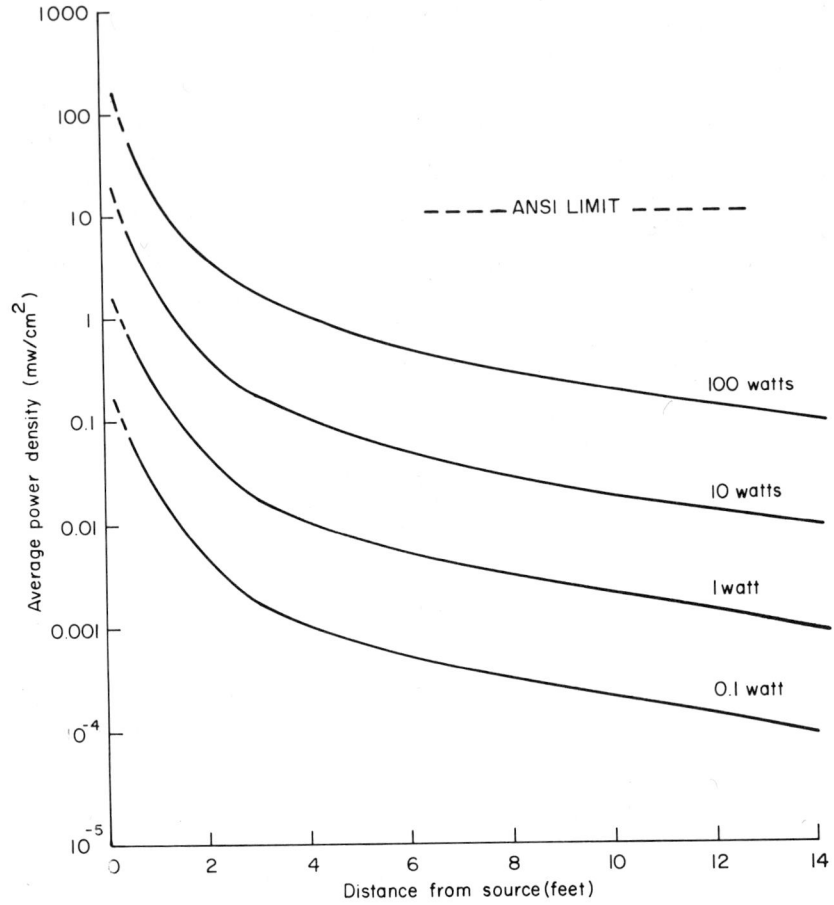

Figure 7. Spatial Dependence of Leakage Power Density for a Range of Radiated Power

EXPANSION OF INDUSTRIAL AND CONSUMER USES

source in a microwave-oven door seal. The HEW oven leakage emission standard of 5(1) mW/cm^2 at 2.0 inches from the external surface is seen to imply at most several microwatts/cm^2 at a distance of several feet.

If careful studies (19,20) of the relationship between microwave oven emission and exposure are made, then it is found that for a typical oven user the worst-case exposure if the oven is leaking at the HEW limit is that shown in Fig. 8. Clearly a very high safety factor is present which justifies the statement in an AMA editorial (5) that "The standards adhered to by industry limits levels of typical exposure in normal use to approximately 1/10,000 the level known to cause harm to humans. Even under conditions of foreseeable misuse, safe margins are very likely maintained under this standard."

Industrial microwave equipment presents microwave leakage hazards similar to those with microwave ovens except for those associated with conveyor tunnels and their entrance and exit ports. A voluntary standard limiting leakage emission to 10 mW/cm^2 at any point two inches from an external surface has been developed by IMPI (11). This standard also calls for adequate warnings to prevent operators from inserting their arms or other objects into accessible ports of the conveyor systems. Adherence to this emission standard should insure fairly straightforward satisfaction by the equipment user of OSHA standards on personnel exposure.

The existing stringent standards on leakage emission and safety interlocks of doors and panels insure a high degree of safety against biological effects of microwave exposure for consumer and industrial microwave equipment. Furthermore, the requirements are satisfied without any severe technical or economic problems. Advances in door seal technology (21) permit suppression of the leakage in the authorized ISM bands well below limits.

In the opinion of the author, interference potentially poses more technical and economic problems. Existing regulations (FCC Regulations, Part 18 in the U.S.) on radiation outside the authorized ISM bands effectively limit emission levels (at 2.0") to a value roughly 60 dB below the HEW leakage limit of 5 mW/cm^2 at 2.0", when measured with a 5 MHz bandwidth. These existing regulations are based on measurement of <u>average</u> power density or signal strength of signals outside the ISM bands and are effective in preventing interference with other authorized services utilizing the spectrum.

There are two potential complications in the status of interference control: (1) "incidental" interference and (2) proposed regulations on impulsive sources.

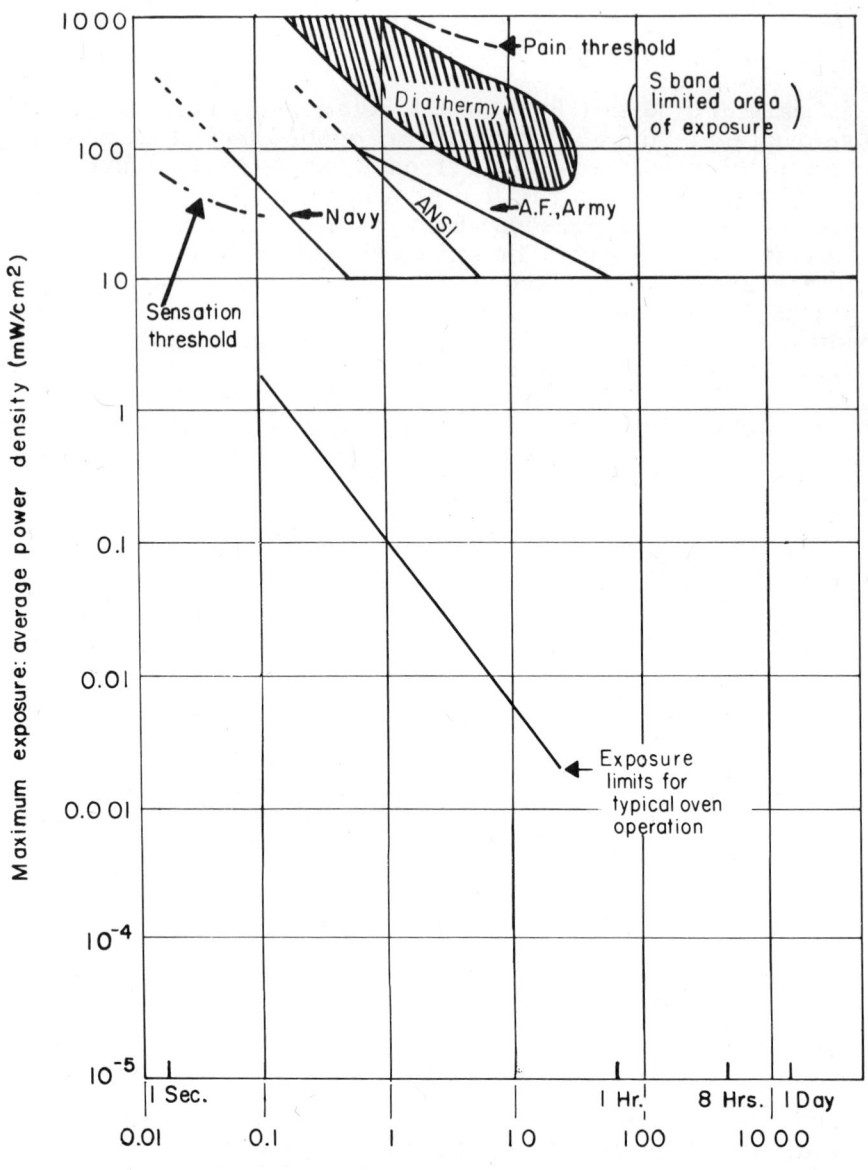

Figure 8. Comparison of Worst-Case Exposure of a Microwave Oven Meeting HEW Leakage Limits with Exposure Standards

"Incidental" interference refers to the fact that authorized radiation sometimes can produce interference in other equipment, particularly those not designed as receivers such as hi-fi sets, tape recorders, instrumentation, electro-explosive devices and cardiac pacemakers. In most cases the source of the problem is most validly described as the susceptibility of the device suffering interference. Susceptibility usually can be suppressed by straightforward techniques of shielding and filtering. Thus cardiac pacemakers which can be susceptible to radiation of broadcast TV (3), amateur radio (24), radar (30), and microwave ovens (16) at power densities as low as $1 \mu W/cm^2$ can be made (23) immune to interference even at a level of $10 mW/cm^2$ by shielding and filtering. It is ineffective and inappropriate to use warning signs around radiation sources in an attempt to protect pacemaker wearers.

The other possible problem of the future is the possibility of new RFI regulations on ISM equipment based on peak limits. These are being proposed (23) with peak limits not far above existing average signal limits. This fails to account for the non-gaussian type of noise peculiar to ISM systems and would be much more stringent than existing standards. A significant economic burden on equipment cost could possibly result and retard the growth of this field of microwaves. It is believed that existing (29) studies on the qualitative nature of impulsive interference support the concept that average and not peak signal strength is a better measure of degree of interference.

The intensified awareness of potential hazards and need for their control will not impede growth in the microwave power field. Users will become more educated in the nature of microwaves and there are potential new applications out of the results of research on biological effects of microwaves. One can predict that in the next decade the microwave oven will become a common major household appliance and that certain industrial applications will become major successes. The challenge to the microwave industry is in foreseeing which applications and when.

REFERENCES

1. BENGTSSON, N. E. and RISMAN, P. O.: Dielectric properties of foods at 3 GHz as determined by a cavity perturbation technique, II. Measurements on food materials. J. Microwave Power 6 (1971) 107.

2. BRIDGES, J. E. and BRUESCHKE, E. E.: Hazardous electromagnetic interaction with medical electronics. Proc. 1970 IEEE EMC Symposium, Anaheim, California (July 14-16) 173.

3. D'CUNHA, G. F. et al.: Syncopal attacks arising from erratic demand pacemaker function in the vicinity of a television transmitter. American J. Cardiology 31 (June 1973) 789.

4. Division of Electronic Products Staff: Survey of selected industrial applications of microwave energy. Document BRH/DEP 70-10, U. S. Dept. of HEW (May 1970).

5. Editorial, J. of the American Medical Assoc., Vol. 215 10 (March 8, 1971) 1661.

6. FREEDMAN, G.: The future of microwave heating equipment in the food industries. J. Microwave Power 8 (July 1973) 161.

7. FREY, J. and BOWERS, R.: What's ahead for microwaves? IEEE Spectrum 9 (1972) 41.

8. GOLDBLITH, S. A.: Applications of microwaves in food processing. Microwave J. (May 1966) 118.

9. GOLDBLITH, S. A. and PACE, W. E.: Some considerations in the processing of potato chips. J. Microwave Power 2 (1967) 95.

10. HEROLD, E. W.: The future of the electron tube. IEEE Spectrum 2 (January 1965) 50.

11. IMPI performance standard on leakage from industrial microwave systems. IMPI Publication IS-1 (Aug. 1973). (Copies may be ordered from IMPI, Box 1556, Edmonton, Alberta, T5J2N7, Canada).

12. Joint Report of DOD and HEW: A partial inventory of microwave towers, broadcasting transmitters and fixed radar by states and regions. Document BRH/DEP 70-15 (June 1970).

13. JOLLY, J.: Industrial microwave power adoption rate and the diffusion of technical information. J. Microwave Power 8 (1973) 337.

14. KAPITSA, P. L.: High power electronics. Uspekhi fizicheskikh nauk 78 2 (1962) 181.

15. McCONNELL, D. R.: Impact of microwaves on the future of the food industry. J. Microwave Power 8 (1973) 123.

16. MICHAELSON, S. M. and MOSS, A. J.: Environmental influence on implanted cardiac pacemakers. J. American Med. Assoc. 216 2006 (1971).

17. O'MEARA, J. P.: Why did they fail. J. Microwave Power 8 (July 1973) 167.

18. OSEPCHUK, J. M.: Trends and limitations of microwave tubes. Digest of 1969 IEEE International Convention.

19. OSEPCHUK, J. M., FOERSTNER, R. A. and McCONNELL, D. R.: Computation of personnel exposure in microwave leakage fields and comparison with personnel exposure standards. Digest of the 1973 Microwave Power Symposium (IMPI).

20. OSEPCHUK, J. M.: A computation of typical operator exposure to microwave oven leakage. Part of the AHAM Presentation to TEPRSSC (HEW Advisory Committee), Nov. 1973, and published in the Microwave Energy Applications Newsletter, Vol. VI (Nov.-Dec., 1973) 11.

21. OSEPCHUK, J. M., SIMPSON, J. E. and FOERSTNER, R.: Advances in choke design for microwave oven door seals. J. Microwave Power 8 (Nov. 1973) 293.

22. Program for control of electromagnetic pollution of the environment: The assessment of biological hazards of non-ionizing electromagnetic radiation. Rept. Office of Telecommunications Policy, Executive Office of the President (March 1973).

23. Recommendation No. 39/1 of the Int. Spec. Committee on Radio Interference (CISPR), Int. Electrotechnical Commission: Limits of interference from ISM RF equipment..... (Aug. 1973).

24. SANCHEZ, S. A.: Danger! When you transmit you can turn off a pacemaker. QST (March 1973) 58.

25. SCHWAN, H. P. and COLE, K. S.: Bioelectricity: Alternating current admittance of cells and tissues. Medical Physics III (Glasser, Ed.) (1960) 52.

26. Special Issue on Industrial, Scientific, and Medical Applications of Microwaves (Voss, W. A. G., Ed.). Proc IEEE 62 1 (Jan. 1974).

27. URBAIN, W.: Some thoughts on the problems of microwave heating and food processing. J. Microwave Power 4 (1969) 59.

28. VOSS, W. A. G.: Advances in the use of microwave power. Paper No. 008, Seminar Series, Bureau of Radiological Health, U. S. Dept. of HEW (Feb. 1970).

29. Working Group 3 Report to the Advisory Committee for Land Mobile Radio Services. Appendix G available from the National Technical Information Service (Report PB174-278).

30. YATTEAU, R. F.: Radar-induced failure of the demand pacemaker. N. E. J. Med. 283:26 (Dec. 24, 1970) 1447.

-DISCUSSION-

DUNN - It was my understanding that the real problem with the microwave oven was assuring that leakage would always remain at this low level that you have indicated, even after the unit has been installed and in use, or when used improperly. I have been told about the wet noodle syndrome, the wet noodle that sort of sticks through the door and acts like a beautiful antenna for radiating microwaves.

OSEPCHUK - A wet noodle wouldn't be a good one; a wire would be much better.

DUNN - But you don't often cook a wire.

OSEPCHUK - No, but you know HEW has a regulation that essentially prevents wires from going into any ventilating port, and it's a very detailed regulation.

DUNN - I understand that, but can you make a comment about assuring that the doors prevent leakage of microwaves.

OSEPCHUK - Assurance of whether or not it will not leak in the future? Well, all of the manufacturers -- and HEW has done a lot of this -- life test these ovens with 100,000 door closings, under various conditions. The Underwriters' Laboratory conducts all types of tests with corn oil, heat tests, drop tests -- in other words there are all types of tests that are professionally established to assure safety. And, in terms of the quality of the goods, one can consider what microwave oven manufacturers give for warranties, I know in our case a five-year warranty is given. So in terms of quality of manufacture, I think that microwave ovens today are meeting high standards. Technically, the reasons why there isn't any great worry about this type of degradation is that more and more people are using what we call the choke seal which is insensitive to putting things in the door region. It is true that a contact seal causes problems if you put paper in it. If you didn't do any of those things, then the contact seal can be very good. As the people from HEW will tell you, as the surveys keep going on the levels get lower and lower. And, I see no practical problem with what's going on.

CZERSKI - I would like to note that we agreed already one year ago in Loughborough that equipment performance standards in the U.S.A. for microwave ovens and the U.S.S.R. personnel exposure standard do conform exactly and we would be only too happy to import any such ovens into Poland. They would be accepted. And this time I am extremely glad that you make such a strong case for controlling the power output of broadcasting and TV stations.

SOLAR POWER VIA SATELLITE

Peter E. Glaser and Owen Maynard

Arthur D. Little, Inc.
Cambridge, Massachusetts

and

Raytheon Company
Waltham, Massachusetts

Abstract

The possibilities for using satellite solar power stations for large-scale power generation on earth, converting solar energy into microwave energy, transmitting it to the earth's surface, and transforming it into electricity have recently been explored. The current state of technology and the necessary developments for accomplishing the four functions, i.e., collection of solar energy, conversion to and transmission of microwaves and rectification to DC on the ground, are reviewed. The requirements for flight control, earth-to-orbit transportation, and orbital assembly are discussed. Environmental issues, including impact of waste heat release, water injection into the upper atmosphere by space vehicle exhaust, noise pollution, and location of antenna sites are listed. Biological effects and radio frequency interference are explored. The time frame for accomplishing the operational system is outlined.

Introduction

Among the different sources of energy, whether they be non-renewable, such as fossil or nuclear fuels, or continuous, such as tidal or geothermal, none have a greater potential than solar energy, on which life in its very essence depends. The question that needs to be answered is: Has society reached the level of sophistication to apply solar energy for its overall long-term benefit consistent with the balance of nature? The answer is not yet obvious. However, efforts required to provide the answer are beginning to be made.

The total influx from solar, geothermal, and tidal energy into the earth's surface environment is estimated to be $173,000 \times 10^{12}$ watts. Solar radiation accounts for 99.98% of it. The sun's contribution to the energy budget of the earth is 5000 times the energy input of other sources combined.

The vast quantities of energy radiated by the sun reach earth in very dilute form. Thus, any attempts to convert solar energy to power on a significant scale will require devices which occupy a large land area as well as locations that receive a copious supply of sunlight. These requirements restrict earth-based solar energy conversion devices which could produce power to a few favorable geographical locations. Even for these locations energy storage must be provided to compensate for the day-night cycle and cloudy weather.

One way to harness solar energy effectively would be to move the solar energy conversion devices off the surface of the earth and place them in orbit away from the earth's active environment and influence and resulting erosive forces.[1,2] The most favorable orbit would be one around the sun, but as a first approximation toward this very long-term goal, an orbit around the earth where solar energy is available nearly 24 hours of every day could be used.

Since the concept of a satellite solar power station (SSPS) (see Fig. 1) was presented as an alternative energy production method,[3] the impending energy crisis has been recognized as one of the major issues facing the nation. Various options to meet the crisis are now being explored.

An assessment of the feasibility of the SSPS concept performed by Arthur D. Little, Inc., Grumman Aerospace Corporation, Raytheon Company, and Textron Inc., with partial support from NASA,[4] has shown that it is worthy of consideration as an alternative energy production method.[5-9] Its development can be realized by building on scientific realities, on an existing industrial capacity for mass production, and on demonstrated technological achievement.

Fig. 1. Design concept for a satellite solar power station

A. Principles of a Satellite Solar Power Station

Figure 2 shows the design principles for an SSPS. Two symmetrically arranged solar collectors convert solar energy directly to electricity by the photovoltaic process while the satellite is maintained in synchronous orbit around the earth. The electricity is fed to microwave generators incorporated in a transmitting antenna located between the two solar collectors. The antenna directs the microwave beam to a receiving antenna on earth where the microwave energy is efficiently and safely converted back to electricity.

An SSPS can be designed to generate electrical power on earth at any specific level. However, for a power output ranging from about 3000 to 15,000 megawatts (MW), the orbiting portion of the SSPS exhibits the best power-to-weight characteristics. Additional solar collectors and antennas could be added to establish an SSPS system at a desired orbital location. Power can be delivered to most desired geographic locations with the receiving antenna placed either on land or on platforms over water near major load centers and tied into a power transmission grid. The status of technology and the advances which will be required to achieve effective operation for an SSPS are described in the following sections.

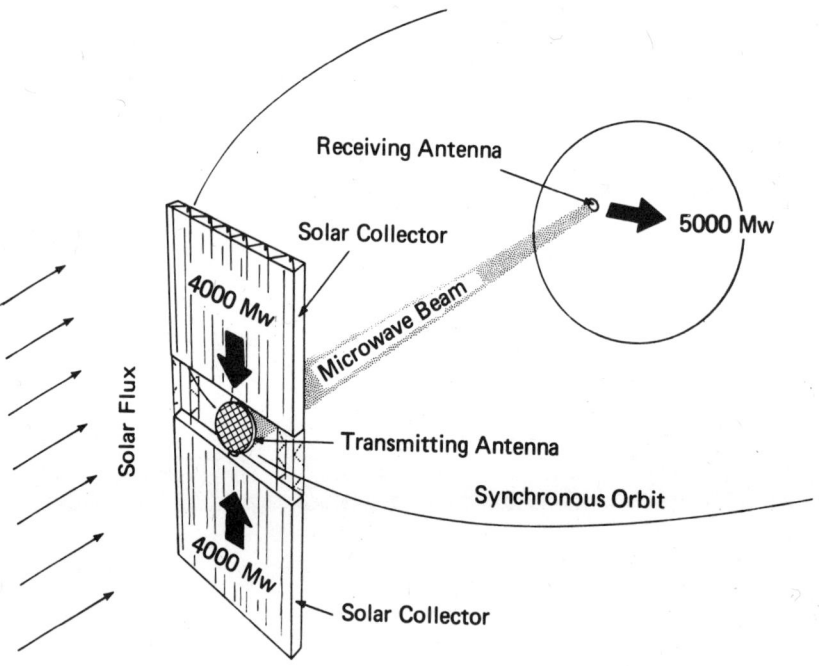

Fig. 2. Design principles for a satellite solar power station.

1. Location of Orbit

The preferred locations for the SSPS are the earth's equatorial synchronous orbit stable nodes which occur near the minor axes at longitudes of about 123° West and 57° East. The SSPS would be positioned so its solar collectors always faced the sun, while the antenna directed a microwave beam to a receiving antenna on earth. In an equatorial synchronous orbit, the satellite can be maintained stationary with respect to any desired location on earth. About 30,000 pounds per year of propellants for attitude control would be required to overcome orbit-disturbing influences, such as the gravitational effects of the sun and moon, solar pressure, and the eccentricity of the earth. The microwave beam would permit all-weather transmission so that full use could be made of the nearly 24 hours of available solar energy.

An SSPS in synchronous equatorial orbit would pass through the earth's shadow around the time of the equinoxes, when it would be eclipsed for a maximum of 72 minutes a day (near midnight at the receiving site). This orbit provides a 6- to 15-fold time advantage over solar energy conversion on earth. A comparison of the maximum allowable costs of solar cells indicates that for terrestrial solar power applications these devices are competitive with other energy conversion methods if they cost about $2.30 per square meter. Because of the favorable conditions for energy conversion that exist in space, these cells would be competitive if they cost about $45 per square meter in an earth-orbit application.[10]

2. Solar Energy Conversion

The photovoltaic conversion of solar energy into electricity is ideally suited to the purposes of an SSPS. In contrast to any process based on thermodynamic energy conversion, there are no moving parts, fluid does not circulate, no material is consumed, and a photovoltaic solar cell can operate for long periods without maintenance. Photovoltaic energy conversion has been developed substantially since the first laboratory demonstration of the silicon solar cell in 1953. Today, such cells are a necessary part of the power supply system of nearly every unmanned spacecraft, and considerable experience has been accumulated to achieve long-term and reliable operations under the conditions existing in space. As a result of many years of operational experience, a substantial technological base exists on which further developments can be based.[11] These developments will be directed toward increasing the efficiency of solar cells, reducing their weight and cost, and maintaining their operation over extended periods.

3. Microwave Power Generation, Transmission, and Rectification

The power generated by the SSPS in synchronous orbit must be transmitted to a receiving antenna on the surface of the earth and then rectified. The power must be in a form suitable for efficient transmission in large amounts across long distances with minimum losses and without affecting the ionosphere and atmosphere. The power flux densities received on earth must also be at levels which will not produce undesirable environmental or biological effects. Finally, the power must be in a form that can be converted, transmitted, and rectified with very high efficiency by known devices.

All these conditions can best be met by a beam link in the microwave part of the spectrum. In this part of the spectrum a frequency of 3.3 GHz was selected because induced radio frequency interference can be limited so that an appropriate internationally agreed upon frequency could be assigned to an SSPS.

As early as 1963, Brown[12] demonstrated that large amounts of power could be transmitted by microwaves. The efficiency of microwave power transmission will be high when the transmitting antenna in the SSPS and the receiving antenna on earth are large. The dimensions of the transmitting antenna and the receiving antenna on earth are governed by the distance between them and the choice of wavelength.[13]

The size of the transmitting antenna is also influenced by the inefficiency of the microwave generators due to the area required for passive radiators to reject waste heat to space and the structural considerations as determined by the arrangement of the individual microwave generators. The size and weight of the transmitting antenna will be reduced as the average microwave power flux density on the ground is reduced by increasing the size of the receiving antenna and as higher-frequency microwave transmission is used. The size of the receiving antenna will be influenced by the choice of the acceptable microwave power flux density, the illumination pattern across the antenna face, and the minimum microwave power flux density required for efficient microwave rectification.

Several power distributions for the transmitting antenna can be identified which concentrate at least 90% of the power at the ground within the main lobe, along with an indication of the nature of side lobes to be expected. Several approaches are available to control the side lobes. The ground site selection criteria will be greatly influenced by results of projected earth resources studies, as well as social and political considerations. System aspects of site selection lend themselves to relatively simple and known analysis techniques.

Highly efficient ground power distribution systems, such as the underground, cryogenic, high-voltage dc system, are anticipated to be well developed in the operation time period of the SSPS. Such systems will permit a rationale for location of power consumption centers, such that they may not necessarily be close to the power supply station. The low power flux density and the efficient receiving antenna will permit a full utilization of land under the antenna. The biological effects of the non-ionizing radiation, which are the subject of investigation now, can be anticipated to become sufficiently well understood so that a straightforward SSPS biological effects program can be implemented to define criteria and limitations for several ground selection and multiple utilization options.

B. SSPS Flight Control

Although the SSPS is orders of magnitude larger than any spacecraft yet designed, its overall design is based on present principles of technology. Thus, its construction and the attainment of a 30-year operating life require not new technology, but substantial advances in the state of the art.

The SSPS structure is composed of high-current-carrying structural elements, the electromagnetic interactions of which will induce loads or forces into the structure. Current stabilization and control techniques are capable of meeting the requirements of spacecraft now under development. Most of these spacecraft have comparatively rigid structures and are amenable to control as a single entity by reaction jets or momentum storage devices. But the large size of an SSPS at first suggests that new structural and control system design approaches may be needed to satisfy orientation requirements. However, results of analyses indicate that present structural analytical techniques and tools are adequate and that an SSPS can be controlled such that the solar collector will point at the sun to ± 1 degree, the transmitting antenna can be mechanically controlled to ± 1 arc min, and the direction of the microwave beam electronically maintained to ± 1 arc sec.[4] Low-thrust, ion propulsion systems appear promising for SSPS control because their performance characteristics are compatible with the potential lifetime required of the SSPS.

C. Earth-to-Orbit Transportation

A high-volume, two-stage transportation system will be required for an SSPS: (1) a low-cost stage capable of carrying high-volume payloads to low-earth orbit (LEO); and (2) a high-performance stage capable of delivering partially assembled elements to synchronous or some intermediate orbit altitude for final assembly and deployment. The factors affecting flight mode selection include payload element size, payload assembly techniques, desirable orbit locations for assembly, time constraints, and requirements for man's participation in the assembly. The choice of transportaion system elements includes currently planned propulsion stages and advanced concepts optimized for an operational SSPS system. Minimum cost transportation combinations will have to be identified which can fulfill the requirements for SSPS delivery, assembly, and maintenance for an operational system. The challenge of an SSPS capable of generating 5000 MW of power on earth is to place into orbit a payload of about 25 million pounds and propellant supplies for station-keeping purposes of about 30,000 pounds per year.

D. Orbital Assembly

Operations in space involving assembly have been limited to docking two actively maneuvering vehicles together. However, studies of modular space stations which dock a number of similar masses to form a large complex in orbit have been performed. In general, to effect a mating, attitude control is required on both target and docking vehicles.

After a docking vehicle contacts a space assembly, several things happen: (1) the assembly experiences loads; (2) modular elements deflect relative to each other; and (3) possibly disruptive control forces and G loads are transmitted throughout the space assembly. In general, the weak link in a space assembly is the docking interface, since it has a smaller cross-sectional area than the prime construction element. Relatively large space assemblies have been analyzed in modular space station studies and found to be controllable during assembly,[14] providing that a prescribed build-up sequence is followed and that grossly asymmetric configurations are avoided. In addition, there is a significant interplay between docking contact velocities, target and docking vehicle flight control, and docking mechanism characteristics. In general, direct vehicle docking contact velocities of about 0.5 fps and manipulator docking contact velocities of 0.1 fps are compatible with currently planned spacecraft systems. The large, more flexible SSPS type spacecraft will require significantly lower contact velocities because of the size and flexibility of the system and its potential damping characteristics. Zero or near-zero contact velocities may be required, together with appropriate docking or joining mechanisms and control techniques during assembly. Because a considerable portion of the SSPS structure may be part of the power distribution system, new joining and assembly techniques may be required.

Assembly sequences and modes and desirable assembly altitudes have to be identified to define the assembly requirements for the SSPS's large area and light-weight structure. Once the framework for the basic SSPS structure has been developed, including the potential sizes and sub-element characteristics of major components, such as the solar collector arrays, the transmitting antenna and structural members, alternative assembly sequences, modes, and attitudes can be evaluated. These could include automatic or operator assembly options, and assessments as to the portions of the assembly process to be carried out at low-earth orbit altitudes, at intermediate or at synchronous altitudes. With this knowledge, the appropriate control techniques during the assembly process can then be identified and developed.

E. Environmental Issues

1. Environmental/Ecological Impact

The SSPS appears to have limited environmental and ecological impacts at the receiving antenna. The following are the potential environmental impacts:[4]

- Waste heat released by natural convection at the receiving antenna does not constitute a significant thermal effect on the atmosphere. If the antenna is located in desert regions where water is limited, there may be some slight modification of the plant community at and near the antenna site.

- With RF shielding incorporated below the rectifying elements, the receiving antenna operation can be compatible with other land uses because there is only a small degree of reduction of solar radiation received on the ground below the antenna. However, installation and maintenance of the antenna has to be planned because extensive activities may be damaging to some ecologically important systems.

- Injection of water into the stratosphere and upper atmosphere by space vehicle exhausts would be small in contrast to the natural abundance, and does not appear to constitute a significant environmental effect. However, a detailed assessment of these effects will have to await a better definition of the nature of the upper atmosphere.

- Noise pollution from the launch operations would be of concern in the immediate vicinity of the launch facility and would have to be reduced by suitable design techniques or location of the launch facilities.

- There is substantial flexibility in choosing a suitable location for the receiving antenna. The area has to be contiguous but need not be flat terrain. The location can be in a region where the land is not suitable for other uses (e.g., desert areas, previously strip-mined land, or near major electrical power users like aluminum smelters).

2. Biological Effects

There exist conflicting interpretations of the effects of microwave exposure throughout the scientific community. Because of the lack of internationally accepted standards, based on experimental data, to place a specific and allowable level on microwave exposure, the SSPS will have to be designed to accommodate a wide range of frequencies and microwave power flux densities. Precise pointing of the microwave beam must be achievable with attitude stabilization and automatic phase control to assure efficient transmission of the power to the receiving antenna. The design approaches already identified indicate that this objective can be met. The transmitting antenna size, the shape of the microwave power distribution across the antenna, and the total power transmitted will determine the level of microwave power flux densities in the beam reaching the earth.

The effects on birds exposed to microwave power flux densities within the beam at the receiving antenna and the effects on aircraft accidentally flying through the beam, even though projected to be negligible, will have to be determined experimentally. The microwave system design requirements of the SSPS must be established so as to result in low microwave exposure levels to assure safe operations which will be acceptable on an international basis.

3. Radio Frequency Interference

Design and development of the microwave generation device with its filters and controls are required to further define criteria for radio frequency spectrum contributions associated with noise and harmonics. High overall efficiency is attributed to circuit efficiency and internal dc-to-rf conversion efficiency which depend on achievable values of magnetic field and low but achievable power losses. The RF spectrum effects are associated with the fundamental frequency and its harmonics, turn-on and shut-down sequence, random background energy, and other superfluous signals resulting from the specific design.

Further studies are required in the areas of filtering technology and the design of the device integrated with its power input and control, and integrated in the higher level of assembly with the transmitting antenna; narrowband operation permitted by the SSPS application in determining and achieving near-optimum gain; and noise measurements specifically associated with the microwave tubes.

The effects of SSPS radio frequency interference on other users will be substantial, unless the frequency is selected so as to avoid the more sensitive services close to the fundamental frequency. It is very likely that current users would be required to relinquish their frequency allocation to the SSPS. Frequency allocation priority may be achieved if the SSPS can be shown to contribute significantly to fulfilling the nation's and the world's power needs.

Most sensitive radio astronomy services require fixed narrow bands for operation spread throughout the spectrum. The 3.3 GHz for SSPS fundamental was selected primarily to maximize the capability of filtering SSPS noise at the existing radio astronomy frequencies because filters can be designed and developed to achieve this objective.

Potential interference with shipborne radar must be investigated unless the advanced technologies in Amplitron device design and filtering are achieved.

Amateur sharing, state police radar, and radio location from high-power defense radar in the 3.23- to 3.37-GHz band will suffer interference. Thus, specific allocation for the SSPS would have to be negotiated with such users in mind. This is not to infer that interference and re-allocation are recommended as acceptable. Rather, detailed and specific effects and impacts on these users in this small band must be determined before the acceptability of specific frequency allocations can be established.

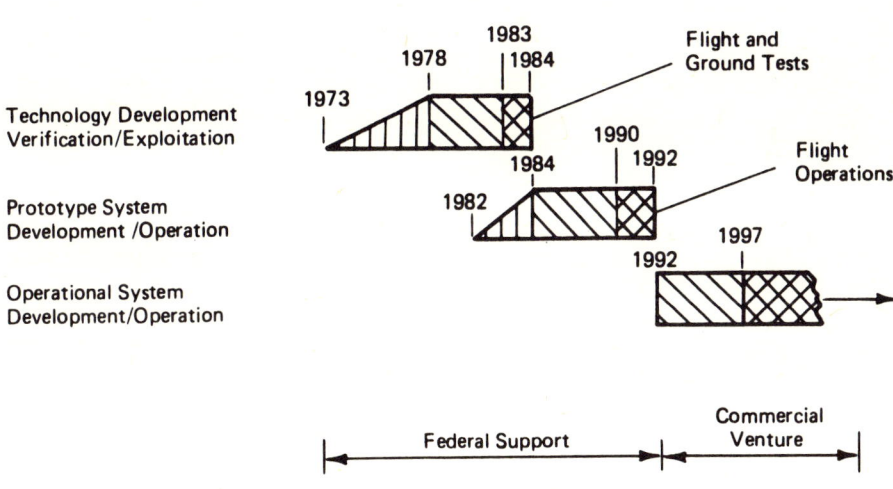

Fig. 3. Program phasing

F. Time Frame

Based on an assessment of the steps required to develop the various technologies, it is reasonable to conclude that a prototype SSPS can be demonstrated in the early 1990's (see Fig. 3). In parallel with the development of technology, environmental effects, economic constraints and social impacts should be assessed so that the overall desirability of an SSPS can be compared with other energy production methods.[15]

The activities and developments specifically related to an operational SSPS system consisting of a network of satellites are assumed to be directed towards a commercial venture with potential for international participation. Such participation would assure that the benefits of the development of the SSPS on a scale which can be of benefit to many nations could be modeled on the successful introduction of communication satellites. The research, development, and verification program is aimed at providing an option for power generation on a time scale which is meaningful in terms of the development of other options whether they be based on solar energy or other energy sources while providing tangible returns even if the SSPS option is not exercised.

Conclusions

The realization of the concept of a Satellite Solar Power Station not only represents a major challenge to technology but also an unparallelled opportunity to apply space technology for the benefit of mankind. No fundamental breakthroughs are required to achieve the objectives identified with the SSPS but rather major advances and improvements over existing technology. Just as only 15 years ago it was unimaginable that a space shuttle could be developed to transport massive payloads into orbit and repeat the mission 100 times, so today the mission to orbit a satellite as large as an SSPS appears to be a formidable undertaking. The questions that need to be answered are not only whether the required technology can be developed - because the answer most likely will be yes - but rather whether the development of an SSPS to meet future energy demands is an option that should be pursued.

In addition to solar energy, several other energy sources have the potential to meet future energy requirements, but only very few are truly in harmony with the environment and conserving the finite resources of the earth. Solar energy applications, such as the SSPS, are still in an early stage of development. Thus, it is too early to tell which of the approaches now being studied will be judged to have the greatest potential to be of overall benefit to society. As more is learned about the operating characteristics of potentially competitive electrical energy-generating systems, the views on what best performance represents in terms of particular operating parameters will continue to evolve. Thus, the criteria for decision-making, whether based on cost, resource conservation, or environmental enhancement, may be quite different in the future, and will continue to change as long as technical developments continue actively on the various energy-production methods.

Alternative energy production methods, such as an SSPS, if successfully developed over the next few decades, will permit society to look beyond the year 2000 with the assurance that future energy requirements could be met without endangering the planet Earth. But even successful development of solar energy conversion alternatives will still require approaches to reduce energy consumption, for the ultimate ceiling to energy production will be the ability to dissipate the heat produced at the point of use of electrical power.

References

1. Glaser, P.E., Solar Power via Satellite, Astronautics and Aeronautics, Vol. 11 (eleven) August, 1973, pp. 60-68.

2. Glaser, P.E., "Power from the Sun: Its Future," Science, Vol. 162 (November, 1968) pp. 857-886.

3. Glaser, P.E., The Future of Power from the Sun, IECEC 1968 Record; IEEE Publication 68C21-Energy, pp. 98-103.

4. Feasibility Study of a Satellite Solar Power Station, NASA Contract No. NAS 3-16804, Arthur D. Little, Inc., April, 1973.

5. Satellite Solar Power Station, Tech. Memoranda, Grumman Aerospace Corporation, Bethpage, N.Y., Jan.-June, 1972.

6. Satellite Solar Power Station, Technical Report, Q-71098, Spectrolab/Heliotek Divisions, Textron, Inc., Nov., 1971.

7. Satellite Solar Power Station, Arthur D. Little, Inc., Jan. 21, 1972.

8. Satellite Solar Power Station, Configuration Status Report, Grumman Aerospace Corporation, ASP-583-R-10, June, 1972, and Satellite Solar Power Station, Master Program Plan Development, Grumman Aerospace Corporation, ASP-611-R-12, June, 1972.

9. Microwave Power Transmission in the Satellite Solar Power Station System, Tech. Report ER72-4038, Raytheon Company, Jan. 27, 1972.

10. Wolf, M., Cost Goals for Silicon Solar Arrays for Large-Scale Terrestrial Applications, Conference Record of the Ninth IEEE Photovoltaic Specialists Conference, Silver Spring, Maryland, May, 1972, pp. 342-350.

11. Berman, P.A., Photovoltaic Solar Array Technology Required for Three Wide-Scale Generating Systems for Terrestrial Applications: Rooftop, Solar Farm, and Satellite, Technical Report 32-1573, California Institute of Technology, Jet Propulsion Laboratory, Pasadena, California, October 15, 1972.

12. Brown, W.C., Experiments in the Transportation of Energy by Microwave Beams, 1964 IEEE Intersociety Conference Record, Vol. 12, Pt. 2, 1964, pp. 8-17.

13. Goubau, G., Microwave Power Transmission from an Orbiting Solar Power Station, Microwave Power, Vol. 5, No. 4, December, 1970, pp. 223-231.

14. Study of Requirements of Assembly and Docking of Spacecraft in Earth Orbit, NASA Contract NAS8-27860, Grumman Aerospace Corp.

15. Glaser, P.E., Space Solar Power: An Option for Power Generation, Presented at the 100th Annual Meeting of the American Public Health Association, Atlantic City, New Jersey, November 14, 1972.

INTERNATIONAL COOPERATION ON NON-IONIZING RADIATION PROTECTION

Michael J. Suess, Sc.D.

World Health Organization Regional Office for Europe

8 Scherfigsvej, 2100 Copenhagen, Denmark

ABSTRACT

The increased use in recent years of non-ionizing radiation (NIR) devices has led to the development by the World Health Organization's Regional Office for Europe (WHO/EURO) of a programme on NIR protection. As part of this programme, WHO/EURO convened two working groups to review the general situation in Europe and to evaluate the results of an international symposium in Warsaw, 1973, on effects from microwaves. A third working group, on health effects from lasers, will take place later in 1974. Other activities of an international nature related to NIR have been undertaken by the Commission of the European Communities and the International Radiation Protection Association. In view of the growing need for inter-country co-operation and agreement on NIR units, definitions and standards, a proposal has been made for the establishment, along the lines of the existing international commissions on ionizing radiation, of a Commission on NIR (CONIR), which could devote itself to these tasks.

The growing use of non-ionizing radiation (NIR) devices in recent years has highlighted the need for a better understanding of the biologic effects which may be caused by NIR and also the need to consider means of protecting occupationally exposed personnel and the general public from possible adverse effects to health. This, in turn, has led in a number of countries to the development of health criteria and the introduction of regulations and standards. However, the time has now come for international co-operation and agreement on aspects related to NIR, along the lines of developments which have already taken place in the field of ionizing radiation.

The conventing of a working group by the World Health Organization's Regional Office for Europe (WHO/EURO) at The Hague in November 1971 could perhaps be considered the beginning of international activities on NIR protection, in contrast to national activities which were of course, already under way in that time. This Working Group was entrusted with the important task of reviewing and assessing the situation prevailing in Europe, discussing which activities were needed and recommending action to be taken. The report and recommendations of this Working Group (2) served as the basis for the development of a sectorial programme on NIR protection as part of the WHO/EURO long-term programme in environmental pollution control.

The objectives of this programme were to assist governments in protecting the general public from NIR in order to eliminate possible adverse effects on human health, some of which may be irreversible. To achieve these objectives, the programme has been developed on the basis of the five following concepts:

(a) collection and evaluation of NIR data in a systematic manner and exchange of information in this field among the countries of the European Region;

(b) study of the effects on health from NIR and support of the relevant investigations needed;

(c) discussion on possible biologic criteria for damage and development of guides and criteria for health protection;

(d) promotion of administrative and legislative measures for the protection of the general public from NIR, and

(e) assistance in the training of national personnel required for the surveillance and control of NIR.

The various activities of this programme have already been described in detail elsewhere (4).

During the meeting of the first Working Group an inconsistency was observed, in that different range limits were being used for the various types of NIR. Therefore, to promote better international discussion and collaboration in future, the various NIR ranges have been fixed and set as give in table I. These ranges, which are being adhered to by WHO/EURO for its own activities, have also been adopted for defining the microwave range of concern to the International Symposium on Biologic Effects and Health Hazards from Microwave Radiation, Warsaw, 15-18 November 1973 (Warsaw Symposium), as well as for the various types of radiation covered in the draft recommendations on health protection now under preparation by the Commission of the European Communities (CEC) (and see below).

Another international activity which was, however, devoted only to the effects of microwaves, was the above-mentioned Warsaw Symposium. Co-sponsored by the WHO, USA and Poland, it brought together for the first time over 50 scientists from 12 different countries. Thirty-nine papers were presented, and the participants exchanged research information and professional experience, and discussed and evaluated current developments and concepts related to the biologic effects and health implications of microwave radiation (1). Undoubtedly one of the major achievements of this Symposium was the initial removal of barriers between scientific workers from countries with different political, social and economic systems, thus constituting an important step forward in promoting international contacts in this field. The participants concluded their meeting with a recommendation that a programme on NIR protection be developed by WHO, which "could exert leadership in this field and facilitate communication among scientists" (1).

The Warsaw Symposium was followed by a special WHO/EURO Evaluation Group in Copenhagen, 22-23 October 1973. This Group discussed the material presented at the Symposium and recognized points for action which could eventually be followed up by the European Regional Office (3).

TABLE 1. Characteristics and sources of electromagnetic type radiation

Type of radiations		Frequency range*	Wave-length range*	Energy range per photon	Typical source
Ionizing		above 30 000 THz	below 10 nm	above 124 eV	Electronic tubes, nuclear decay, nuclear fission
Ultraviolet		30 000 THz to 790 THz	10 nm to 380 nm	124 eV to 3.3 eV	Sun, gas discharge tubes
	vacuum	3 000 THz to 1 600 THz	100 nm to 190 nm	12.4 eV to 6.5 eV	
	far	1 600 THz to 1 000 THz	190 nm to 300 nm	6.5 eV to 4.1 eV	
	near	1 000 THz to 790 THz	300 nm to 380 nm	4.1 eV to 3.3 eV	
	non-ionizing portion	1 800 THz to 750 THz	170 nm to 400 nm	7.3 eV to 3.1 eV	
Visible		790 THz to 400 THz	380 nm to 750 nm	3.1 eV to 1.6 eV	Sun, thermally excited atoms
Infrared		400 THz to 300 GHz	750 nm to 1 mm	1.6 eV to 1.2 meV	Sun, hot bodies
	near	400 THz to 100 THz	750 nm to 3 μm	1.6 eV to 0.4 eV	
	middle	100 THz to 10 THz	3 μm to 30 μm	0.4 eV to 41 meV	
	far	10 THz to 300 GHz	30 μm to 1 mm	41 meV to 1.2 meV	
	Laser	1 500 THz to 15 THz	200 nm to 20 μm	6.2 eV to 62 meV	
Microwaves		300 GHz to 300 MHz	1 mm to 1 m	1.2 meV to 1.2 μeV	Klystron, Magnetron
	EHF**	300 GHz to 30 GHz	1 mm to 10 mm	1.2 meV to 0.1 meV	
	SHF**	30 GHz to 3 GHz	10 mm to 100 mm	0.1 meV to 12 μeV	
	UHF**	3 GHz to 300 MHz	100 mm to 1 m	12 μeV to 1.2 μeV	
	Radar	56 GHz to 220 MHz	5.4 mm to 1.3 m	0.2 meV to 0.9 μeV	
Radio-frequencies		300 MHz to 300 kHz	1 m to 1 km	1.2 μeV to 1.2 neV	Tubes, transistors and tuned circuits

* The given ranges are only approximations, as no exact end-point can be defined.
** Extremely high frequencies, Super-high frequencies, Ultra-high frequencies

The next inter-country activity planned by WHO/EURO is a
Working Group on Health Effects from Lasers to be held in Dublin,
21-24 October 1974. This Group will comprise nearly 40 participants from 14 countries and will discuss, evaluate and prepare
recommendations on effects to the eye and the skin; risks from
the use of lasers in engineering works, industry and medicine;
dose measurements; international standards, and necessary legislation.

It may be assumed that the participation of the CEC representatives in the WHO/EURO Working Group at The Hague in 1971 (2)
led to an increased awareness on the part of the CEC of the importance of NIR protection. Since then, a number of activities have
taken place, at which the health implications of NIR have been
examined. One such activity was the convening in Paris in April
1974 of a sub-group, set up by the CEC Group of Experts on Basic
Standards, to discuss and prepare proposals for draft recommendations for protection of individuals against the hazards arising
from laser and microwave radiation. The recommendations, after
disucssion by the Group of Experts on Basic Standards, will be
reviewed by the nine CEC Member States before being finalized.

The International Radiation Protection Association (IRPA),
which for many years has concentrated on protection from ionizing
radiation only, has for the first time included in its international meetings   papers on NIR and has devoted a whole session
solely to this new and growing field. Thus, its Third
International Conference in Washington, 9-14 September 1973,
provided a good forum for the presentation of introductory and
review papers on NIR to an audience of workers in the ionizing
radiation field. The first step having been made, it may be
expected that IRPA will play an important role in drawing the
attention of its members, many of whom are practising health physicists, to the health hazards involved in, and the need for protection from NIR devices, the use of which is constantly growing.
An important step in this direction has also been taken by <u>Health
Physics</u>, the official journal of the American Health Physics
Society, which has decided to publish material on NIR which could
be beneficial to the large international body of its readers (3, 4).

No one will question the fact that the introduction of the
concepts of Standard Man and maximum persmissible dose, and the
classification of dose definition and dose measurement by The
International Commission on Radiological Protection (ICRP) and
the International Commission on Radiation Units and Measurements
(ICRU), were fundamental in the control of ionizing radiation.

Moreover, to develop an understanding of the hazards of NIR, a body with similar terms of reference is urgently needed in this field. In this connexion, the first WHO/EURO Working Group has already proposed in its recommendation No.8 that "the WHO should initiate the setting up of such an organ, either within its own organization or through international collaboration " (2).

Moreover, the need for the standardization of nomenclature in the field of NIR has been emphasized by the WHO/EURO Evaluation Group (3) following the Warsaw Symposium. When summarizing their deliberations, the Symposium participants expressed the need for an extensive effort leading to the establishment of "internationally acceptable nomenclature and definitions of physical quantities and units, and to standardize measurement techniques and dosimetry". Furthermore, they recommended that "an international group be established to work out procedures for achieving these objectives" (1). It became evident at the Symposium that numerous uncertainties resulted from linguistic difficulties that arose not only in translating specific words but also in conveying shades of meaning. This would concern the description of effects, organism response, techniques and criteria used in the assessment of the physiopathologic reaction of the body. It is important, therefore, that the terminology and definitions used by different countries and organizations should be understood and agreed upon. The CEC sub-group convened in Paris in April 1974 found it necessary to discuss and agree on a list of basic definitions as part of its proposals for recommendations on laser and microwave radiation protection.

As the appropriate international bodies which deal with ionizing radiation have rejected any involvement with NIR, the way is now open for the establishment of a new body which could perhaps be known as the Commission on Non-ionizing Radiation (CONIR). In view of the repeated recommendations to WHO to initiate such a body, and of the recent and growing interest expressed by IRPA and other international and national organizations in joining forces, it is very much to be hoped that the urgently needed CONIR will soon be established and begin its work.

REFERENCES

1. Biologic Effects on Health Hazards of Microwave Radiation. Proceedings of an International Symposium, Warsaw, 15-18 October 1973. Polish Medical Publishers. Warsaw (1974). (In print).

2. Health Effects of Ionizing and non-ionizing Radiation. WHO Regional Office for Europe. Document EURO 4701. Copenhagen (1972).

3. Health Hazards from Exposure to Microwaves. WHO Regional Office for Europe. Document EURO 3170. Copenhagen (1974). (This report will be reprinted in Health Physics (1974)).

4. SUESS, M.J.: The development of a long-term programme on non-ionizing radiation protection. Health Physics (1974). (In print).

Note: This paper expresses the views of the writer and does not necessarily represent the decisions or stated policies of the World Health Organization.

-DISCUSSION-

DUNN - Could I ask if there are any plans in any of these organizations to increase the involvement of, say, mechanical radiation as regards ultrasound, infrasound, and ordinary audible acoustics? Or is this something that you don't consider to be very important?

SUESS - I didn't mention all the details, but first of all our working group in The Hague in 1971 covered everything from ionizing radiation to ultrasound. At that meeting priority was given to non-ionizing radiation. Earlier today my colleagues have indicated that while we know the problems we also know our limitations of what we can do, and we can do only so much. So, the working group gave priority to microwaves and lasers. However, I can say that in preparation of our activities to 1982, as part of our long-term program on environmental pollution control, we have already inserted activities on ultrasound. Moreover, I think that this meeting now surely encourages me to formulate detailed proposals.

After the Stockholm Conference on the Human Environment, our headquarters has been assigned the task of developing international environmental health criteria and standards

SUESS – also for physical factors, including non-ionizing radiation. Now, I know that in their program they have ultrasound at a rather high priority, with other non-ionizing radiation types at lower priorities. The reason, to the best of my understanding, is the concern of radiologists with ultrasound being intensively used for medical purposes. Lasers would follow. I think microwaves are less applied in the medical field in comparison with the other two, and thus have a lower priority. In any way the answer to your question is "yes."

HILL – The question I was going to ask is whether you really believe that non-ionizing radiation exists as a concept? The reason I ask this is that it seems to me that a large part of the reason that non-ionizing radiation has come into being as a term is for administrative convenience and partly I think arising out of the situation that you mentioned, that ionizing radiation has become looked after by organizations like ICRP and ICRU, to some extent. And there is an extent to which non-ionizing radiation is sort of a rag-bag of things that don't quite fit into this other pattern. And I think the comment I wanted to make then was that it does seem important that this situation shouldn't go too far. Because there is tremendous benefit to be gained in the ultrasound field, and I am sure in the microwave field, and others, from relationship with ionizing radiation. And it would be a great pity I think if the world became divided into radiations that were ionizing and radiations that were thought to be non-ionizing.

SUESS – I am very glad, Dr. Hill, that you have raised this point. I should perhaps mention here that Dr. Hill was one of the members of The Hague meeting in 1971, and I think that he has really the answer to his question. One point that has appeared at the very beginning of the meeting's report is the indication that for the sake of the non-ionizing radiation field we should learn from all our experience in the ionizing radiation field. Therefore, I definitely agree that we should not make a very clear-cut separation between the two -- between ionizing and non-ionizing radiation.

With respect to the IAEA (International Atomic Energy Agency) I have only indicated that it is not getting involved, at least for the time being, in non-ionizing radiation. It is involved in the nuclear field, be it environmental studies for siting of a reactor or isotope-type problems. WHO is similarly involved in ionizing radiation and radioisotopes with respect to health, but, this again, is a different field.

As far as the WHO Regional Office for Europe is concerned

SUESS - if I didn't make it clear before I will try to do it now. We are very limited with manpower and, of course, also with funds. Therefore, if we can do only so much, the question is whether we should spread ourselves thin and do a superficial job with very little, if any, impact; or should we concentrate on one or a few aspects, going into depth, and have an impact with our work and really serve the governments by giving them something that is new or needed.

As to the situation in Europe, many countries still have areas that are in an early stage of development with regard to even ionizing radiation. I am sure it is still a problem everywhere. The knowledge, however, is available, and most governments in Europe should be able to help themselves if they so wish, and, in our opinion, they don't need the office at this stage. In comparison, they may need assistance on non-ionizing radiation. That is really the only difference.

OSEPCHUK - I want to ask what you mean or what Dr. Czerski means by non-stationary fields and stationary fields?

CZERSKI - A field from a moving antenna or beam is non-stationary. A point in a specific location is exposed intermittently. This would perhaps in effect correspond to the concept of an effective radiation time at a point in the vicinity of an installation, where you have a moving beam or antenna. From this the concept of a stationary field is self-evident - a point in a specific location is continuously irradiated, i.e., exposed to a stationary field.

MICHAELSON - Dr. Czerski has agreed to spend a few minutes to tell us the major conclusions of the Polish symposium (International Symposium on Biologic Effects and Health Hazards of Microwave Radiation, Warsaw, October 15-18, 1973, Polish Medical Publishers, Warsaw, 1974).

CZERSKI - The symposium was held under the sponsorship of the World Health Organization, the U. S. Department of Health, Education and Welfare, as well as the Polish Minister of Health and Social Welfare. The participants, as much as possible, were a good representation of the international community interested in microwave biologic effects. Dr. Michaelson, Dr. Suess, who represented the WHO during the symposium, and I feel that perhaps you would be interested in hearing simply the verbatim text of the conclusions and recommendations reached at the end of this symposium and approved by all the participants:

"The widespread and increasing use of microwave power has greatly increased the possibility of exposure of both occupational and the general population groups in many countries of the world. Protection measures and health and safety standards have varied widely in different parts of the world mainly because of differences in research approaches, findings, and interpretations. In so far as possible, it is imperative to resolve or remove obstacles to a common understanding of the scientific basis for protection measures. The accomplishments of this International Symposium in terms of meaningful exchange of data, frank discussion of viewpoints and enthusiastic interest and opportunities for collaborative undertakings and the recommendations that follow constitute a significant step toward the advancement of the knowledge in the field. The following recommendations were made:

(1) to promote international coordination of research on the biologic effects of microwave radiation; there should be a continuing exchange of information, improved efficiency of translation services, exchange visits, and closer collaboration in research projects, and publications.

(2) a program concerned with non-ionizing radiation should be developed by an international health agency that could exert leadership in this field and facilitate communication among scientists. It was hoped that the World Health Organization would assume this responsibility.

(3) thought should be given to establish internationally accepted nomenclature and definitions of physical quantities and units, and to standardize measurements, techniques, and dosimetry. An international group should be established to work out procedures for achieving these objectives.

(4) to achieve a more uniform approach to the discussion of mechanisms underlying biologic effects, it is proposed that the microwave intensities be considered as divided into three approximate ranges as follows:

(a) the range above 10 milliwatts per cm^2 in which distinct thermal effects predominate.

(b) the range below 1 milliwatt per cm^2 in which thermal effects are improbable.

(c) an intermediate range in which weak but noticeable thermal effects occur as well as direct field effects and other effects of a microscopic or macroscopic nature, the details of which have not yet been clarified.

The limits of these ranges have not yet been determined. They may differ for various species of animals, and may also depend on a variety of parameters, such as modulation and frequency.

(5) in view of the importance of electrophysiological recording for studies of microwave effects, there is need for the development of new electrode systems and integrated electro-signal amplifying systems, capable of use and full operation during microwave exposure.

(6) further biologic, medical, epidemiological, and biophysical studies are needed to improve understanding of the interaction of microwave radiation with biologic systems, and clarify the risks that may be associated with microwave exposure. Specific attention should be given to the following:

(a) investigation into the occurrence of accumulated effects and delayed effects.

(b) study of low intensity effects.

(c) determination of possible threshold values.

(d) study of combined effects of radiation and other environmental factors.

(e) investigation of differential radiation sensitivity as a function of organs, system, and age or intrauterine development.

(f) study of effects related to cellular transformation.

(g) study of effects occurring at the molecular level.

(h) determination of absorbed energy and its spatial distribution.

The desirability of conducting similar investigations in the radio frequency range was emphasized."

SPEAKERS

Tadeusz M. Babij, M.Sc., Ph.D.
Technical University of Wroclaw
Wroclaw, Poland

Ronald R. Bowman, Ph.D.
National Bureau of Standards
Boulder, Colorado 80302

John Bligh, Ph.D.
Institute of Animal Physiology
Babraham, Cambridge, England

P. Czerski, M.D.
National Research Institute of
  Mother & Child
Warsaw, Poland

Floyd Dunn, Ph.D.
University of Illinois
Urbana, Illinois   61801

Thomas Ely, M.D.
Eastman Kodak Company
Rochester, New York   14650

Owen Maynard
Raytheon Company
Waltham, Massachusetts   02154

James Lin, Ph.D.
University of Washington
Seattle, Washington   98195

C. R. Hill, Ph.D.
Institute of Cancer Research
Belmont, Sutton, Surrey
  United Kingdom

Justus F. Lehmann, M.D.
University of Washington
Seattle, Washington   98195

Padmakar P. Lele, M.D., Ph.D.
Massachusetts Institute of
  Technology
Cambridge, Massachusetts   02139

Sol M. Michaelson, D.V.M.
The University of Rochester
Rochester, New York   14642

Mr. John C. Mitchell, M.S.
USAF School of Aerospace Medicine
Brooks Air Force Base
San Antonio, Texas   78235

Wesley Nyborg, Ph.D.
The University of Vermont
Burlington, Vermont   05401

Lawrence T. Odland, M.D., Ph.D.
Radiological Health Laboratory
Wright-Patterson AFB
Ohio   45433

John M. Osepchuk, Ph.D.
Raytheon Company
Waltham, Massachusetts   02154

Adolfo Portela, Ph.D.
National Council of Scientific
  & Technical Investigations
Buenos Aires, Argentina

John R. K. Savage, D.Phil.
MRC Radiobiology Unit
Harwell, Didcot, Berks
  England

Harold F. Stewart, Ph.D.
Bureau of Radiological Health
Rockville, Maryland   20852

Michael J. Suess, D.Sc.
World Health Organization
Regional Office for Europe
Copenhagen, Denmark

K. J. W. Taylor, Ph.D.
The Royal Marsden Hospital
Sutton, Surrey, England

SESSION CHAIRMEN

SESSION I: BIOPHYSICS AND DOSIMETRY

>Dr. Dietrich E. Beischer
>Naval Aerospace Medical Research Laboratory
>Naval Aerospace Medical Center
>Pensacola, Florida   32512

SESSION II: ENERGY ABSORPTION

>James D. Hardy, Ph.D.
>John B. Pierce Foundation Lab.
>Yale University
>New Haven, Connecticut   06519

SESSION III: MICROWAVES - BIOLOGICAL EFFECTS

>Karl Lowy, M.D.
>Center for Brain Research
>The University of Rochester
>Rochester, New York   14642

SESSION IV: ULTRASOUND - BIOLOGICAL EFFECTS

>George W. Casarett, Ph.D.
>The University of Rochester
>Rochester, New York   14642

SESSION V: MEDICAL APPLICATIONS

>Raymond Gramiak, M.D.
>The University of Rochester
>Rochester, New York   14642

SESSION VI: OCCUPATIONAL ASPECTS

>Commander Paul E. Tyler, Jr. (MC, USN)
>National Naval Medical Center
>Bethesda, Maryland   20014

SESSION VII: FUTURE APPLICATIONS AND CONTROLS

>Edythalena Tompkins
>Environmental Protection Agency
>Research Triangle Park
>North Carolina   27711

PARTICIPANTS

OBSERVERS

Frank Alfano
The University of Rochester
Rochester, New York 14642

Kurt Altman, Ph.D.
The University of Rochester
Rochester, New York 14642

Kym Arcuri
The University of Rochester
Rochester, New York 14642

Edward Aslan
The Narda Microwave Corp.
Plainview, L.I., N.Y. 11803

Robert Baxter
The University of Rochester
Rochester, New York 14642

M. Michael Brady
Norconsult A.S.
Hovik, Norway

Dr. C. A. Cain
University of Illinois
Urbana, Illinois 61801

Sally Child
The University of Rochester
Rochester, New York 14627

Christopher L. Christman
Bureau of Radiological Health
Rockville, Maryland 20852

Fred Eames
The University of Rochester
Rochester, New York 14642

Dr. Zorach Glaser
National Naval Medical Center
Bethesda, Maryland 20014

Major R. B. Graham
Wright-Patterson AFB
Ohio   45433

Winborn Gregory
The University of Rochester
Rochester, New York 14642

Henry Han
The University of Rochester
Rochester, New York 14627

Mr. L. Heynick
Stanford Research Institute
Menlo Park, California 94025

Deborah Hollister
The University of Rochester
Rochester, New York 14642

Dr. William Houk
Naval Aerospace Medical Center
Pensacola, Florida 32512

Dr. David Janes
Environmental Protection Agency
Washington, D. C.

Dr. Curtis C. Johnson
University of Utah
Salt Lake City, Utah

Gary E. Kaufman, Ph.D.
The University of Rochester
Rochester, New York 14642

Edwin Kinnen, Ph.D.
The University of Rochester
Rochester, New York 14627

Dr. Frederick Kremkau
Bowman Gray School of Medicine
Winston Salem, N. C. 27103

Don E. Lee
3M Center
St. Paul, Minnesota 55101

Hy Lisman
The University of Rochester
Rochester, New York 14642

Observers (Continued)

Abelardo Lopez
The University of Rochester
Rochester, New York 14642

W. Gregory Lotz
The University of Rochester
Rochester, New York 14642

Shin-Tsu Lu
The University of Rochester
Rochester, New York 14642

Robert A. McAfee, Ph.D.
Veterans' Administration Hospital
New Orleans, Louisiana 70126

H. David Maillie, Ph.D.
The University of Rochester
Rochester, New York 14642

Dr. George Mickey
Richfield, Connecticut

Douglas Miller
The University of Vermont
Burlington, Vermont 05401

William C. Milroy, USN, MC
Naval Weapons Laboratory
Dahlgren, Virginia 22448

Mr. Kenneth Oscar
Department of the Army
Fort Belvoir, Va. 22060

Robert Pizzutiello
The University of Rochester
Rochester, New York 14642

Dr. Peter Polson
Stanford Research Institute
Menlo Park, California 94025

Elliott Postow, Ph.D.
National Naval Medical Center
Bethesda, Maryland 20014

William Quinlan
The University of Rochester
Rochester, New York 14642

Mr. Paul Roney
Bureau of Radiological Health
Rockville, Maryland 20852

Dr. Thomas C. Rozzell
Office of Naval Research
Arlington, Virginia 22217

Richard M. Schreck
General Motors Technical Center
Warren, Michigan 48090

Dr. Y. T. Seto
Tulane University
New Orleans, Louisiana 70118

Kathy Smachlo
The University of Rochester
Rochester, New York 14627

Vernon Steele
The University of Rochester
Rochester, New York 14642

Al Thomson
The University of Rochester
Rochester, New York 14642

Dr. Joseph H. Vogelman, P.E.
Vogelman Development Company
Roslyn, New York 11576

Susan M. Voorhees
The University of Rochester
Rochester, New York 14642

Dr. Robert Waag
The University of Rochester
Rochester, New York 14627

Dr. Bernard Weiss
The University of Rochester
Rochester, New York 14642

PARTICIPANTS

Observers (Continued)

Dr. Seymour Zigman
The University of Rochester
Rochester, New York 14642

Conference Committee

Solomon M. Michaelson, Chairman
George G. Berg
Edwin L. Carstensen
Richard Magin
Morton W. Miller

The University of Rochester
Rochester, New York   14642

SUBJECT INDEX

Absorbed power, 174-177, 182, 207
Absorption coefficient
    microwave, 412, 413
    ultrasound, 25, 26, 345
Action potential, 116
    amplitude, 106, 197, 198
    excitability, 197, 198
    propagation, 106, 197, 198
    threshold, 10
Acoustic
    impedance, 24
    intensity, 24
    pressure, 24
    spectrum, 22
    streaming, 140, 278, 284-287
    wave intensity, 24
Acoustics
    non-linear, 37, 278, 282-284
Antenna, 368
    dipole, 41
    loop, 50
    radio, 422-426
    television, 373, 422-426
Audition, 217, 225
Auditory
    nerve stimulation, 229
    threshold, 211, 218, 219
Auditory effects
    continuous wave, 183-190
    pulsed wave, 190-195, 210, 211

Blood flow, 155, 306

Bradycardia, 209, 371
Brain
    grey matter, 332-334
    pacemaker, 362
    white matter, 332-334
Broadcast station
    radio, 422-426
    television, 422-426

Calorimeter, 68
Cancer, 256
    ascites tumor cells, 253-255
    therapy, 322
Cardiazole tolerance, 371, 375
Cataracts, 314, 321, 371, 396-399
    calcium deposits, 405, 406
    incidence in exposed groups, 406, 407
    lens protein precipitation, 405
Cavitation, 140, 248, 250, 251, 254, 258, 259, 264-266, 274, 279-282, 318, 319, 322, 323, 345
    nuclei, 275, 279
    stable, 261, 276, 278, 280
    transient, 260, 278, 297
Cell
    binucleated, 298
    culture, 236, 262
    death, 252, 254
    kinetics, 252
    structure, 253, 255
    survival, 252, 262

Chromosome aberration, 234, 236-238, 243, 253, 255, 259, 260
    breaks, 235, 247, 248
    chromatid, 235, 246
    coiling, 245, 247
    gaps, 235, 245
    resolution, 240
Chromosome anomaly, 243, 244
Circadian rhythm
    adrenalin/noradrenalin, 376, 377
Commission of the European Communities (CEC), 449-452
Circulation, local, 144, 162, 332
Conductivity, 4-9
Craniotomy, 329, 335
Densitometry, 52
Detector
    diode, 44
    linear, 44
    square law, 44
Diagnostic ultrasound, 85, 348-350, 342, 343
    equipment safety, 264, 341-347
    doppler, 264, 265, 343
    holographic, 264, 265
    pacemaker, 360
    procedures, 343
    sonar, 264, 265
Diathermy, 155, 308-319
    microwave, 309-314
    short-wave, 309
    ultrasound, 314-319
Dielectric constant, 4-9
Dielectrophoresis, 11
Diffraction, 39, 71
Dipole orientation, 45
Dispersion ($\alpha, \beta, \delta, \gamma$), 4-9
Dosimetry, 17, 52, 54, 209
    microwave, 56, 57, 168
    personnel, 401
    ultrasound, 85, 343

EEG abnormalities, 370, 371

Effects
    electrophonic, 229
    linear, 10, 25, 37
    non-linear, 25, 37
    non-thermal, 94, 95, 131, 132
    resonance, 11
    thermal, 11, 94, 95, 133, 134
Electromagnetic radiation compatibility, 352
Electronic cardiac pacemakers, 351-358, 384, 385
    nonionizing radiation compatibility, 354-358
    frequency sensitivity, 355, 362
    interference susceptibility, 361-363, 429
Energy density, 24
Equilibrium
    keto-enol, 33
    solvation, 33
Epidemiology, 346, 370-373
Evoked potential
    amplitude, 183-190
    latency, 183-190, 208, 209

Fatigue, 370
Fever
    pyrogen induced, 157, 158, 208
Field
    forces, 11, 278
    non-stationary, 368, 455
    threshold, 14
Flagellation function, 253, 256
Free radical, 248, 251, 258, 279
Fresnel zone, 41, 42

Gas discharge lamp, 383
Genetic damage, 240, 241, 252, 253, 345

INDEX

Head exposure, 169-176
Headache, 370
Hearing, 14, 15, 210, 211, 226
Heat, 94, 303, 304, 345
Heat loss, 144, 159
Helix coil transition, 32
Hydration layer, 30, 31
Hydrophones, 76, 87, 88, 264
Hypothalamus, 155
Hypotonia, 370
Hypoxia, 267

Incandescent lamps, 382
Interferometer, 74
International Atomic Energy Agency (IAEA), 454
International Commission on Radiological Protection (ICRP), 451
International Commission on Radiation Units and Measurement (ICRU), 54, 451
International Radiation Protection Association (IRPA), 451, 452
International Symposium on the Biological Effects and Health Hazards of Microwave Radiation, Warsaw, Poland, Oct. 15-18, 1973, Summary of major conclusions, 17, 456, 457
In vitro exposure
    saphenous nerve, 195-198
    sartorius muscle, 97-102
    sciatic nerve, 195-198
    superior cervical ganglion, 198-202, 209
    tibialis anticus, 103-105
Ionic currents, 116, 119, 126, 127
Ionizing radiation, 454, 455
ISM bands, 381, 382

Joints, 305, 306, 314-319

Karyotype instability, 236

Lasers, 383, 384
Lysosomes, 256

Macromolecules, 30-33
Malignancy, 306, 322
Maxwell-Wagner effect, 4
Measurements
    absorption, 37, 38
    attenuation, 39
    intensity distribution, 67
    internal electric field, 51, 52
    scattering, 38
Membrane
    electric properties, 96, 106, 116
    permeability, 9
    resting potential, 10
    structure, 253
Metabolic rate, 207
Methacrylate medium, 330
Micronuclei, 244, 245
Microstreaming, 251, 279, 323
Microwave
    consumer applications, 422-429
    hearing threshold, 219-221, 225-227
    imaging, 53
    industrial applications, 419-422
    oven, 382, 416-419, 427-429, 432
    sickness, 371
    warning sign, 385
Microwave exposure
    free field, 352-354
    general public, 372, 373, 432, 442, 448
    military, 389-402
    occupational, 367-373, 448-452
    simulated chest implant, 352, 353
    waveguide, 180-182

Microwave sources
    electron tube, 413-416
    solid state, 414-416
Mitotic index, 261, 262
Mode conversion, 37
Modulation, 225
Muscle
    sartorius, 96
    tibialis anticus, 102
Muscle cell water, 102-105, 117, 163
Mutation, 233, 243, 252, 253
    ultrasound induced, 254, 255, 297-299
    UV induced, 386

Nerve-muscle preparation, 96-102, 209
Nervous system
    central, 95, 167, 168
    peripheral, 95, 167, 168
Neuronal thermoregulator, 149-152, 157
Nonionizing radiation
    frequency ranges, 450
Non-thermal microwave effects, 94, 95
    hearing, 14, 15, 183-195, 217-221
Non-thermal ultrasonic effects, 269-271
    abnormal embryonic development, 266
    focal brain lesions, 136
    liver centrilobular necrosis, 267, 268
    red blood cell stasis, 274, 275, 268, 269
    spinal cord hemorrhage and paraplegia, 266, 267
Nystagmus, 228

Obstetrics, 325, 342
Ophthalmology, 325
Ozone, 385-387

Pain, relief, 303, 304
Particle displacement, 24
Pearl-chain formation, 11, 19, 291, 296
Pellicle, 74, 84-87
Perception
    electromagnetic radiation, 213, 214, 227, 228
Peripheral vasomotor tone, 144, 155
Power measurements
    acousto-optical, 70-76
    microphone, 76, 77
    optical, 71
    radiation force, 60-67
    thermal, 68, 84
Probe
    electric field, 41, 50, 51
    magnetic field, 50, 51
    parabolic thermal, 70
    three orthogonal dipole, 42
Protonema, 298
Proton transfer, 32, 33

Radar
    development of, 391, 392
Radiation forces, 278, 292, 293
Radiation force methods
    analytical balance, 62-65
    float, 60-62
    portable, 65-67
Radiation pressure, 292, 293
Radiofrequency heaters, 381, 382
Raman-Nath theory, 71, 74
Random coil, 32
Rectified diffusion, 282
Reflection coefficient, 27
Relaxation
    counter ion, 6, 9
    electrophoretic, 6
    membrane, 6
    process, 26
    regions, 5

Resonance
      dipole/diode, 49
Reversible lesion, 336, 337

Satellite solar power generation, 434-445
Schlieren visualization, 71
Screening, 345-347
      occupational, 392, 393
      ophthalmological, 396
Seasonal variation, 95, 116-119
Sensation, 213, 214
      cutaneous thermal, 214-217, 225
      thermal, 16, 227, 228
Servo-mechanism, 147-149, 156
Set-point, 145-147, 156, 158
Shear wave, 37
Shivering, 144
Solar energy, 434
Solvent-solute interaction, 32, 33
Somatosensory effects, 183-190
Sonic torque, 287-291
Spinal cord exposure, 178-180, 195, 196
Standards
      ANSI C95.1 (microwave), 381, 382
      ANSI Z-136.1 (laser), 383, 384
      general public, 15, 16
      occupational, 324, 367-373, 448-452
      OSHA 1910.97 (microwave), 382
      military, 389-402
Stress
      classical response, 376
      psychological, 375
Sweating, excessive nightly, 370

Tendons, 305, 306
Teratogenesis, 345
Thermal
      gradients, 16, 161
      hypothesis, 130, 131

Thermistor, 68
Thermocouple, 68, 86
Thermogenesis, non-shivering, 144
Thermography, 55, 56
Thermoregulation, 143-152, 156
Thermoregulatory effectors, 151
Thermosensors, 149, 155, 162
Threshold temperature hypothesis, 129
Tissue
      collagenous, 303-307
      conductivity, 4
      dielectric constant, 4
      metabolism, 144
      rectification, 228
      regeneration, 264
      temperature distribution, 307-319
      vascularity, 304-308, 321, 322
Transducer
      calibration, 77-80, 88
      ceramic, 327, 328
      lens, 328, 329
      near-field, 86
      quartz, 327, 328

Ultrasonic
      absorption, 26-33
      diagnostic devices, 264,
      focused lesions, 134-136, 326, 327, 331-335
      power measurements, 59-80
      tissue damage, 129, 249-257, 318, 319, 324
      velocity, 24
Ultrasound
      doppler-CW, 343
      pulse-echo, 336, 343
      safety, 264, 342, 345, 348-350, 453
      surgical tool, 326
Ultraviolet light, 385-387

Wave
　　equation, 23
　　progressive, 297
　　standing, 296
World Health Organization
　Regional Office for Europe
　(WHO/EURO), 448-452

X-ray sensitivity, 253, 256